RECEPTORS IN THE DEVELOPING NERVOUS SYSTEM

VOLUME 2

RECEPTORS IN THE DEVELOPING NERVOUS SYSTEM

VOLUME 2
Neurotransmitters

Edited by

*Ian S. Zagon and
Patricia J. McLaughlin*

*Department of Neuroscience and Anatomy
The Milton S. Hershey Medical Center
The Pennsylvania State University
Hershey, Pennsylvania, USA*

 SPRINGER-SCIENCE+BUSINESS MEDIA, B.V.

First edition 1993

© 1993 Springer Science+Business Media Dordrecht
Originally published by Chapman & Hall in 1993
Softcover reprint of the hardcover 1st edition 1993

Typeset in 10/12 Pt Palatino by Expo Holdings, Malaysia

ISBN 978-94-010-4674-9   Vols. 1 and 2 (set) 0 412 54520 9

A catalogue record for this book is available from the British Library

Library of Congress Cataloging-in-Publication data

Receptors in the developing nervous system / edited by Ian S. Zagon and
   Patricia J. McLaughlin.—1st ed.
        p. cm.
   Includes bibliographical references and index.
      Contents: v. 1. Receptors related to growth factors and hormones
   —v. 2. Neurotransmitters.
      ISBN 978-94-010-4674-9   ISBN 978-94-011-1544-5 (eBook)
      DOI 10.1007/978-94-011-1544-5
   alk. paper)
      1. Developmental neurology. 2. Neurotransmitter receptors. I. Zagon,
   Ian S. II. McLaughlin, Patricia J.
   [DNLM: 1. Growth Substances. 2. Nervous System—embryology.
   3. Neuroregulators. 4. Receptors, Endogenous Substances.
   5. Receptors, Sensory, WL 101 R295]
   QP363. 5.R43 1993
   599' .0333—dc20
   DNLM/DLC   92-49100
   for Library of Congress   CIP

For Eileen
And my grandparents,
    Mary and Abraham Shaffer
                            I. S. Z.

For my parents,
        Katharine and Charles
                            P. J. M.

CONTENTS

CONTRIBUTORS

JOHN D. ALVARO
Laboratory of Molecular Psychiatry,
   Department of Psychiatry and Program in
   Neuroscience, Abraham Ribicoff Research
   Facility, Yale University School of
   Medicine, 34 Park St, New Haven, CT
   06508, USA.

LUCIO G. COSTA
Department of Environmental Health, SC-34,
   University of Washington, Seattle, WA
   98195, USA.

THAN-VINH DAM
Douglas Hospital Research Centre and
   Department of Psychiatry, Faculty of
   Medicine, McGill University, Verdun,
   Quebec, Canada H4H IR3.

ANGEL LUIS DE BLAS
Division of Molecular Biology and
   Biochemistry, School of Biological Sciences,
   University of Missouri-Kansas City,
   Kansas City, MO 64110-2499, USA.

RONALD S. DUMAN
Laboratory of Molecular Psychiatry,
   Department of Psychiatry and Program in
   Neuroscience, Abraham Ribicoff Research
   Facility, Yale University School of
   Medicine, 34 Park St, New Haven, CT
   06508, USA.

F. JAVIER GARCIA-LADONA
Institute of Pathology, Department of
   Neuropathology, University of Basel,
   Switzerland.

GAIL E. HANDELMANN
Department of Pharmacology and
   Toxicology, University of Utah, Salt Lake
   City, UT 84112, USA.

FRANCES M. LESLIE
Department of Pharmacology, University of
   California at Irvine, Irvine, CA 92717, USA.

EDYTHE D. LONDON
Addiction Research Center, National
   Institute on Drug Abuse, Baltimore, MD
   21224, USA.

SANDRA E. LOUGHLIN
Department of Anatomy and Neurobiology,
   University of California at Irvine, Irvine,
   CA 92717, USA.

GUADALUPE MENGOD
Department of Neurochemistry, Centro
   Investigacion y Desarrollo, Consejo
   Superior de Investigaciones Cientificas
   (CSIC), Jordi Girona, 18–26, Barcelona,
   Spain.

TIMOTHY H. MORAN
Department of Psychiatry, Johns Hopkins
   University School of Medicine, Baltimore,
   MD 21205, USA.

JOSÉ M. PALACIOS
Department of Neurochemistry, Centro
   Investigacion y Desarrollo, Consejo
   Superior de Investigaciones Cientificas
   (CSIC), Jordi Girona, 18–26, Barcelona,
   Spain, and Research Institute, Laboratorios
   Almirall, Barcelona, Spain.

REBECCA M. PRUSS
Marion Merrell Dow Research Institute, 2110 E. Galbraith Road, Cincinnati, OH 45215, USA.

RÉMI QUIRION
Douglas Hospital Research Centre and Department of Psychiatry, Faculty of Medicine, McGill University, Verdun, Quebec, Canada H4H 1R3.

PAUL H. ROBINSON
Westminster Hospital, London SW1, UK.

THOMAS ROTHE
University of Leipzig, Paul Flechsig Institute for Brain Research, Department of Neurochemistry, Leipzig, Germany.

REINHARD SCHLIEBS
University of Leipzig, Paul Flechsig Institute for Brain Research, Department of Neurochemistry, Leipzig, Germany.

JOHN D. STEPHENSON
Institute of Psychiatry, London SE5, UK.

ANN TEMPEL
Hillside Hospital, Division of Long Island Jewish Medical Center, The Long Island Campus for the Albert Einstein College of Medicine, Glen Oaks, New York 11004, USA.

PATRICIA M. WHITAKER-AZMITIA
Department of Psychiatry, State University of New York, Stony Brook, NY 11794, USA.

STEPHEN R. ZUKIN
Albert Einstein College of Medicine, Yeshiva University, Bronx, New York, NY 10461, USA.

CONTENTS TO VOLUME ONE

CONTRIBUTORS TO VOLUME ONE

MARTIN ADAMO
Section on Molecular and Cellular
    Physiology, Diabetes Branch, National
    Institute of Diabetes and Digestive and
    Kidney Diseases, National Institutes of
    Health, Bethesda, MD 20892, USA

SEEMA BHATNAGAR
Developmental Neuroendocrinology
    Laboratory, Douglas Hospital Research
    Centre, Department of Psychiatry and
    Neurology and Neurosurgery, McGill
    University, Montreal, Canada H4H 1R3

GERARD J. BOER
Netherlands Institute for Brain Research,
    Meibergdreef 33, 1105AZ Amsterdam Z0,
    The Netherlands

CAROLYN A. BONDY
Developmental Endocrinology Branch,
    National Institute of Child Health and
    Human Development, National Institutes
    of Health, Bethesda, MD 20892, USA

JEAN-GUY CHABOT
Douglas Hospital Research Centre and
    Department of Psychiatry, Faculty of
    Medicine, McGill University, Verdun,
    Québec, Canada H4H 1R3

KWEN-JEN CHANG
Department of Anesthesiology and
    Pharmacology, Duke University Medical
    Center, and Division of Cell Biology,
    Wellcome Research Laboratories,
    Burroughs Wellcome Co.
    Research Triangle Park, NC 27709, USA

MOSES V. CHAO
Department of Cell Biology and Anatomy,
    Hematology/Oncology Division, Cornell
    University Medical College, 1300 York
    Avenue, New York, NY 10021, USA

ERROL B. DE SOUZA
Central Nervous System Diseases Research,
    The DuPont Merck Pharmaceutical Co.,
    Wilmington, DE 19880, USA

JEAN DE VELLIS
Departments of Anatomy and Psychiatry,
    Mental Retardation Research Center, Brain
    Research Institute, Laboratory of
    Biomedical and Environmental Sciences,
    UCLA School of Medicine, University of
    California, Los Angeles, CA 90024, USA

ARACELI ESPINOSA DE LOS MONTEROS
Departments of Anatomy and Psychiatry,
    Mental Retardation Research Center, Brain
    Research Institute, Laboratory of
    Biomedical and Environmental Sciences,
    UCLA School of Medicine, University of
    California, Los Angeles, CA 90024, USA

DIMITRI E. GRIGORIADIS
Central Nervous System Diseases Research,
    The DuPont Merck Pharmaceutical Co.,
    Wilmington, DE 19880, USA

JEFFREY A. HEROUX
Central Nervous System Diseases Research,
    The DuPont Merck Pharmaceutical Co.,
    Wilmington, DE 19880, USA

THOMAS R. INSEL
Laboratory of Neurophysiology NIMH
Poolesville, MD 20837, USA

SATYABRATA KAR
Douglas Hospital Research Centre and
Department of Psychiatry, Faculty of Medicine, McGill University, Verdun, Québec,
Canada H4H 1R3

STEPHEN L. KINSMAN
The Kennedy Krieger Institute, 707 North
Broadway, Baltimore, MD 21205, USA

DEREK LEROITH
Section of Molecular and Cellular
Physiology, Diabetes Branch, National
Institute of Diabetes and Digestive and
Kidney Diseases, National Institutes of
Health, Bethesda, MD 20892, USA

PATRICIA J. MCLAUGHLIN
Department of Neuroscience and Anatomy,
The Pennsylvania State University, The
M.S. Hershey Medical Center, Hershey, PA
17033, USA

MICHAEL J. MEANEY
Developmental Neuroendocrinology
Laboratory, Douglas Hospital Research
Centre, Departments of Psychiatry and
Neurology and Neurosurgery, McGill
University, Montreal, Canada H4H 1R3

DAJAN O'DONNELL
Developmental Neuroendocrinology
Laboratory, Douglas Hospital Research
Centre, Departments of Psychiatry and
Neurology and Neurosurgery, McGill
University, Montreal, Canada H4H 1R3

LUIS F. PARADA
Molecular Embryology Group, ABL-Basic
Research Program, NCI-Frederick Cancer
Research, P O Box B, Frederick, MD 21701,
USA

DANIEL H. POLK
UCLA School of Medicine, Perinatal
Laboratories, Harbor-UCLA Medical
Center, Torrance, CA 90509, USA

RÉMI QUIRION
Douglas Hospital Research Centre and
Department of Psychiatry, Faculty of
Medicine, McGill University, Verdun,
Québec, Canada H4H 1R3

MOHAN K. RAIZADA
Department of Physiology, University of
Florida, Gainesville, FL 32610, USA

CHARLES T. ROBERTS
Section on Molecular and Cellular
Physiology, Diabetes Branch, National
Institute of Diabetes and Digestive and
Kidney Diseases, National Institutes of
Health, Bethesda, MD 20892, USA

ALAIN SARRIEAU
Developmental Neuroendocrinology
Laboratory, Douglas Hospital Research
Centre, Departments of Psychiatry and
Neurology and Neurosurgery, McGill
University, Montreal, Canada H4H 1R3

NOLA SHANKS
Developmental Neuroendocrinology
Laboratory, Douglas Hospital Research
Centre, Departments of Psychiatry and
Neurology and Neurosurgery, McGill
University, Montreal, Canada H4H 1R3

SAMUEL A. SHOLL
Wisconsin Regional Primate Research Center,
University of Wisconsin, Madison,
WI53715–1299, USA

JAMES SMYTHE
Developmental Neuroendocrinology
Laboratory, Douglas Hospital Research
Centre, Departments of Psychiatry and
Neurology and Neurosurgery, McGill
University, Montreal, Canada H4H 1R3

YING-FU SU
Department of Anesthesiology and Pharma-
cology, Duke University Medical Center,
Durham, NC 27710, USA

VICTOR VIAU
Developmental Neuroendocrinology Labora-
tory, Douglas Hospital Research Centre,
Departments of Psychiatry and Neurology
and Neurosurgery, McGill University,
Montreal, Canada H4H 1R3

CLAIRE-DOMINIQUE WALKER
Department of Physiology, University of
California at San Francisco, San Francisco,
CA 94143, USA

HAIM WERNER
Section on Molecular and Cellular
Physiology, Diabetes Branch, National
Institute of Diabetes and Digestive and
Kidney Diseases, National Institutes of
Health, Bethesda, MD 20892, USA

IAN S. ZAGON
Department of Neuroscience and Anatomy,
The Pennsylvania State University, The
M.S. Hershey Medical Center, Hershey, PA
17033, USA

PREFACE

Receptors for cell hormones, growth factors, and neurotransmitters are involved in the control and modulation of an enormous array of biological processes. The development of these receptors has distinct spatial and temporal arrangements, and alterations in this pattern during embryogenesis can have significant consequences for the well-being of the fetus, infant, child and adult. The developing nervous system is particularly dependent on receptors because its period of structural and functional organization extends through both prenatal and postnatal phases. Moreover, receptors are a key element in neural communication in both the developing and adult organism, so that the ontogeny of receptors is crucial in determining the myriad connections forming the circuitry of the nervous system.

The purpose of these two volumes is to provide a comprehensive review of receptors in the developing nervous system, placing basic and clinical information into perspective in order to formulate future scientific inquiry. A number of themes are maintained throughout the books. First, the receptors discussed in these books are part of a unit, and both the receptor and neurotransmitter, hormone, or growth factor must be considered. Therefore, the authors have spent some time in each chapter introducing the compound(s) that interacts with each receptor. Second, some receptors appear transiently and are responsible for participating in a biological process related to ontogeny, disappearing by maturation. Third, a receptor in the process of evolving into a structure important in the mature animal may also be vital in establishing the framework of particular pathways or the architecture of the brain.

Fourth, alterations in the development of neural receptors may have profound implications for the structure and function of the organism. As much as possible, the repercussions of disrupting the orchestration of receptor development in the nervous system are discussed. In many instances, however, we are just beginning to learn about some receptors and the authors may not be in a position to discuss the consequences of receptor dysfunction.

In designing these two volumes, we have asked major figures in each field to review the literature, to apprise the audience of their latest findings, and to provide a perspective on the role of receptors in the developing nervous system. These books are intended to summarize not only where we have been, but also to map directions for future research efforts, with the ultimate goals of understanding processes of normal development and the prevention or treatment of receptor dysfunction. The book is intended to be part of a continuing dialog about an important and emerging field. Given the broad scope of the subject matter, it is our expectation that the information provided by these experts will be of interest to basic and clinical researchers and graduate students in the field of developmental neurobiology, as well as to individuals in psychology, cellular and molecular biology, endocrinology, embryology, neuroscience, pharmacology, and in clinical professions such as pediatrics, neonatology, and neurology. The style adopted by the contributors should also ensure that the contents of the two volumes will be accessible to undergraduate students enrolled in advanced courses concerned with neuroscience and cell and molecular biology.

The subject matter with respect to receptors and the developing nervous system has been divided into two volumes. The editors have attempted to group central themes and topics within three major areas: receptors related to growth factors and hormones (Volume I) and to neurotransmitters (Volume II). The reader should be aware that these are rather arbitrary divisions, and that a particular compound may have multiple functions. For example, opioid receptors known to be involved in neurotransmission may also have a trophic influence on development, and could also be considered a growth factor. An introductory chapter on the biology of receptors has been included in Volume I. For convenience, an index of the subject matter in each volume is included. Finally, since Volumes I and II should be viewed as a single, integral entity, we have included the chapter titles and authors of both volumes in each book.

We thank all the authors for their thoughtful and stimulating chapters, and for the prompt attention shown to editorial requests. We have learned a great deal from reviewing these chapters and trust that the reader will have an equally enjoyable experience.

Ian S. Zagon
Patricia J. McLaughlin

DEVELOPMENTAL EXPRESSION OF ADRENERGIC RECEPTORS IN THE CENTRAL NERVOUS SYSTEM

Ronald S. Duman and John D. Alvaro

1.1 INTRODUCTION

The norepinephrine (NE) neurotransmitter receptor system is involved in numerous processes in the nervous system, including vigilance, attention, arousal, memory, neuroendocrine function, drug addiction, anxiety, depression, mania and stress-related disorders. The receptors for NE and epinephrine have been divided into α and β adrenergic receptors (αAR and βAR, respectively) based on functional and ligand binding studies. The development of more selective pharmacological agents has resulted in the characterization of α_1AR, α_2AR, β_1AR, and β_2AR subtypes, and molecular cloning has isolated further subtypes of each pharmacological type. Adrenergic receptors belong to the G protein-coupled receptor superfamily, a group of proteins which share a common feature of having seven hydrophobic transmembrane domains and which include dopaminergic, serotonergic, neuropeptide and other receptors. Members of this family couple with specific G proteins which then interact with effector systems such as enzymes or ion channels. Adrenergic receptor coupled effector systems regulate intracellular pathways that result in immediate or short-term changes in neuronal function as well as more long-term adaptive changes which may be dependent on regulation of gene expression.

The formation and function of adrenergic receptor systems during development ultimately influence a number of vital neuronal and behavioral processes. A clear understanding of the ontogeny of adrenergic receptors and the factors which regulate their development will help define the normal physiological processes regulated by these receptors. Moreover, the mechanisms which underlie psychiatric abnormalities associated with adrenergic receptors will be elucidated. The focus of this chapter is to review studies of the development of adrenergic receptors and their coupling factors/effector systems in the nervous system. The studies discussed in this review will focus on the development of adrenergic receptors in rat since the majority of the work has been conducted with this species.

1.2 DEVELOPMENT OF THE NOREPINEPHRINE NEUROTRANSMITTER SYSTEM IN THE NERVOUS SYSTEM

Both NE and epinephrine are effective neurotransmitter ligands at αAR and βAR, although NE is more widely distributed and more abundant than epinephrine in the central nervous system. For this reason the majority of

Receptors in the Developing Nervous System Vol. 2: Neurotransmitters. Edited by Ian S. Zagon and Patricia J. McLaughlin. Published in 1993 by Chapman & Hall. ISBN 0 412 49400 0. Vols. 1 and 2 (set) ISBN 0 412 54520 9.

studies have focused on the NE system. The distribution of NE- and epinephrine-containing cells and their projections has been reviewed by Moore and Bloom (1979) and Foote *et al.* (1983) and is summarized below. The majority of NE-containing cells are located in the pons and medulla of the brainstem. The nucleus locus coeruleus is a nearly homogeneous population of cells which contains approximately 40% of the NE cell bodies of the brain. The noradrenergic neurons located in the lateral tegmentum are more diffusely organized than those of the locus coeruleus and are spread throughout the lateral tegmental fields from the most rostral aspect of the medulla through the pons. Epinephrine-containing cell bodies are often intermingled with NE-containing cells in the lateral tegmentum and dorsal medulla.

NE projections throughout the brain are extensive and in many brain regions follow a very uniform plexus pattern of innervation. The locus coeruleus noradrenergic neurons project to many areas of the central nervous system including the cerebral cortex, thalamus, amygdala, striatum, hippocampus, hypothalamus, olfactory bulb, brainstem, cerebellum, and spinal cord; locus coeruleus is the primary source of NE to most of these brain regions. The lateral tegmental NE neurons project to many of the same brain areas and provide the primary NE input to some of these regions, specifically amygdala, septum and some hypothalamic nuclei such as the paraventricular nucleus and the supraoptic nucleus. Epinephrine-containing cells send projections primarily to the hypothalamus and midbrain regions.

Norepinephrine neurons differentiate and begin to express catecholamine during embryogenesis, and levels of NE continue to increase after birth (for reviews see Coyle, 1977; Foote *et al.*, 1983). In the rat, locus coeruleus neurons appear between embryonic day (ED) 10 and 13. The presence of catecholamine histofluorescence and the immuno-chemical staining of NE biosynthetic enzymes (tyrosine hydroxylase and dopamine β-hydroxylase) in these neurons appear as early as ED14. Catecholamine projections to the thalamus and hypothalamus appear around ED14, and ascending NE fibers to the cerebral cortex and subcortical regions appear around ED16. Neurons in the hippocampus appear around ED13–ED18 and in the cerebellar Purkinje neurons around ED14–ED15. This time course for development of LC neurons precedes the development of neurons in terminal field brain regions of this neurotransmitter. During the last week of gestation in the rat (ED15–ED22) there is a linear increase in levels of NE biosynthetic enzymes, and by ED18 high affinity uptake of NE can be demonstrated in synaptosomes.

At birth the levels of NE and its biosynthetic enzymes are between 10 and 20% of adult levels (Fig. 1.1). The innervation of cerebral cortex progresses to the adult pattern, determined by NE histofluorescence and tyrosine hydroxylase immunocytochemistry, by the end of postnatal day (PD) 7 in the rat. Although the adult pattern of NE innervation is present throughout the brain by the first week after birth, the arborization of NE terminals continues; adult levels of NE and tyrosine hydroxylase are not attained until 4–5 weeks after birth.

1.3 DEVELOPMENT OF β_1 AND β_2 ADRENERGIC RECEPTORS IN THE NERVOUS SYSTEM

The physiology, pharmacology and function of βAR subtypes have been widely studied. These receptors play a role in the central control of cardiovascular function, are involved in memory, and have been implicated in psychiatric illnesses, particularly depression (Minneman, 1981; Heninger and Charney, 1987). β_1AR and β_2AR subtypes have been defined based on physiological and pharmacological studies, and individual cDNA clones

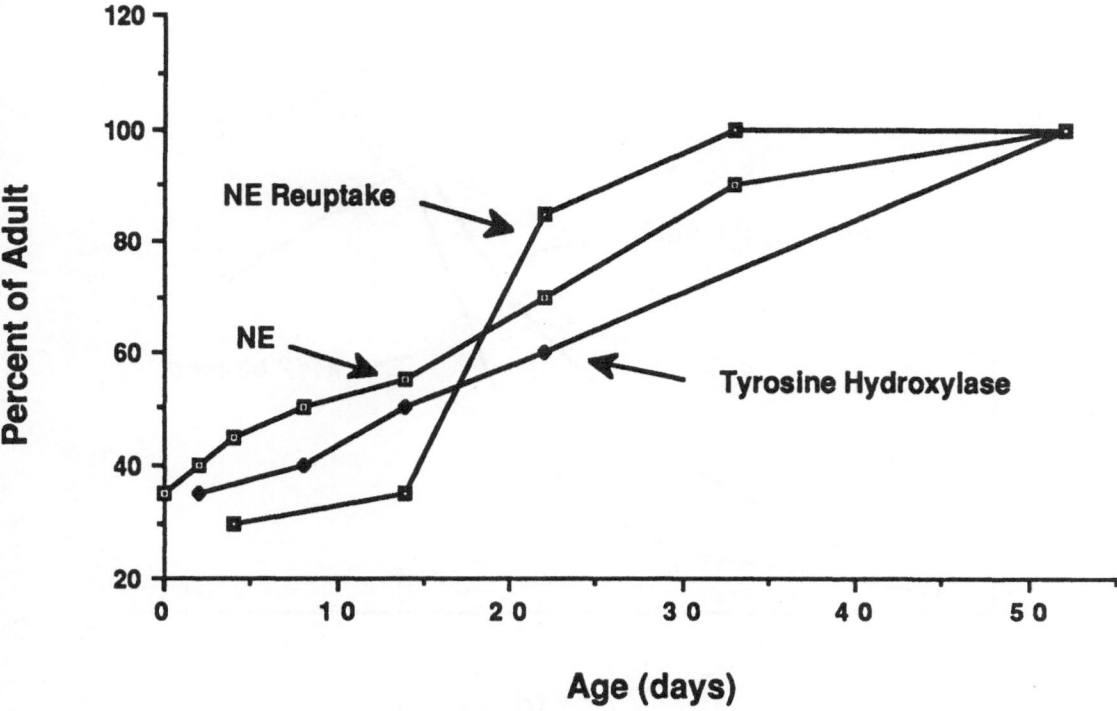

Fig. 1.1. Development of NE, tyrosine hydroxylase and NE re-uptake in rat brain. Levels of endogenous NE, tyrosine hydroxylase enzyme activity, and uptake of [³H]NE were determined at different times after birth in lateral cerebral cortex. The results are expressed as percentages of adult values and are derived from Johnston and Coyle (1980).

have been isolated for each receptor subtype (Dixon *et al.*, 1986; Frielle *et al.*, 1987). The predicted amino acid sequence for each receptor contains seven putative hydrophobic transmembrane domains and the human β_1AR and β_2AR are 54% homologous at the amino acid level. Both βAR subtypes are positively coupled to adenylate cyclase via the stimulatory G protein, $G_{s\alpha}$, and activation of the cyclic AMP system represents the primary second messenger pathway through which βARs regulate cellular function.

A number of studies have examined the development of βAR in fetal and postnatal rat brain. Many of these studies utilized the non-selective ligands [¹²⁵I]-labeled hydroxyben-zylpindolol ([¹²⁵I]HYP) and [³H]dihydroalprenolol (DHA) which, under the assay conditions used, label both β_1AR and β_2AR subtypes. In fetal forebrain [³H]DHA binding was observed as early as ED16 (Bruinink and Lichtensteiger, 1984). On PD1 levels of [¹²⁵I]HYP binding in cerebral cortex were reported to be approximately 10–20% of adult levels (Fig. 1.2), increased slightly during the first week, and then increased four- to five-fold to adult levels by PD16–PD21 (Harden *et al.*, 1977). The increase was shown to be due to an increase in the total number of binding sites (B_{max}) with no change in receptor affinity (K_d). Other studies have reported a similar pattern for the development of total βAR binding but with a time course which was either slightly faster (Keshles and Levitski, 1984) or slower (Waddington and Banks, 1981; Bruinink and Lichtensteiger, 1984). Levels of NE, re-uptake

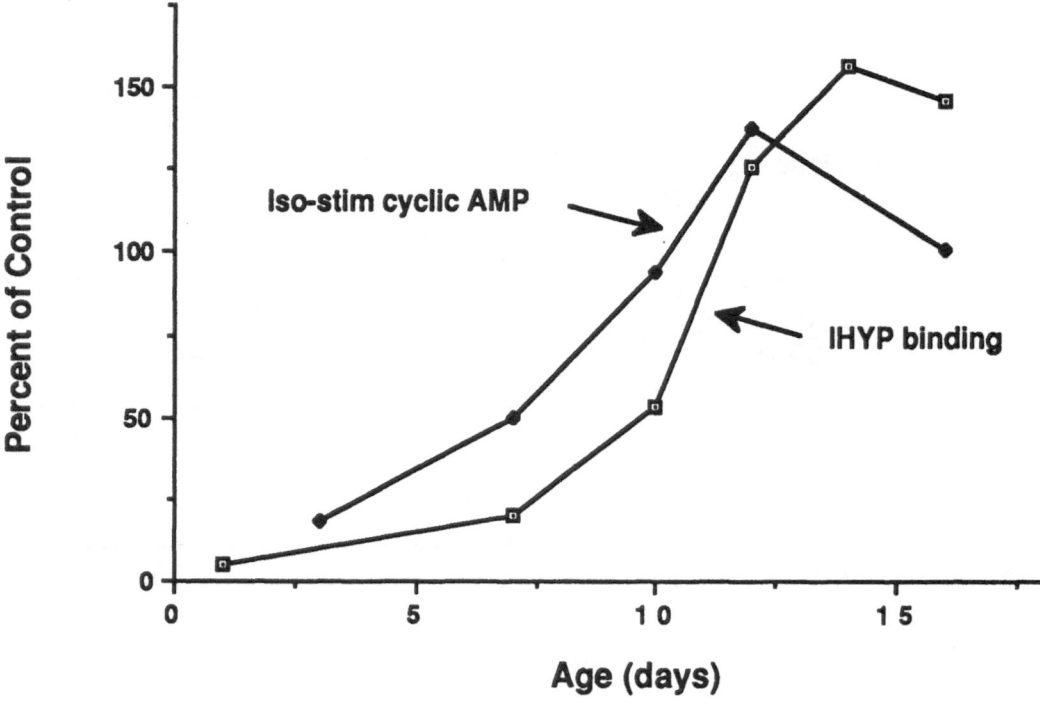

Fig. 1.2. Development of [^{125}I]HYP binding and isoproterenol-stimulated cyclic AMP in rat cerebral cortex. Levels of [^{125}I] HYP binding were determined in homogenates of cerebral cortex, and isoproterenol-stimulated cyclic AMP accumulation (Iso-stim cyclic AMP) was determined in slices of cerebral cortex. Results are expressed as percentages of control values which were either adult or P16 levels of binding or cyclic AMP, respectively. The data are derived from Harden *et al.* (1977).

sites, and catecholamine synthetic enzymes develop more slowly and do not reach adult levels until 5–6 weeks after birth. Because βAR reach adult levels much sooner than does NE, it is believed that the developmental expression of βAR is independent of expression of neurotransmitter. The latter point is supported by a later study demonstrating the presence of βAR in rats which had been treated at birth with 6-hydroxydopamine (6-OHDA), a neurotoxin which destroys catecholamine terminals (Minneman *et al.*, 1979). However, levels of NE are not completely eliminated by this treatment and it is conceivable that exposure to NE during gestation is sufficient to stimulate expression and continued development of βAR.

The development of β_1AR and β_2AR subtypes has also been measured by utilizing subtype-selective antagonists. Pittman *et al.* (1980) examined the development of β_1AR and β_2AR in cerebral cortex and cerebellum, two brain regions which express the subtypes at measurably different levels in the adult. In adult cerebral cortex approximately 65% of βAR ligand binding is to β_1ARs whereas in adult cerebellum about 80% of βAR ligand binding is to β_2ARs. In both cerebral cortex and cerebellum Pittman *et al.* (1980) found that levels of β_1AR increased four- to fivefold by PD21, a time course similar to that reported previously for total βAR binding. In contrast, the development of β_2AR followed a much slower pattern of development in both brain regions, reaching adult levels by PD28 in cerebral cortex and by PD42 in cerebellum. Results of a recent study on the development of β_1AR and β_2AR in rat forebrain support these findings (Erdtsieck-Ernste *et al.*, 1991). In this study β_1AR and β_2AR ligand binding are also reported to be present in forebrain as early as ED13 and increase severalfold before birth. This study also demonstrates that the amount of β_2AR in forebrain is greater than 50% of total βAR ligand binding before PD4 (including gestation) but less than 30% of total βAR ligand binding after PD4.

The slower development of β_2AR binding in brain together with the reports that levels of β_2AR are not influenced by treatments which alter synaptic levels of NE (i.e. 6-OHDA or desipramine treatment) lead to the hypothesis that β_2ARs are located on glia which do not have direct noradrenergic synaptic connections (Minneman *et al.*, 1979). However, autoradiographic studies have demonstrated that β_2AR ligand binding is regulated by 6-OHDA and desipramine treatments in certain brain regions (Ordway *et al.*, 1988; Johnson *et al.*, 1989). These findings suggest that β_2ARs may be located on either neurons or glia. In contrast, levels of β_1AR ligand binding are clearly regulated by treatments which alter synaptic levels of catecholamine and therefore are thought to be located on neurons receiving direct noradrenergic synaptic input.

In addition to the use of binding assays, another approach for studying the development of βAR subtypes is to measure levels of β_1AR and β_2AR mRNA by either hybridization blot analysis (Northern blot) or solution hybridization. The advantage of this approach is that the DNA and RNA probes used for these studies are extremely specific and under stringent conditions hybridize only with mRNA of the appropriate receptor subtype. We have used hybridization blot analysis to study the development of β_1AR and β_2AR mRNA (Duman *et al.*, 1989). For these studies an enriched mRNA fraction was used since β_1AR and β_2AR mRNAs are expressed at very low levels in brain as well as in other tissues. Enrichment was achieved by subjecting total RNA to oligo(dT) selection which results in the isolation of polyadenylated RNA (most but not all mRNA is polyadenylated). The mRNA was fractionated on an agarose gel and transferred to nitrocellulose. Human β_1AR and hamster β_2AR cDNA clones (provided by Dr Lefkowitz, Duke University) were labeled with ^{32}P by nick translation and hybridized with the nitrocellulose filters containing mRNA. At birth, β_1AR and β_2AR mRNA were present at 20% and 50% of adult levels respectively, and both then gradually increased to adult levels by PD25. This time course was parallel to the development of β_1AR and β_2AR ligand binding determined in the same study and is approximately the same as reported by others. These results indicate that development of both βAR subtypes is dependent on activation of gene transcription, expression of mRNA and the resultant *de novo* receptor synthesis.

A number of studies have examined the development of βAR-stimulated cyclic AMP formation as a measure of the ontogeny of functional βAR. βAR-stimulated cyclic AMP formation in brain slices can be observed as early as ED17 (Walton *et al.*, 1979). After birth the development of βAR-stimulated cyclic AMP formation in slices of cerebral cortex (Fig. 1.2) corresponds to the expression of βAR binding sites (Perkins and Moore, 1973; Harden *et al.*, 1977). These studies also examined both the development of cyclic AMP accumulation in brain slices and the ontogeny of levels of adenylate cyclase activity in brain homogenates. Basal levels of cyclic AMP formation and adenylate cyclase activity on PD1 are present at approximately 30–40% of adult levels and reach adult levels by PD21. Stimulation of adenylate cyclase activity by Mn^{2+}, which directly activates the catalytic unit, follows a similar time course of development (Keshles and Levitski, 1984). Stimulation of adenylate cyclase by fluoride or Gpp(NH)p, which activate $G_{s\alpha}$, also follows a similar time course. These results indicate that the level of functional $G_{s\alpha}$ and adenylate cyclase present at birth are sufficient to support βAR-stimulated cyclic AMP formation and therefore suggest that the development of functional βAR is dependent on the expression of the receptors and not on the development of the effector system.

1.4   DEVELOPMENT OF α_1AR IN THE NERVOUS SYSTEM

α_1ARs appear to mediate the actions of NE on locomotor activity and behavioral excitation (Bylund and U'Prichard, 1983; Berridge and Dunn, 1989). α_{1A} and α_{1B} subtypes have been defined on the basis of physiological and pharmacological studies (Morrow and Creese, 1986; Minneman et al., 1988), and molecular cloning studies have confirmed these subtypes by isolation of α_{1A} and α_{1B} cDNA clones (Cotecchia et al., 1988; Lomasney et al., 1991). Like the other adrenergic receptor subtypes α_{1A} and α_{1B} receptor proteins have the characteristic seven transmembrane spanning domains.

Both subtypes appear to be localized on postsynaptic elements in brain and display a regional heterogeneity in the nervous system (Table 1.1; Cotecchia et al., 1988; Wilson and Minneman, 1989; Lomasney et al., 1991).

Activation of α_1ARs stimulates phospholipase C and the phosphatidylinositol second messenger pathway which has two components: (1) formation of inositol triphosphate which stimulates the release of intracellular calcium (Berridge and Irvine, 1989), and (2) formation of diacylglycerol which together with calcium stimulates protein kinase C (Nishizuka, 1988). In addition, α_1ARs appear indirectly to regulate the cyclic AMP system in brain by enhancing the response to other stimulatory ligands (Duman et al., 1986). Finally, α_1ARs have also been reported to regulate voltage-dependent calcium channels independent of an effect on phospholipase C (Han et al., 1987). The α_{1B} subtype has been proposed to couple to phospholipase C whereas the α_{1A} may be coupled to Ca^{2+} influx. G_q appears to mediate receptor stimulation of the phosphatidylinositol pathway, although the specific G protein subtypes which mediate the effects of α_1ARs on this second messenger

Table 1.1.   Adrenergic receptor subtypes

Receptor subtype	G protein	Effector system	Regional distribution
α_1 subtypes			
α_{1A}	G_i/G_o	Ca^{2+} influx	CTX>HP>BSM>CB
α_{1B}	G_q	IP_3/DG ↑	CTX>BSM>>>CB>>HP
α_{1C}	G_q	IP_3/DG ↑	Bovine tissues only
α_2 subtypes			
$\alpha_{2A}(C10)^a$	G_i/G_o	Cyclic AMP ↓ K^+ channel ↑	CTX=BSM>MID>NS>CB>HP
$\alpha_{2B}(C2)$	G_i/G_o	Cyclic AMP ↓	Very low levels in brain
$\alpha_{2C}(C4)$	G_i/G_o	Cyclic AMP ↓	NS>CTX>CB>HP>MID>BSM
β subtypes			
β_1	G_S	Cyclic AMP ↑	CTX>HP>NS>CB
β_2	G_S	Cyclic AMP ↑	CB>NS>CTX>HP
β_3	G_S	Cyclic AMP ↑	Fat tissue

[a] C2, C4 and C10 refer to the cDNA clones which appear to encode for each receptor subtype; these cDNA clones or their rat homologues were also used for studies of the regional distribution of mRNA. Brain regions examined include: CTX, cerebral cortex; HP, hippocampus; BSM, brainstem; CB, cerebellum; MID, midbrain; NS, neostriatum.

system and on ion channels have not been identified.

The development of α_1AR ligand binding and NE-stimulated inositol phosphate accumulation have been examined in rat brain. Using [³H]prazosin, a selective α_1AR ligand, the levels of α_1AR ligand binding in rat whole brain (minus cerebellum) are reported to be approximately 10–20% of adult levels during the first week after birth; by PD14 the levels of ligand binding are 40–50% of adult levels and then increase to adult levels by approximately PD21 (Morris *et al.*, 1980). Similar results have been observed using [³H]WB4101 as a ligand although some regional differences were noted; in hippocampus, frontal cortex, and hypothalamus levels of [³H]WB4101 ligand binding do not reach adult levels until PD30 (Hartley

and Seeman, 1983; Bylund and U'Prichard, 1983).

The appearance of α_1AR ligand binding appears to correlate qualitatively with the development of NE-stimulated inositol phosphate formation in rat cerebral cortex in that there is a general increase to adult levels in both measures during the first 3 weeks of postnatal development (Fig. 1.3). However, there are some differences (Fig. 1.3). Between PD1 and PD14 NE-stimulated inositol phosphate formation in cerebral cortex is approximately 20% of adult levels (whereas ligand binding has reached 40–50%) and then increases to adult levels by PD21–PD24 (Schoepp and Rutledge, 1985). Similar results were observed in hippocampal slices in a separate study (Nicolleti *et al.*, 1986). One other study also

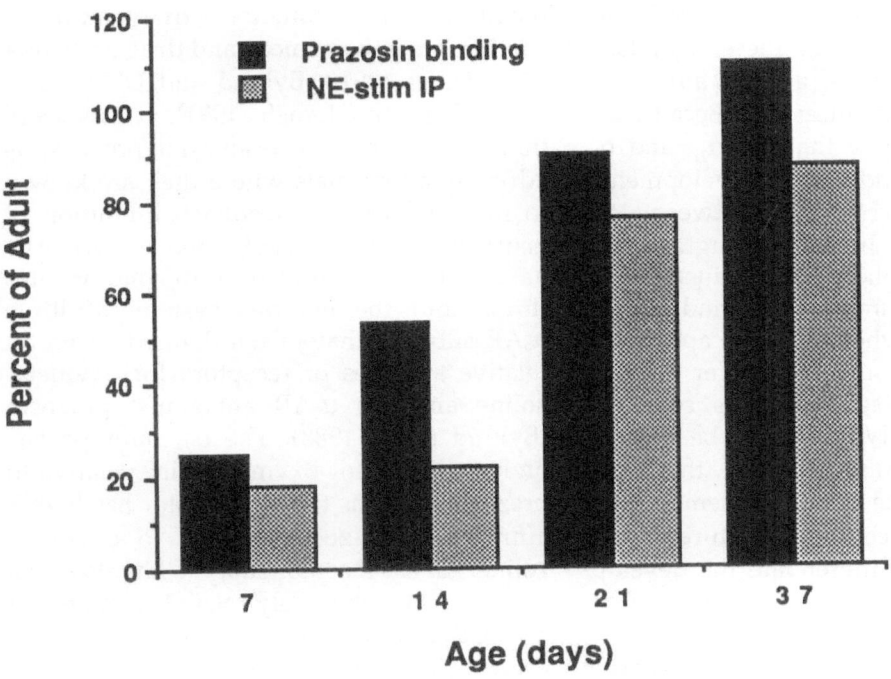

Fig. 1.3. Development of [³H]prazosin binding and NE-stimulated inositol phosphate production in rat cerebral cortex. Levels of [³H]prazosin ligand binding were determined in homogenates of cerebral cortex; K_d = 0.15 nM; B_{max} = 125 fmol/mg protein. NE-stimulated inositol phosphate accumulation was determined in slices of cerebral cortex. Results are expressed as percentage of adult values. The data are derived from Schoepp and Rutledge (1985).

reported a complete lack of correlation between levels of ligand binding and NE-stimulated inositol phosphate accumulation (Rooney and Nahorski, 1987). These discrepancies cannot be explained by the time course for development of the effector enzyme as basal and glutamate- or acetylcholine-stimulation of inositol phosphate accumulation are expressed at adult levels or higher at birth. The development of the α subunit of G_q has not been examined.

Taken together the results suggest that the ontogeny of α_1AR coupling to and stimulation of inositol phosphate formation lags behind the developmental expression of the receptors themselves. However, it should be noted that the ligand binding studies used to measure the receptor levels were carried out either with [^{3}H]prazosin, which does not distinguish between α_{1A} and the α_{1B}, or [^{3}H]WB4101, which only has a 10–20-fold selectivity for these subtypes. By using these ligands, a combination of both α_{1A} and α_{1B} adrenergic receptor subtypes was measured. Because it is conceivable and likely that the α_{1A} and α_{1B} subtypes display individual developmental profiles, the discrepancy between the expression of α_1AR ligand binding and NE-stimulated inositol phosphate production may be due to the measurement of ligand binding to both subtypes when only α_{1B} appears to stimulate this second messenger system. Under the appropriate conditions, some ligands, such as [^{3}H]WB4101, can be used to distinguish one subtype from the other (Morrow and Creese, 1986; Minneman *et al.*, 1988). Using these conditions future studies should be able to differentiate the developmental time courses for the expression of α_{1A} and α_{1B} ligand binding sites.

As with the βAR subtypes, an alternative approach to studying the development of α_1AR subtypes is to examine mRNA levels using subtype-specific cDNA clones. Using an α_{1B} cDNA probe, McCune and Voigt (1991) examined the developmental expression of α_{1B} mRNA. Levels of α_{1B} mRNA increase severalfold during the first week of postnatal development and reach adult levels by about PD21 (McCune and Voigt, 1991). This time course corresponds to that of total α_1AR ligand binding. The development of α_{1A} mRNA in brain remains to be examined. By combining studies of α_1AR subtype mRNA levels with binding assays that use more selective ligands, the time course for the developmental expression of the α_{1A} and α_{1B} subtypes may be revealed. Such studies should elucidate the apparent discrepancies between α_1AR ligand binding and NE-stimulation of the phosphatidylinositol system in brain.

1.5   DEVELOPMENT OF α_2AR IN THE NERVOUS SYSTEM

The α_2ARs influence a variety of physiological functions including regulation of blood pressure, locomotor activity, antinociception, anxiety, memory and drug addiction (Aghajanian, 1978; Bylund and U'Prichard, 1983; Zeng and Lynch, 1991). α_2ARs were initially thought to be localized to presynaptic adrenergic terminals where they are known to mediate negative feedback inhibition of neurotransmitter release. Subsequently, α_2ARs have also been found on postsynaptic sites throughout the nervous system. Multiple α_2AR subtypes have been defined based on relative affinities of receptors for oxymetazoline and the α_1AR antagonist prazosin (Bylund *et al.*, 1988). The α_{2A} subtype has higher affinity for oxymetazoline relative to prazosin whereas the α_{2B} receptor has higher affinity for prazosin relative to oxymetazoline. An α_{2C} receptor subtype has also been defined which is closely related to α_{2B} based on relative affinities for these two compounds (Murphy and Bylund, 1988).

Molecular cloning studies have confirmed in part this subdivision of α_2AR subtypes (Table 1.1). An α_{2A} subtype has been isolated and is also referred to as C10 because it is located on chromosome number 10 (Kobilka *et al.*, 1987). Additional α_2AR subtypes have been cloned but their relationship to pharmacologically defined receptor subtypes is not clear. One subtype referred to as α_2-C4 (located on chromosome number 4) was isolated from human kidney and appears to have ligand binding characteristics most similar to that of an α_{2B} receptor (Regan *et al.*, 1988). However, based on its regional distribution in peripheral tissues (it is not present in neonatal rat lung) and that it is glycosylated, this receptor appears to be more closely related to the α_{2C} subtype (Lorenz *et al.*, 1990; Zeng and Lynch, 1991). A third subtype was subsequently isolated and is referred to as either RNGα_2 (Zeng *et al.*, 1990) or α_2-C2 (Lomasney *et al.*, 1990). This receptor also has ligand binding properties characteristic of an α_{2B} subtype, and this assignment is supported by its presence in neonatal rat lung and by the fact that this receptor is not glycosylated.

Activation of α_2ARs has been shown to regulate cellular function in a number of ways depending on the neuronal system (Table 1.1). Activation of α_{2A} receptors inhibits the adenylate cyclase second messenger system in brain (Duman and Enna, 1986). In addition, the indirect effect of αAR activation to enhance the neurotransmitter receptor-stimulation of cyclic AMP in brain appears to be mediated in part by an α_2AR (Duman *et al.*, 1986). Finally, stimulation of α_2ARs activates K$^+$ channels in locus coeruleus neurons (Williams *et al.*, 1985) and inhibits Ca^{2+} channels (Bean, 1989). In most cases α_2ARs couple with pertussis toxin-sensitive G proteins (G$_i$/G$_o$) to influence both ion channels and adenylate cyclase.

A number of studies have described the developmental expression of α_2AR binding but in all cases [^{3}H]clonidine was used. [^{3}H]clonidine is a non-selective ligand for α_2AR subtypes and also labels nonadrenergic imidazoline binding sites (for discussion of imidazoline binding sites see Michel *et al.*, 1988 and Michel and Insel, 1989). In addition, clonidine is an agonist and therefore its affinity is influenced by the interaction of the receptor with G proteins. A more selective ligand would be [^{3}H]rauwolscine which only labels α_2AR and displays some selectivity for the subtypes.

The multiple limitations of [^{3}H]clonidine ligand binding complicate the interpretation of developmental studies using this ligand. However, with these problems in mind it is useful to examine the results of these studies. In whole rat brain minus cerebellum, levels of [^{3}H]clonidine binding at birth are approximately 35% of adult levels, increase to about 50% of the adult level by PD10, and reach adult levels by PD20 (Morris *et al.*, 1980). This study reports that [^{3}H]clonidine labels a single class of binding sites with a K_d of 2.3 nM and that the developmental increase reflects an increase in the total number of binding sites. The same study reports that levels of NE do not reach adult levels until 5–6 weeks after birth. The results indicate that the expression of [^{3}H]clonidine binding sites precedes the expression of adult levels of neurotransmitter, a trend that has also been observed for the other adrenergic receptor subtypes.

Subsequent studies reported a similar pattern of development for [^{3}H]clonidine binding in rat cerebral cortex (Nomura *et al.*, 1982) and more specifically in mesolimbic and hypothalamic regions (Hartley and Seeman, 1983). In hippocampus, levels of [^{3}H]clonidine binding increase to adult levels by PD10, and in frontal cortex, adult levels of ligand binding are reached 5–6 weeks after birth (Fig. 1.4). The latter study also reported that [^{3}H] clonidine labels a single class of sites with a K_d in the nanomolar range (1.5 nM). However, Nomura *et al.* (1982) reported that in neonates [^{3}H]clonidine binding displays a subnanomolar K_d (0.27 nM) whereas on PD7 and in adults the K_d is 1.6 nM.

In subsequent studies Nomura *et al.* (1984) investigated the nature of the high and low affinity [^{3}H]clonidine binding sites observed at different times of development. The presence of a single class of low affinity site in neonatal rat cortex was confirmed, but after the first week of development a low as well as the high

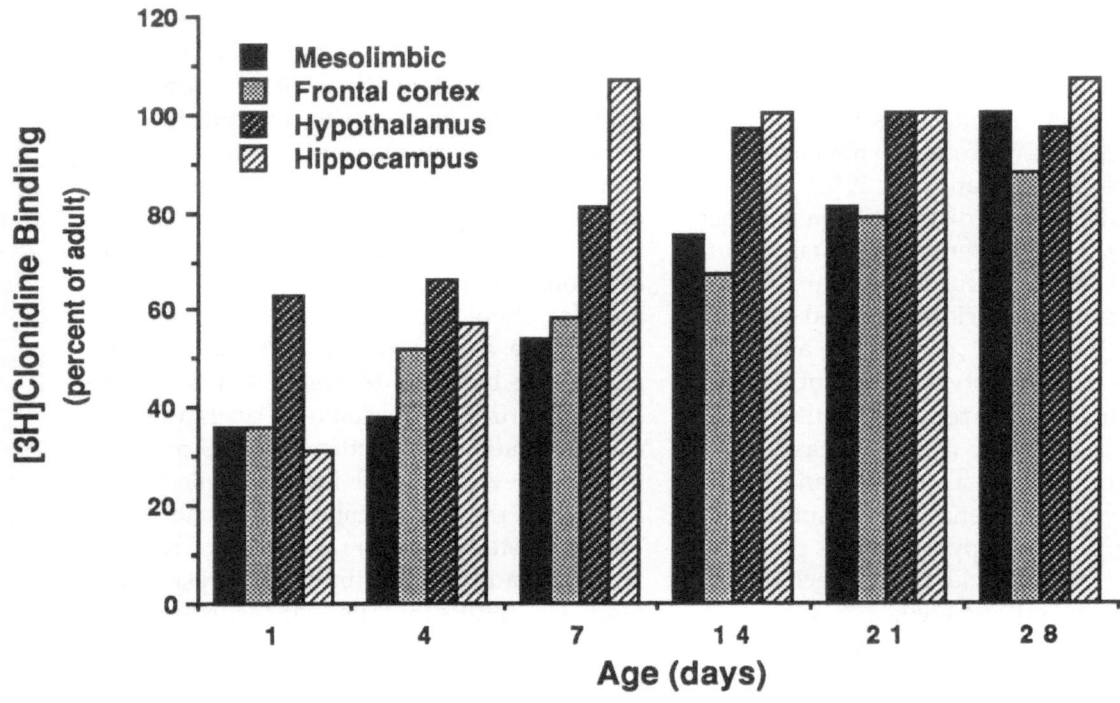

Fig. 1.4.   Development of [³H]clonidine binding in different rat brain regions. The K_d was approximately 1.5 nM in all brain regions. The B_{max} for each brain region was (in fmol/mg protein): mesolimbic, 160; frontal cortex, 120; hypothalamus, 115; hippocampus, 75. The results are expressed as percentages of adult values and are derived from Hartley and Seamon (1983).

affinity [³H]clonidine binding site (0.7 and 7.0 nM, respectively) was seen. Both sites are present at approximately 60% of adult levels on PD7 and reach adult levels by PD30. In a later study (Kitamura *et al.*, 1989) it was shown that in the presence of GTP, which converts the majority of receptors into low affinity binding sites, there is little or no change in levels of [³H]clonidine binding during development. In contrast, in the presence of Mn^{2+}, which increases the level of high affinity binding, there is an increase in the expression of [³H] clonidine binding during development. From these results the authors concluded that expression of the high affinity ligand–receptor–G protein ternary complex is the rate limiting step in the development of functional clonidine binding sites. This pattern of expression may be related to the expression of G protein subunits which couple with clonidine binding sites although the G protein subunits have been reported to be present at 35–50% of adult levels at birth (section 1.6).

One study has examined the development of functional clonidine binding sites by measuring the expression of K⁺-induced [³H]NE release from cerebral cortical slices and inhibition of neurotransmitter release by clonidine (Nomura *et al.*, 1982). High K⁺ induced release of [³H]NE is observed on ED18 and increases to near adult levels by PD7. Clonidine inhibition of neurotransmitter release is not observed on PD1 but by PD7 clonidine inhibition of [³H]NE release is observed and is approximately 70% of that seen in adult brain slices. These results indicate that clonidine receptor binding sites are functionally expressed at this time during development which corresponds to the appearance of low affinity [³H]clonidine binding sites. Moreover, these results also demonstrate the developmental expression of the presynaptic α_2AR which regulates the release of NE.

While measurement of α_2AR subtypes by ligand binding poses some difficulty due to the lack of selective ligands for these receptors, expression of receptor mRNA, specifically α_2-C10 and α_2-C4 receptor mRNA, has been examined using specific cDNA probes (McCune and Voigt, 1991). The developmental time course of α_2-C4 and α_2-C10 has been confirmed and extended in our laboratory (Fig. 1.5). Levels of α_2-C4 are present, but are very low on ED18, increase several fold by PD8, and increase to adult levels by PD16. These results are similar to the time course for development of [³H]clonidine ligand binding described above. In contrast, levels of α_2-C10

are found to be highest at birth and then gradually decrease to adult levels by PD16. The high levels of α_2-C10 mRNA suggest that levels of α_{2A} receptor protein and ligand binding are present at higher levels at birth than in adults. However, it is possible that turnover of mRNA and/or receptor protein is higher in neonates and that levels of functional receptor protein are at low levels. Analysis of α_{2A} ligand binding or immunoblot analysis of α_{2A} receptor protein will be required to determine the nature of this discrepancy.

1.6 DEVELOPMENT OF G PROTEINS AND EFFECTOR SYSTEMS IN THE NERVOUS SYSTEM

The regulation of cellular function by adrenergic receptors and other members of this receptor family occurs through coupling of

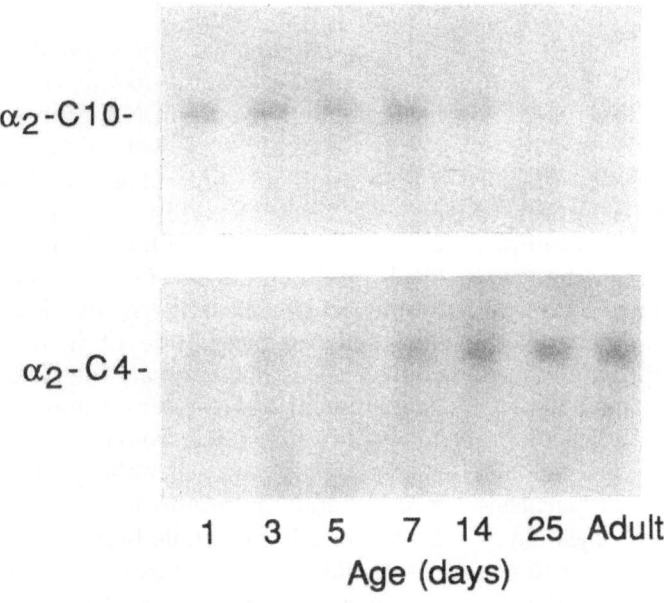

Fig. 1.5. Development of α_2-C4 and α_2-C10 mRNA in rat whole brain. Total RNA was isolated from rat whole brain at different times of development and subjected to hybridization blot analysis. The RNA (20 µg) was fractionated on an agarose gel, transferred to nitrocellulose, and then hybridized with ³²P-labeled α_2-C4 or α_2-C10 cDNA (probes were kindly provided by S. Lanier). The filters were then exposed for 3 days without an intensifying screen.

receptors to transmembrane signalling proteins known as GTP binding proteins or G proteins (Stryer and Bourne, 1986; Gilman, 1987; Simon *et al.*, 1991). The interaction of ligand bound receptors with G proteins results in the hydrolysis of GTP and then coupling of G proteins with effector systems. The effector systems regulated by receptor-coupled G proteins include second messenger enzymes such as adenylate cyclase, phospholipase C and phospholipase A_2, and K^+ and Ca^{2+} channels (Table 1.1). Activation of these second messengers results in regulation of intracellular pathways, such as protein kinases, which have acute or immediate effects on cellular function as well as long-term effects including regulation of gene transcription. Developmental expression of receptors which are functional depends on the expression of the G proteins and effector systems which couple these receptors to intracellular second messenger pathways. For this reason it is useful to review studies of the development of specific G proteins and effector systems.

G proteins are heterotrimers composed of α, β and γ subunits (Stryer and Bourne, 1986; Gilman, 1987; Simon *et al.*, 1991). Specific α subunits appear to exist for each effector system whereas the $\beta\gamma$ subunits appear to subserve a number of α subunits. For example, receptor stimulation and inhibition of adenylate cyclase is mediated by $G_{S\alpha}$ and $G_{i\alpha}$, respectively, while regulation of phospholipase C is mediated by $G_{o\alpha}$. Receptor regulation of ion channels is also thought to be mediated by subtypes of $G_{i\alpha}$ and/or $G_{o\alpha}$. Studies of the structure and function of G proteins have been aided by the use of cholera and pertussis toxins which catalyze the ADP-ribosylation of the α subunits of G_s and G_i/G_o, respectively. Molecular cloning studies have identified at least 16 G protein α subunits, at least three β subunits, and four types of γ subunits, all encoded by separate genes (Simon *et al.*, 1991). However, with the exception of $G_{S\alpha}$, the receptor/effector system regulated by each of these subtypes has not yet been determined.

The development of G protein subunits in brain has been examined using a variety of approaches. Levels of $G_{o\alpha}$ and G_{β} immunoreactivity in cerebral cortex were reported to increase approximately two-fold between ED16 and birth, at which time levels of these G proteins subunits were equal to the levels observed in adults (Kitamura *et al.*, 1989). A somewhat different time course was reported in forebrain (Milligan *et al.*, 1987); levels of $G_{o\alpha}$, G_{β} and $G_{i\alpha}$ immunoreactivity were reported to be present at 25, 35 and 50%, respectively, of adult levels and then increased to adult levels by PD30. A third study reported a similar developmental pattern for $G_{o\alpha}$ immunoreactivity in whole brain (Chang *et al.*, 1988). Levels of pertussis toxin-catalyzed ADP-ribosylation are reported to be approximately 35% of adult levels at birth and increase to adult levels by PD16, a somewhat slower pattern of development compared with levels of $G_{i\alpha}$ and $G_{o\alpha}$ immunoreactivity.

The developmental expression of mRNA for these G protein subunits has also been studied using subunit specific cDNA probes (Fig. 1.6, Duman *et al.*, 1989). Levels of $G_{o\alpha}$ and G_{β} in whole brain are present at or above adult levels at birth, increase to 150–200% of adult levels by PD3–PD7, and then decrease to adult levels by PD14. Levels of $G_{i1\alpha}$ and $G_{i2\alpha}$ mRNA follow a similar pattern of development although the magnitude of increase above adult levels is not as great. The differences in development of G protein immunoreactivity and mRNA may result from the different brain regions being examined, although levels of $G_{o\alpha}$ mRNA and immunoreactivity were both determined in rat whole brain. Alternatively, the stability or rate of turnover of mRNA and protein may change during development. One other possibility is that the antibodies used recognize more than one G protein subtype whereas the cDNA probes are relatively more selective. Thus, although sufficient levels of some G protein subtypes are present during early stages it is possible that some of the

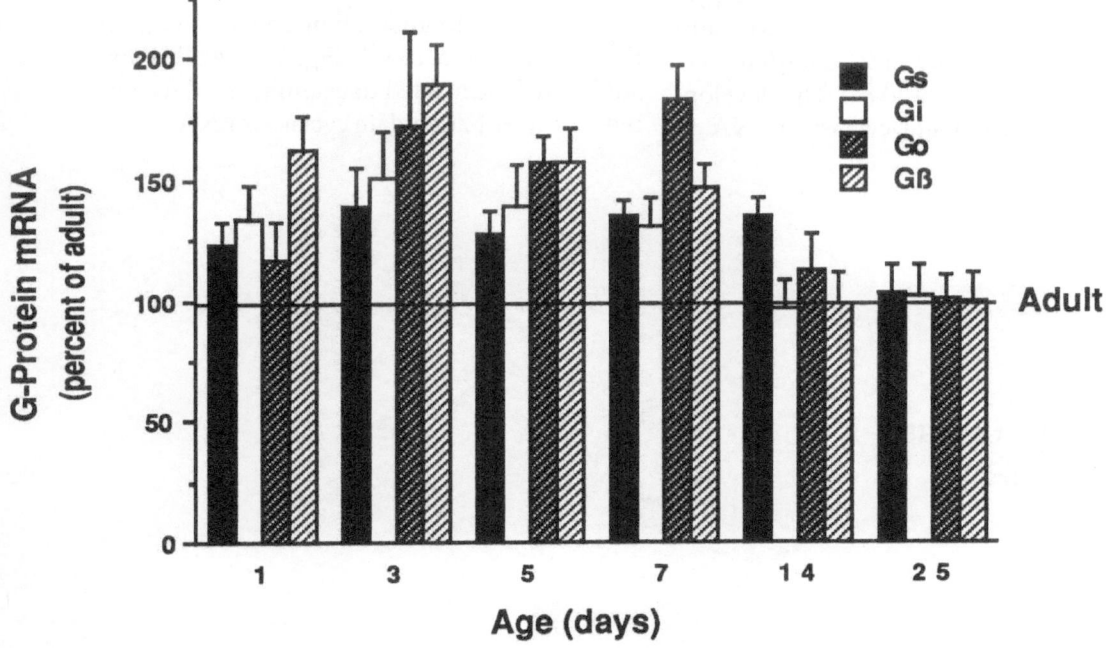

Fig. 1.6. Development of G protein subunit mRNA in rat brain. Total RNA was isolated from whole brain at different times of development and submitted to hybridization blot analysis for levels of $G_{s\alpha}$, $G_{i2\alpha}$, $G_{o\alpha}$, and G_β. The results are expressed as percentages of adult values and are derived from Duman *et al.* (1989).

many other subtypes are not expressed until later periods of development.

The measurement of guanine nucleotide-stimulated adenylate cyclase has been used as a measure of the development of functional $G_{s\alpha}$. These studies have demonstrated that levels of Gpp(NH)p-stimulated adenylate cyclase are approximately 35% of adult levels at birth and then gradually increase to adult levels by PD21 (section 1.3). The development of $G_{s\alpha}$ has been further studied by measurement of cholera toxin catalyzed ADP-ribosylation of $G_{s\alpha}$ in homogenates of cerebral cortex and by measuring levels of mRNA for $G_{s\alpha}$. Levels of cholera toxin-stimulated ADP-ribosylation of $G_{s\alpha}$ at birth are reported to be present at adult levels, increase about two-fold by day 12 and then gradually decrease to adult levels (Kitamura *et al.*, 1989). Similarly, levels of $G_{s\alpha}$ mRNA in whole brain are found

to be expressed at adult levels on PD1, increase slightly, and then return to adult levels by PD21 (Fig. 1.6, Duman *et al.*, 1989). These results indicate that adult levels of $G_{s\alpha}$ are present at birth and support the conclusion that development of functional βAR is dependent on expression of receptors.

Although little information is available on the development of most of the effector systems, the development of adenylate cyclase has been studied. At least two forms of the enzyme have been characterized in brain, calmodulin-sensitive (type 1) and calmodulin-insensitive (type 2) adenylate cyclase, and cDNA clones for each have been identified (Krupinski *et al.*, 1989; Feinstein *et al.*, 1991). A number of studies have examined the development of adenylate cyclase and under the conditions used, the level of enzyme activity measured would have been a combination of both enzyme types. Basal and Mn²⁺-stimulated enzyme activity are reported to be present at 30–40% of adult levels at birth and increase to adult levels by PD21 (Harden *et al.*, 1977; Keshles and Levitski, 1984). The development of type 2 adenylate cyclase mRNA in whole brain has been studied in our laboratory. Expression of levels of type 2 adenylate cyclase mRNA follows a time course similar to that of enzyme activity (Fig. 1.7). Additional studies will be needed to examine the development of type 1 adenylate cyclase mRNA.

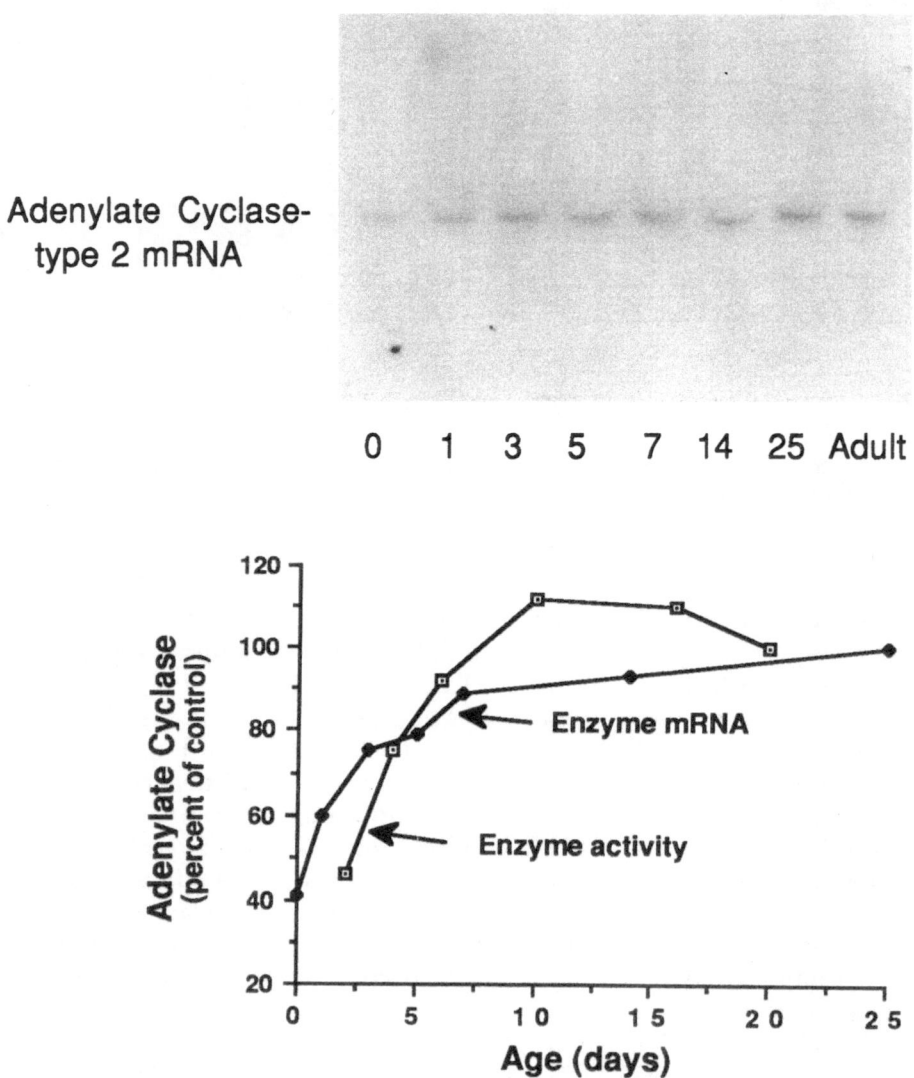

Fig. 1.7. Development of adenylate cyclase enzyme activity and type 2 mRNA in rat brain. Levels of Mn²⁺ plus forskolin-stimulated adenylate cyclase activity were determined in homogenates of cerebral cortex and are expressed as percentages of levels at PD21 (derived from Keshles and Levitski, 1984). Levels of adenylate cyclase type 2 mRNA were determined in rat whole brain and are expressed as percentages of adult levels. A 20 μg sample of total RNA was subjected to hybridization blot analysis using ³²P-labeled adenylate cyclase type 2 cDNA (kindly provided by R. Reed, Johns Hopkins).

Although the development of receptor-stimulated inositol phosphate accumulation in brain has been examined (section 1.4), the ontogeny of phospholipase C activity and mRNA has not been studied. Similarly, the expression of other phospholipases as well as K^+ and Ca^{2+} channels have not been examined. In addition, it is conceivable that adrenergic receptors as well as other neurotransmitter receptors regulate additional, as yet unidentified, effectors, some of which are expressed only during early stages of development. Examination of these possibilities must await future studies.

1.7 ADRENERGIC RECEPTORS AND IMPLICATIONS FOR DEVELOPMENT

The early appearance of NE during gestation and the extensive innervation of the NE neurotransmitter system throughout the nervous system have led to the hypothesis that NE may act as a neurotrophic factor. Consistent with this hypothesis are reports that NE enhances the growth of neurons and glia in cerebellar explants (Vernadakis and Gibson, 1974). Destruction of noradrenergic nerve terminals in neonatal rats leads to significant morphological changes in both the cerebral cortex and the cerebellum, and the extent of these changes correlates with the reduction in levels of norepinephrine (Konkol *et al.*, 1978; Felten *et al.*, 1982; Lovell, 1982). Such changes may underlie the influence of neonatal 6-OHDA on locomotor activity (Pappas *et al.*, 1975) and attention deficits (Mason and Iverson, 1979). Additional studies will be needed to test this hypothesis further and to elucidate the role of NE in central nervous system development.

Although these findings suggest that the noradrenergic system may have neurotrophic actions, the mechanisms by which NE could regulate neuronal growth have not been identified. Ornithine decarboxylase (ODC), which has been shown to be obligatory for neuronal development, is under βAR regulation during development (Morris *et al.*, 1983;

Morris and Slotkin, 1985). ODC catalyzes the initial reaction in the synthesis of polyamines which have been shown to play a significant role in the development and maturation of cells. βAR-stimulation of ODC expression appears to be mediated by $β_2AR$ and activation of the cyclic AMP second messenger system. Maximal βAR-stimulated ODC activity correlates with major periods of replication, differentiation and outgrowth during development. Although these results support the hypothesis that ODC activity is critical to neuronal development and maturation, that βAR-stimulated ODC is required for such development is unclear. Additional studies will be required to further elucidate the role of βAR-stimulated ODC in neuronal development.

1.8 SUMMARY AND FUTURE PROSPECTS

The developmental expression of most adrenergic receptor subtypes matures to adult levels by the end of the third week after birth, whereas adult levels of NE are not attained until approximately 6 weeks after birth. This has led investigators to conclude that the expression of adrenergic receptors precedes that of the NE neurotransmitter system. However, it has also been noted that although adult levels of NE, its biosynthetic enzymes, and re-uptake sites do not attain adult levels until a later postnatal time, the adult pattern of noradrenergic innervation is achieved much earlier, by the first to second week after birth. The development of adrenergic receptors appears to correspond more closely with this latter time course. According to this interpretation, the expression of neurotransmitter and receptors may develop in parallel although it is difficult to identify whether NE or adrenergic receptors appears first during development. One possibility is that expression of NE and adrenergic receptors are independent events and that continued expression of receptors is dependent on the presence of neurotransmitter. Alternatively, expression of adrenergic receptors may be turned on earlier so that the cells expressing these receptors are fully functional when NE innervation occurs.

The embryonic expression and further development of both adrenergic receptors and the norepinephrine neurotransmitter system are controlled by regulation of gene expression. The development of brain and other tissues is controlled by turning on and off sets of genes encoding proteins needed for each period of development. Thus, to understand fully the development of adrenergic receptors, the gene transcription factors which control expression of these receptors must be identified. Regulation of β_1AR and β_2AR gene transcription has been examined in cultured cells. The genes for both receptors have been shown to be under the control of thyroid hormone (Bahouth, 1991), and β_2AR gene expression is also regulated by glucocorticoid hormone (Hadcock and Malbon, 1988). It is possible that expression of fetal hormones plays a role in the ontogeny of these receptors. Alternatively, maternal levels of these hormones may regulate fetal gene transcription since these hormones are lipophilic and therefore can penetrate the placental barrier.

In addition, both the β_1AR and β_2AR genes contain a cyclic AMP response element (CRE), and β_2AR gene expression has been shown to be regulated by the cyclic AMP system, presumably through activation of a CRE binding protein or CREB (Collins *et al.*, 1990). The regulation and function of these gene transcription factors and a variety of other classes of transcription factors are being studied in cultured cells although the role of cyclic AMP in the developmental expression of adrenergic receptors has not been examined. In addition, rapid advances are being made in studies of the expression and function of a variety of nerve growth factors which may also participate in the development of the central nervous system. Identification of the transcription factors, growth factors, and other determinants of gene expression will contribute to a more comprehensive understanding of both the development of adrenergic recept-

ors and regulation of these receptors in adults. These studies will also help elucidate the mechanisms which underlie perturbations of adrenergic receptor function that occur during development, some of which could underlie abnormalities of receptor function in adults.

REFERENCES

Aghajanian, G.K. (1978) Tolerance of locus coeruleus neurons to morphine and suppression of withdrawal response by clonidine. *Nature*, **276**, 186–8.

Bahouth, S. (1991) Thyroid hormones transcriptionally regulate the β1-adrenergic receptor gene in cultured ventricular myocytes. *J. Biol. Chem.*, **266**, 15863–9.

Bean, B.P. (1989) Neurotransmitter inhibition of neuronal calcium currents by changes in channel voltage dependence. *Nature*, **340**, 153–6.

Berridge, C.W. and Dunn, A.J. (1989) Restraint-stress-induced changes in exploratory behavior appear to be mediated by norepinephrine-stimulated release of CRF. *J. Neurosci.*, **9**, 3513–21.

Berridge, M.J. and Irvine, R.F. (1989) Inositol phosphates and cell signalling. *Nature*, **341**, 197–205.

Bruinink, A. and Lichtensteiger, W. (1984) β-Adrenergic binding sites in fetal rat brain. *J. Neurochem.*, **43**, 578–81.

Bylund, D.B. and U'Prichard, D.C. (1983) Characterization of α_1- and α_2-adrenergic receptors. *Int. Rev. Neurobiol.*, **24**, 343–431.

Bylund, D.B., Ray-Prenger, C. and Murphy, T.J. (1988) Alpha-2A and alpha-2B adrenergic receptor subtypes: antagonist binding in tissues and cell lines containing only one subtype. *J. Pharmacol. Exp. Ther.*, **245**, 600–7.

Chang, K.-J., Puch, W., Blanchard, S.G. *et al.* (1988) Antibody specific to the α subunit of the guanine nucleotide-binding regulatory protein G_o: developmental appearance and immunocytochemical localization in brain. *Proc. Natl. Acad. Sci. USA*, **85**, 4929–33.

Collins, S., Altschmied, J., Herbsman, O. *et al.* (1990) A cAMP response element in the β_2-adrenergic receptor gene confers transcriptional autoregulation by cAMP. *J. Biol. Chem.*, **265**, 19330–5.

Cotecchia, S., Schwinn, D.A., Randall, R.R. *et al.* (1988) Molecular cloning and expression of the cDNA for the hamster α_1-adrenergic receptor. *Proc. Natl. Acad. Sci. USA*, **85**, 7159–63.

Coyle, J.T. (1977) Biochemical aspects of neurotransmission in the developing brain. *Int. Rev. Neurobiol.*, **20**, 65–103.

Dixon, R.A.F., Kobilka, B.K., Strader, D.J. *et al.* (1986) Cloning of the gene and cDNA for mammalian β-adrenergic receptor and homology with rhodopsin. *Nature*, **321**, 75–9.

Duman, R.S. and Enna, S.J. (1986) A procedure for measuring α_2-adrenergic receptor-mediated inhibition of cyclic AMP accumulation in rat brain slices. *Brain Res.*, **384**, 391–4.

Duman, R.S., Karbon, E.W., Harrington, C. and Enna, S.J. (1986) An examination of the involvement of phospholipases A_2 and C in the α-adrenergic and γ-aminobutyric acid receptor modulation of cyclic AMP accumulation in rat brain slices. *J. Neurochem.*, **47**, 800–10.

Duman, R.S., Saito, N. and Tallman, J.F. (1989) Development of β-adrenergic receptor and G protein messenger RNA in rat brain. *Mol. Brain Res.*, **5**, 289–96.

Erdtsieck-Ernste, B.H.W., Feenstra, M.G.P. and Boer, G.J. (1991) Pre- and postnatal developmental changes of adrenoceptor subtypes in rat brain. *J. Neurochem.*, **57**, 897–903.

Feinstein, P.G., Schrader, K.A., Bakalyar, H.A. *et al.* (1991) Molecular cloning and characterization of a Ca^{2+}/calmodulin insensitive adenylyl cyclase from rat brain. *Proc. Natl. Acad. Sci. USA*, **88**, 10173–7.

Felten, D.L., Hallman, K. and Johsson, G. (1982) Evidence for a neurotrophic role of noradrenaline neurons in the postnatal development of rat cerebral cortex. *J. Neurocytol.*, **11**, 119–35.

Foote, S.L., Bloom, F.E. and Aston-Jones, G. (1983) Nucleus locus ceruleus: new evidence of anatomical and physiological specificity. *Physiol. Rev.*, **63**, 844–914.

Frielle, T., Collins, S., Daniel, K.W. *et al.* (1987) Cloning of the cDNA for the human β_1-adrenergic receptor. *Proc. Natl. Acad. Sci. USA*, **84**, 7920–4.

Gilman, A.G. (1987) G-proteins: transducers of receptor-generated signals. *Annu. Rev. Biochem.*, **56**, 615–49.

Hadcock, J.R. and Malbon, C.C. (1988) Regulation of β-adrenergic receptors by 'permissive' hormones: glucocorticoids increase steady-state levels of receptor mRNA. *Proc. Natl. Acad. Sci. USA*, **85**, 8415–19.

Han, C., Abel, P.W. and Minneman, K.P. (1987) α_1-adrenoceptor subtypes linked to different mechanisms for increasing intracellular Ca^{2+} in smooth muscle. *Nature*, **329**, 333–5.

Harden, T.K., Wolfe, B.B., Sporn, J.R. *et al.* (1977) Ontogeny of β-adrenergic receptors in rat cerebral cortex. *Brain Res.*, **125**, 99–108.

Hartley, E. and Seeman, P. (1983) Development of receptors for dopamine and noradrenaline in rat brain. *Eur. J. Pharmacol.*, **91**, 391–7.

Heninger, G.R. and Charney, D.S. (1987) Mechanism of action of antidepressant treatments: implications for the etiology and treatment of depressive disorders, in *Psychopharmacology, the Third Generation of Progress* (ed. H.Y. Meltzer), Raven Press, New York, pp. 535–44.

Johnson, E.W., Wolfe, B.B. and Molinoff, P.B. (1989) Regulation of subtypes of β-adrenergic receptors in rat brain following treatment with 6-hydroxydopamine. *J. Neurosci.*, **9**, 2297–305.

Johnston, M.V. and Coyle, J.T. (1980) Ontogeny of neurochemical markers for noradrenergic, GABAergic, and cholinergic neurons in neocortex lesioned with methylazoxymethanol acetate. *J. Neurochem.*, **34**, 1429–41.

Keshles, O. and Levitzki, A. (1984) The ontogenesis of β-adrenergic receptors and of adenylate cyclase in the developing rat brain. *Biochem. Pharmacol.*, **33**, 3231–3.

Kitamura, Y., Mochii, M., Kodama, R. *et al.* (1989) Ontogenesis of α_2-adrenoceptor coupling with GTP-binding proteins in the rat telencephalon. *J. Neurochem.*, **53**, 249–57.

Kobilka, B.K., Matsui, H., Kobilka, T.S. *et al.* (1987) Cloning, sequencing, and expression of the gene coding for the human platelet α_2-adrenergic receptor. *Science*, **238**, 650–6.

Konkol, R.J., Bendeich, E.G. and Breese, G.R. (1978) A biochemical and morphological study of the altered growth pattern of central catecholamine neurons following 6-hydroxydopamine. *Brain Res.*, **140**, 125–35.

Krupinski, J., Coussen, F., Bakalyar, H.A. *et al.* (1989) Adenylyl cyclase amino acid sequence: possible channel- or transporter-like structure. *Science*, **244**, 1558–64.

Lomasney, J.W., Lorenz, W., Allen, L.F. *et al.* (1990) Expansion of the α_2-adrenergic receptor family: cloning and characterization of a human α_2-adrenergic receptor subtype, the gene for which is located on chromosome 2. *Proc. Natl. Acad. Sci. USA*, **87**, 5094–8.

Lomasney, J.W., Cotecchia, S., Lorenz, W. *et al.* (1991) Molecular cloning and expression of the cDNA for the α_{1A}-adrenergic receptor. *J. Biol. Chem.* **266**, 6365–9.

Lorenz, W., Lomasney, J.W., Collins, S. *et al.* (1990) Expression of three α_2-adrenergic receptor subtypes in rat tissues: implications for α_2 receptor classification. *Mol. Pharmacol.*, **38**, 599–603.

Lovell, J. (1982) Effects of 6-hydroxydopamine-induced norepinephrine depletion on cerebellar development. *Dev. Neurosci.*, **5**, 359–68.

Mason, S.T. and Iversen, S.D. (1979) Theories of the dorsal bundle extension effect. *Brain Res. Rev.*, **1**, 107–37.

McCune, S.K. and Voigt, M.M. (1991) Regional brain distribution and tissue ontogenic expression of a family of alpha-adrenergic receptor mRNAs in the rat. *J. Mol. Neurosci.*, **3**, 29–37.

Michel, M.C. and Insel, P.A. (1989) Are there multiple imidazoline binding sites? *Trends Pharmacol. Sci.*, **10**, 342–5.

Michel, M.C., Brodde, O.-E., Schnepel, B. *et al.* (1988) [^{3}H]Idazoxan and some other α_2-adrenergic drugs also bind with high affinity to a nonadrenergic site. *Mol. Pharmacol.*, **35**, 324–30.

Milligan, G., Streaty, R.A., Gierschik, P. *et al.* (1987) Development of opiate receptors and GTP-binding regulatory proteins in neonatal rat brain. *J. Biol. Chem.*, **262**, 8626–30.

Minneman, K.P. (1981) Adrenergic receptor molecules, in *Neurotransmitter Receptors*, Part 2 (*Receptors and Recognition*. Series B, Vol. 10) (eds H.I. Yamamura and S.J. Enna), Chapman & Hall, London.

Minneman, K.P., Dibner, M.D., Wolfe, B.B. and Molinoff, P.B. (1979) β_1- and β_2-adrenergic receptors in rat cerebral cortex are independently regulated. *Science*, **204**, 866–8.

Minneman, K.P., Han, C. and Abel, P.W. (1988) Comparison of α_1-adrenergic receptor subtypes distinguished by chlorethylclonidine and WB 4101. *Mol. Pharmacol.*, **33**, 509–14.

Moore, R.Y. and Bloom, R.E. (1979) Central catecholamine neuron systems: anatomy and physiology of the norepinephrine and epinephrine systems. *Annu. Rev. Neurosci.*, **2**, 113–68.

Morris, G. and Slotkin, T.A. (1985) Beta-2 adrenergic control of ornithine decarboyxlase activity in brain regions of the developing rat. *J. Pharmacol. Exp. Ther.*, **233**, 141–7.

Morris, G., Seidler, F.J. and Slotkin, T.A. (1983) Stimulation of ornithine decarboyxlase by histamine or norepinephrine in brain regions of the developing rat: evidence for biogenic amines as trophic agents in neonatal brain development. *Life Sci.*, **32**, 1565–71.

Morris, M.J., Dausse, J.-P., Devynck, M.-A. and Meyer, P. (1980) Ontogeny of α_1 and α_2-adrenoceptors in rat brain. *Brain Res.*, **190**, 268–71.

Morrow, A.L. and Creese, I. (1986) Characterization of α_1-adrenergic receptor subtypes in rat brain: a reevaluation of [^{3}H]WB4104 and [^{3}H]prazosin binding. *Mol. Pharmacol.*, **29**, 321–30.

Murphy, T.J. and Bylund, D.B. (1988) Characterization of α_2-adrenergic receptors in the OK cell, an opossum kidney cell line. *J. Pharmacol. Exp. Ther.*, **244**, 571–8.

Nicoletti, F., Iadarola, M.J., Wroblewski, J.T. and Costa, E. (1986) Excitatory amino acid recognition sites coupled with inositol phospholipid metabolism: developmental changes and interaction with α_1-adrenoceptors. *Proc. Natl. Acad. Sci. USA*, **83**, 1931–5.

Nishizuka, Y. (1988) The molecular heterogeneity of protein kinase C and its implications for cellular regulation. *Nature*, **334**, 661–5.

Nomura, Y., Yotsumoto, I. and Nishimoto, Y. (1982) Ontogeny of influence of clonidine on high potassium-induced release of noradrenaline and specific [^{3}H]clonidine binding in the rat brain cortex. *Dev. Neurosci.*, **5**, 198–204.

Nomura, Y., Kawai, M., Mita, K. and Segawa, T. (1984) Developmental changes of cerebral cortical [^{3}H]clonidine binding in rats: influences of guanine nucleotide and cations. *J. Neurochem.*, **42**, 1240–5.

Ordway, G.A., Gambarana, C. and Frazer, A. (1988) Quantitative autoradiography of central beta adrenoceptor subtypes: comparison of the effects of chronic treatment with desipramine or centrally administered l-isoproterenol. *J. Pharmacol. Exp. Ther.*, **247**, 379–89.

Pappas, B.A., Peters, D.A.V., Sobrian, S.K. *et al.* (1975) Early behavioural and catecholaminergic effects of 6-hydroxydopamine and guanethidine in the neonatal rat. *Pharmacol. Biochem. Behav.*, **3**, 683–5.

Perkins, J.P. and Moore, M.M. (1973) Characterization of the adrenergic receptors mediating the rise in cyclic 3',5'-adenosine monophosphate in rat cerebral cortex. *J. Pharmacol. Exp. Ther.*, **185**, 371–8.

Pittman, R.N., Minneman, K.P. and Molinoff, P.B. (1980) Ontogeny of β_1- and β_2-adrenergic receptors in rat cerebellum and cerebral cortex. *Brain Res.*, **188**, 357–68.

Regan, J.W., Kobilka, T.S., Yang-Feng, T.L. *et al.* (1988) Cloning and expression of a human kidney cDNA for an α_2-adrenergic receptor subtype. *Proc. Natl. Acad. Sci. USA*, **85**, 6301–5.

Rooney, T.A. and Nahorski, S.R. (1987) Postnatal ontogeny of agonist and depolarization-induced phosphoinositide hydrolysis in rat cerebral cortex. *J. Pharmacol. Exp. Ther.*, **243**, 333–41.

Schoepp, D.D. and Rutledge, C.O. (1985) Comparison of postnatal changes in alpha1-adrenoceptor binding and adrenergic stimulation of phosphoinositide hydrolysis in rat cerebral cortex. *Biochem. Pharmacol.*, **34**, 2705–11.

Simon, M.I., Strathmann, M.P. and Gautam, N. (1991) Diversity of G proteins in signal transduction. *Science*, **252**, 802–8.

Stryer, L. and Bourne, H.R. (1986) G proteins: a family of signal transducers. *Annu. Rev. Cell Biol.*, **2**, 391–419.

Vernadakis, A. and Gibson, D.A. (1974) Role of neurotransmitter substances in neural growth, in *Perinatal Pharmacology: Problems and Priorities* (eds J. Dancis and J.C. Hwang), Raven Press, New York, pp. 65–76.

Waddington, G. and Banks, P. (1981) The development of pre- and postsynaptic components of the noradrenergic system in the rat cerebellum. *J. Neurochem.*, **37**, 576–81.

Walton, K.G., Miller, E. and Baldessarini, R.J. (1979) Prenatal and early postnatal β-adrenergic receptor-mediated increase of cyclic AMP in slices of rat brain. *Brain Res.*, **177**, 515–22.

Williams, J.T., Henderson, G. and North, R.A. (1985) Characterization of alpha 2-adrenoceptors which increase potassium conductance in rat locus coeruleus neurons. *Neuroscience*, **14**, 95–101.

Wilson, K.M. and Minneman, K.P. (1989) Regional variations in α_1-adrenergic receptor subtypes in rat brain. *J. Neurochem.*, **53**, 1782–6.

Zeng, D. and Lynch, K.R. (1991) Distribution of α_2-adrenergic receptor mRNAs in the rat CNS. *Mol. Brain Res.*, **10**, 219–25.

Zeng, D., Harrison, J.K., D'Angelo, D.D. *et al.* (1990) Molecular characterization of a rat α_{2B}-adrenergic receptor. *Proc. Natl. Acad. Sci. USA*, **87**, 3102–6.

MUSCARINIC RECEPTORS AND THE DEVELOPING NERVOUS SYSTEM

Lucio G. Costa

2.1 INTRODUCTION

The cholinergic system plays a primary role in a number of important behaviors, and cholinergic dysfunctions may result in several neurological disorders (e.g. dementias, affective disorders; Singh *et al.*, 1985). In the central nervous system, muscarinic receptors constitute the majority of receptors of the cholinergic type. Though the muscarinic action of acetylcholine has been known since the beginning of the century, a significant advancement in our knowledge of the biochemical and molecular characteristics of muscarinic receptors has occurred only in the last decade. A number of excellent reviews on muscarinic receptors give a comprehensive discussion of the subject (Nathanson, 1987; Mitchelson, 1988; Brown, 1989; Goyal, 1989; Mei *et al.*, 1989; van Delft *et al.*, 1989; Hulme *et*

al., 1990). The most striking findings of the last decade are probably the discovery of the existence of several muscarinic receptor subtypes and the characterization of their second messenger responses. This chapter will briefly review some basic concepts of muscarinic receptor structures and functional responses and will then discuss in more detail our current knowledge of their ontogeny and their potential roles in brain development. A discussion of the development of muscarinic receptors in other tissues, such as the heart, can be found in Nathanson (1989).

2.2 MUSCARINIC RECEPTOR SUBTYPES

The first strong evidence of muscarinic receptor heterogeneity came from studies with

Table 2.1. Subtypes of muscarinic receptors

Molecular sequence	m_1	m_2	m_3	m_4	m_5
Amino acid residues					
Human	460	466	590	479	532
Rat	460	466	589	478	531
Pharmacological subtype	M_1	M_2	M_3	—	—
Selective antagonists	Pirenzepine (+) Telenzepine	AFDX-384 Methoctramine	*p*-Fluorohexahydrosiladiferidol	—	—

Adapted from Mei *et al.* (1989) and Levine and Birdsall (1989).

Receptors in the Developing Nervous System Vol. 2: Neurotransmitters. Edited by Ian S. Zagon and Patricia J. McLaughlin. Published in 1993 by Chapman & Hall. ISBN 0 412 49400 0. Vols. 1 and 2 (set) ISBN 0 412 54520 9.

the muscarinic antagonist pirenzepine, which revealed a unique regional distribution of binding, different from that of other antagonists, such as quinuclidinyl benzylate (QNB; Hammer *et al.*, 1980). This finding led to the classification of muscarinic receptors into two subtypes: a high affinity pirenzepine site, found primarily in the CNS, which was named M_1, and a site with low affinity for pirenzepine, present in heart and smooth muscle, which was named M_2. A third type of receptor, named M_3, was soon identified pharmacologically in exocrine glands and smooth muscle. New tools in molecular biology have since led to the cloning of five distinct rat and human muscarinic receptors (Table 2.1). Muscarinic receptors from mouse, *Drosophila* and chick have also been cloned (Shapiro *et al.*, 1988; Tietje *et al.*, 1990). The rat genes are designated m_1 to m_5 and their products correspond to the pharmacologically defined subtypes (M_1, M_2, etc.; Kubo *et al.*, 1986; Peralta *et al.*, 1987; Buckley *et al.*, 1988; Bonner, 1989).

Sequence analysis of cloned muscarinic receptors has indicated that the receptor protein is composed of seven transmembrane domains (α helical segments), which are highly conserved among subtypes. However, the third inner cytoplasmatic loop differs among subtypes, and this portion of the protein is believed to infer different functional characteristics in the receptor. Each muscarinic receptor subtype, unlike other receptors coupled to ion channels (e.g. $GABA_A$ or cholinergic nicotinic receptors) which consist of several subunits, but similar to other G protein-coupled receptors, appears to be the product of a single gene. The distribution of muscarinic receptors in the CNS has been determined by measuring receptor recognition sites by radioligand binding assays, as well as subtype mRNAs by *in situ* hybridization (Potter *et al.*, 1983; Buckley *et al.*, 1988). These studies have shown that regions such as the cerebral cortex and the hippocampus contain mostly M_1, and to a minor extent, M_2 and M_3 receptors; the striatum has a preponderance of M_4, M_1 and possibly M_2 subtypes, whereas cerebellum and medulla pons contain almost exclusively M_2 receptors. Low levels of the M_5 subtype have been found in the hippocampus and substantia nigra (Vilarò *et al.*, 1990). The pre- and postsynaptic location of receptor subtypes has also received much attention. For example, in the cerebral cortex and the hippocampus M_1 and M_3 receptors are believed to be located postsynaptically, whereas M_2 receptors might be located presynaptically in cholinergic terminals, and appear to be involved in the modulation of acetylcholine release (Kilbinger, 1987). In the striatum, on the other hand, an M_3 muscarinic autoreceptor appears to regulate the release of acetylcholine (De Boer *et al.*, 1990).

2.3   SECOND MESSENGER SYSTEMS

Activation of muscarinic receptors in the nervous system can lead to several biochemical responses, the most important or, at least, the most studied of which, are an increase in the levels of cyclic GMP, a decrease in the level of cyclic AMP and an increased turnover of membrane phospholipids, particularly phosphoinositides (Nathanson, 1987).

2.3.1   REGULATION OF CYCLIC AMP METABOLISM

In several tissues, activation of muscarinic receptors leads to an inhibition of adenylate cyclase and to a decrease in the rate of formation of cyclic AMP (adenosine 3',5'-monophosphate) (Olianas *et al.*, 1983; Harden, 1989). Similar to activation of adenylate cyclase by beta adrenergic agonists, muscarinic receptor-mediated inhibition of this enzyme involves a G protein, Gi. Inhibition of adenylate cyclase leads to a decrease in the intracellular levels of cyclic AMP. Although a direct association of cyclic AMP with muscarinic receptor-mediated responses has been unambiguously demonstrated in only a few cases, in several tissues, particularly those rich in the M_2 receptor subtype, physiological effects of acetylcholine involve alterations of cyclic AMP metabolism (Harden, 1989). In addition to causing inhibition of adenylate cyclase through an interaction with Gi, there is limited evidence that activation of muscarinic receptors may lead to activation of phosphodiesterase. This also will lead to a decrease in intracellular levels of cyclic AMP (Harden, 1989). The observation that muscarinic receptors can stimulate phosphodiesterase, together with the observation that activation of alpha$_1$-adrenoceptors can produce a similar event, has led to a proposed mechanism for such stimulation. Since both alpha$_1$-adrenoceptors and muscarinic receptors are coupled to phospholipase C, this model considers the interaction between the phosphoinositide and the cyclic AMP pathway (Harden, 1989). Elevation of cytoplasmatic Ca^{2+} levels occurs as a consequence of the increased inositol 1,4,5-trisphosphate (Ins-1,4,5-P_3) levels, and an activation of a Ca^{2+} calmodulin-regulated phosphodiesterase ensues. Thus, whereas inhibition of cyclic AMP accumulation through interaction with Gi can be seen as a direct mechanism, the muscarinic receptor-mediated attenuation of cyclic AMP accumulation probably should be considered more as a modulatory mechanism that follows receptor-stimulated phosphoinositide hydrolysis (Harden, 1989). Although detailed pharmacological studies on muscarinic receptor-mediated activation of phosphodiesterase have not been carried out, it would appear that all subtypes of muscarinic receptor are capable of affecting cyclic AMP levels, the M_2 subtype, directly, via inhibition of adenylate cyclase, and the M_1 and M_3 subtypes indirectly, as a consequence of an increase of intracellular calcium which follows phosphoinositide hydrolysis. It is becoming apparent, however, that several muscarinic receptor subtypes couple to each of the second-messenger response systems with varying degrees of efficiency.

2.3.2 REGULATION OF CYCLIC GMP METABOLISM

Activation of muscarinic receptors also causes the elevation of intracellular cyclic GMP (guanosine 3',5'-monophosphate) levels. This effect has generally been considered to be indirect, involving an additional second messenger, since muscarinic agonists cannot activate guanylate cyclase (the enzyme which converts GTP into cyclic GMP) in broken cell preparations (McKinney and Richelson, 1989). The most likely candidates for this role are calcium ions or a metabolite of arachidonic acid. Evidence for a role of calcium comes from experiments showing that extracellular calcium ions are necessary for the muscarinic receptor stimulation of cyclic GMP (Schultz et al., 1973), and that calcium channel antagonists block this effect (El-Fakahany and Richelson, 1983). The role of a metabolite of arachidonic acid is suggested by the finding that inhibitors of lipoxygenase (but not cyclo-oxygenase) block the cyclic GMP response (Snider et al., 1984; McKinney and Richelson, 1986). The muscarinic receptor involved in the cyclic GMP response appears to be the M_1 subtype (McKinney and Richelson, 1989), with a pharmacological profile similar to that of stimulation of phosphoinositide metabolism. Some evidence has suggested a link between these two biochemical responses (e.g. Ohsako and Deguchi, 1981), but this is not always the case (e.g. Kendall, 1986).

Increases in intracellular levels of cyclic GMP contribute to some of the effects of muscarinic receptor stimulations in various tissues such as smooth muscle, heart and brain (McKinney and Richelson, 1989). Most of these effects are probably mediated by protein phosphorylation regulated by cyclic GMP-dependent protein kinases. The cyclic GMP pathway has recently been accorded much attention, since it has been discovered that nitric oxide (NO), synthesized from L-arginine by NO synthase, binds to a heme moiety which is attached to guanylate cyclase and stimulates the accumulation of cyclic GMP (Ignarro, 1991; Snyder and Bredt, 1991). It has been shown that NO mediates glutamate-stimulation of cyclic GMP formation in the cerebellum (Garthwaite et al., 1989), as well as norepinephrine-induced increases in cyclic GMP in astrocytes (Agullo and Garcia, 1991). It has been recently reported that the accumulation of cyclic GMP stimulated by activation of muscarinic receptors is mediated by the NO pathway (Castoldi et al., 1993).

2.3.3 ACTIVATION OF THE PHOSPHOINOSITIDE/PROTEIN KINASE C PATHWAY

Interaction of acetylcholine with the appropriate subtype of muscarinic receptor activates a phosphoinositidase (phosphoinositidase C) which hydrolyzes phosphatidylinositol 4,5-bisphosphate (PtdIns 4,5-P_2) to Ins 1,4,5-P_3 and 1,2-diacylglycerol (DG) (Gonzales and Crews, 1984; Berridge, 1987; Costa, 1990). A GTP-binding protein is believed to couple the receptor to phosphoinositidase C (Chiu et al., 1988; Harden, 1990; Fain, 1990; Candura et al., 1992b). The exact nature of such protein (possibly Gp or Gq) has not been fully elucidated, although the current view is that there might be more than one protein (Lo and Hughes, 1987; Smrcka et al., 1991).

Ins 1,4,5-P_3 binds to specific and saturable receptor sites in, or near, the endoplasmic reticulum (Worley et al., 1989) and causes the mobilization of calcium ions in the cytosol (Stundermann et al., 1988). Ins 1,4,5-P_3 is then dephosphorylated by phosphatases to generate Ins 1,4-P_2, Ins 1-P and inositol. This latter reaction, catalyzed by an Ins 1-P phosphatase, is inhibited by lithium ions (Drummond et al., 1987). Ins 1,4,5-P_3 can also be phosphorylated by a 3-kinase to form inositol 1,3,4,5-tetrakisphosphate (Ins 1,3,4,5-P_4), which in turn is dephosphorylated to Ins 1,3,4-P_3 (Irvine et al., 1988). It has been suggested that

Ins 1,3,4-P_3 might be involved in long-term cellular effects (e.g. regulation of gene transcription or cell division), but evidence for this is still lacking. Ins 1,3,4,5-P_4 might play a role in modulating the mobilization of intracellular Ca^{2+} by Ins 1,4,5-P_3 and/or in regulating Ca^{2+} entry from outside the cell (Irvine et al., 1988). All these inositol phosphates are formed, in different amounts and over different time-courses, following activation of muscarinic receptors by agonists (Stephens and Logan, 1989). Ins 1,3,4,5,6-P_5 and Ins-P_6 may also be formed, and it has been suggested that they serve as a storage for phosphates (the same role they have in higher plants). Both have been shown to induce dose-dependent changes in heart rate and blood pressure in the rat (Vallejo et al., 1987) and specific binding sites for Ins-P_6 have been identified in mammalian brain (Hawkins et al., 1990). The cyclic inositol (1:2 cyclic),4,5-trisphosphate is also capable of mobilizing Ca^{2+} and is formed and metabolized at a slower rate; it could therefore act as a long-term messenger, although it is still of doubtful physiological significance. Thus, a major consequence of the activation of the phosphoinositide system by muscarinic agonists is a change in the intracellular concentration of Ca^{2+} which plays a pivotal role in synaptic transmission and in cell functions (Llinas, 1982; Kudo et al., 1988; Boess et al., 1990).

The other product of phosphoinositide hydrolysis, DG, is capable of activating a novel protein kinase, PKC (Nishizuka, 1988). In most tissues, PKC is present in its inactive form in the cytosol and is translocated to membranes when cells are stimulated; activation of PKC requires Ca^{2+} and phospholipids, particularly phosphatidylserine. It is now evident that at least seven subspecies of PKC exist in nerve tissue, and their structures, deduced by analysis of their DNA sequences, have been elucidated (Nishizuka, 1988; Kikkawa et al., 1988). These different PKCs, one of which, the gamma, is expressed only in the brain and spinal cord, but not in other tissues, including peripheral nerves, differ in their specific activities in different brain areas, in their developmental profile, and in their activation requirements. In addition to DG, the gamma form can also be activated by arachidonic acid and by other eicosanoids (Kikkawa *et al.*, 1988; Rana and Hokin, 1990), suggesting that production of these compounds upon receptor stimulation might have differential intracellular effects depending on the forms of PKC present in a certain cell, and that PKC might be regulated independently of DG and Ca^{2+}, i.e. in a manner not necessarily linked to phosphoinositide turnover. In recent years, it has been shown that activation of muscarinic receptors can also cause the hydrolysis of membrane phosphatidylcholine; this hydrolysis appears to be mediated both by phospholipase C and by phospholipase D (Martinson *et al.*, 1989; Qian and Drewes, 1989; Diaz-Meco *et al.*, 1989; Horwitz, 1990; Sandmann and Wurtman, 1991). Hydrolysis by phospholipase C generates phosphorylcholine and DG, whereas hydrolysis by phospholipase D leads to the formation of choline and phosphatidic acid. The latter can be metabolized to DG by the action of a phosphohydrolase. Thus, significant amounts of DG, available for activation of PKC, can be formed by hydrolysis of phosphoinositides and phosphatidylcholine on activation of muscarinic receptors.

PKC is the receptor for a class of tumor promoters, the phorbol esters (Ashendel, 1985) and the use of these compounds as direct activators of PKC is proving extremely useful for investigating the role of PKC in cellular functions. A large number of proteins have been shown to be substrates for PKC; these include such diverse molecules as receptors, including the muscarinic receptors (Haga *et al.*, 1990), ion channels, cytoskeletal proteins and enzymes (Nishizuka, 1988). PKC has been shown to provide both a 'positive forward action' leading to a synergistic interaction with the Ca^{2+} pathway (e.g. in control of gene expression), and a 'negative feedback control' over various steps of the cell-signalling process (Berridge, 1987; Nishizuka, 1988). Muscarinic receptor-induced phosphoinositide hydrolysis has been shown to be inhibited by phorbol esters in brain slices and neurons in cultures (Labarca *et al.*, 1984; Gonzales *et al.*, 1987). Furthermore, activation of PKC is also believed to be involved in muscarinic receptor desensitization and down regulation (Liles *et al.*, 1986; Jia *et al.*, 1989; however, see also Lai *et al.*, 1990; Hoover and Toews, 1990).

It is now well established that phorbol esters enhance the release of various neurotransmitters in a number of neuronal preparations (Kikkawa and Nishizuka, 1986), an effect believed to be due to activation of PKC, since it is not induced by inactive phorbol esters and is inhibited by PKC antagonists, such as H-7. The involvement of PKC in neurotransmitter release might also be linked to its role in the maintenance of long-term potentiation (LTP), a rapidly induced, persistent increase in synaptic efficacy (Brown *et al.*, 1988). Since LTP is a leading candidate for a synaptic mechanism of rapid learning in mammals, a role of PKC in memory formation has also been suggested (Chiarugi *et al.*, 1989). There is also increasing evidence that phosphoinositol metabolism plays a relevant role in the control of cell proliferation (Vicentini and Villareal, 1986). For example, in the nervous system, proliferation of astrocytes has been shown to be associated with activation of the phosphoinositide/PKC system by acetylcholine (Ashkenazi *et al.*, 1989). Stimulation of phosphoinositide metabolism differs in different brain areas, the highest effect being found in cerebral cortex, hippocampus and striatum (Gonzales and Crews, 1984; Balduini *et al.*, 1990a). The m_1, m_3 and m_5 receptor subtypes induce a strong phosphoinositide response, whereas m_2 and m_4 receptors weakly activate phosphoinositide hydrolysis (Peralta *et al.*, 1988; Forray and El-Fakahany, 1990). Ongoing studies with chimeric receptors are attempting to define the specific structural domain(s) conferring selectivity of coupling to phosphoinositide hydrolysis (Wess *et al.*, 1990).

2.4 REGULATION OF MUSCARINIC RECEPTORS

The sensitivity of muscarinic receptors is regulated by the prolonged presence of an agonist or an antagonist at the receptor site (Nathanson, 1989; El-Fakahany and Cioffi, 1990). Activation of muscarinic receptors by an agonist leads, in a very short time, to receptor desensitization; persistent activation results in a down-regulation of muscarinic receptors. Prolonged exposure to muscarinic antagonists, on the other hand, causes a supersensitivity of muscarinic receptors, often accompanied by receptor up-regulation, a phenomenon similar to that observed after denervation. A large number of studies have investigated the modulation of muscarinic receptors in several tissues and model systems, from cell cultures to intact animals. These findings have been summarized in several recent reviews (Nathanson, 1987; 1989; Costa, 1988; El-Fakahany and Cioffi, 1990). Prolonged exposure to direct muscarinic agonists such as oxotremorine, and indirectly to agonists, such as cholinesterase inhibitors, has been shown to cause a decrease in the sensitivity of muscarinic receptors. This alteration of muscarinic receptor sensitivity is manifested in decreases in receptor density measured by radioligand binding, in receptor mRNA levels (Wang *et al.*, 1990), in muscarinic receptor-mediated cellular responses and in various behavioral effects elicited by cholinergic agonists. Conversely, repeated treatments with muscarinic antagonists cause an increase in the density of muscarinic receptors and a supersensitivity to the behavioral effects of muscarinic agonists.

In addition to muscarinic agonists (including cholinesterase inhibitors) and antagonists, a number of other agents can modify muscarinic receptor density, affinity or sensitivity following *in vivo* administration. These include drugs, such as barbiturates (Nordberg *et al.*, 1980), hormones, such as estrogens (Dohanick *et al.*, 1982), environmental neurotoxicants, such as mercury (Aronstam and Eldefrawi, 1979), or physical agents, such as repeated noise (Lai *et al.*, 1989). A loss of muscarinic receptors, particularly of the M_2 subtype, has been found to occur with aging (Araujo *et al.*, 1990); a severe loss of M_2 autoreceptors has also been shown in brain of patients with Alzheimer's disease (Mash *et al.*, 1985).

2.5 MUSCARINIC RECEPTORS IN THE DEVELOPING NERVOUS SYSTEM

2.5.1 RECEPTOR BINDING

A large number of studies have examined the ontogenesis of muscarinic receptors in the CNS. Most of these studies have been carried out in rodents, particularly in rats, although there is some limited information for other species, including human; data also exist for the development of muscarinic receptors in neurons in primary cultures. One early study (Coyle and Yamamura, 1976) measured the binding of the specific but non-selective antagonist [^{3}H]quinuclidinyl benzilate (QNB) to rat brain membranes. At 15 days of gestation, the whole brain exhibited low, but detectable, levels of binding, approximately 1% of adult levels. Whole brain [^{3}H]QNB binding increased in a relatively linear fashion during the last week of gestation and first month *post partum*, when the concentration of specific binding sites for [^{3}H]QNB reached adult levels. An analysis of muscarinic receptors in brain regions revealed a similar developmental pattern, with adult levels of [^{3}H]QNB binding being reached at 28 days of age. The rate of muscarinic receptor development was, however, different in different regions; for example, at two weeks after birth, the medulla pons and hypothalamus had nearly 80% concentration of the adult muscarinic receptors, whereas the parietal cortex and striatum had less than 40% of adult levels (Coyle and Yamamura, 1976). The region-specific development of muscarinic receptors was later confirmed by Kuhar *et al.* (1980), who found a faster development of [³H]methyl scopolamine binding in medulla pons than in cortex or diencephalon. A differential development of high and low affinity agonist binding sites was also found, with the high affinity site appearing only 6–7 days after the low affinity sites (Kuhar *et al.*, 1980).

Several additional studies have examined the ontogenesis of muscarinic receptors, labeled by non-selective antagonists such as [³H]QNB, in rat or mouse brain or brain areas, by means of receptor binding or receptor autoradiography, with similar results (Aronstam *et al.*, 1979; Rotter *et al.*, 1979; Egozi *et al.*, 1980; Dudai *et al.*, 1980; East and Dutton, 1980; Mallol *et al.*, 1984; Hohman *et al.*, 1985; Large *et al.*, 1986; Balduini *et al.*, 1987; Represa *et al.*, 1989). Similar developmental profiles were also observed in rat cerebral and hippocampal cells in primary culture (Dudai and Yavin, 1978; Fernandez-Tomè and Segal, 1987), and in rabbit or chick brain (Yavin and Harel, 1979; Enna *et al.*, 1976).

A number of studies have also investigated the development of muscarinic receptors in the human fetal and neonatal brain (Brooksbank *et al.*, 1978). Egozi *et al.* (1986) measured the binding of *N*-[³H]methyl-4-piperidyl benzilate in several regions of human fetal brain from week 14 to week 24 of gestation, and found a gradual increase in binding. Longer gestational ages (up to 40 weeks) were studied by Ravikumar and Sastry (1985) using [³H]QNB. These investigators found an increase in receptor density from week 16 to 20, a lag period between 20 and 24 weeks, and a rapid increase of receptors during the third trimester reaching a maximum at birth and declining thereafter. Some of these observations were also made by Nordberg and Winblad (1981). On the other hand, Gremo *et al.* (1987) reported a decrease in receptor density between prenatal weeks 28–32 and birth in

most brain areas. It should be noted that in all these animal and human studies the affinity of the antagonist ligands did not change significantly during development.

Recent attention has focused on the development of muscarinic receptor subtypes in brain tissue. In human fetal frontal cortex the density of M_1 and M_2 sites increased from gestational week 8 to week 22 (Perry *et al.*, 1986). The ratio of M_1/M_2 sites also increased with age, as might be expected if presynaptic elements, those containing M_2 receptors, are established prior to functional neurotransmission, including postsynaptic activity. On the other hand, an autoradiographic study in the rat indicated that M_1 receptors appear early (gestational day 18 in striatum and 20 in hippocampus), whereas M_2 receptors develop more slowly and increase significantly only during the second and third weeks of development (Aubert *et al.*, 1990). Evans *et al.* (1985) measured the postnatal development of muscarinic receptors labeled with either [³H]QNB or [³H]pirenzepine in mouse cerebral cortex. Binding of both ligands increased with age and reached adult levels at about postnatal day 21. Interestingly, in this study the density of binding sites was found to be almost identical for both ligands, suggesting that the murine cerebral cortex contains predominantly M_1 sites. These findings were confirmed by an *in vitro* autoradiographic study (Miyoshi *et al.*, 1987). No data are available, to date, on the development of mRNA levels for the five muscarinic receptor subtypes.

In summary, most receptor binding studies have shown that muscarinic receptor binding sites increase with development in all species and, in the rat, they reach maximal density at approximately 30 days of age.

2.5.2 DEVELOPMENT OF MUSCARINIC RECEPTOR-COUPLED SECOND MESSENGER SYSTEMS

In recent years, the development of muscarinic receptor-coupled second messenger systems, particularly the phosphoinositide/ protein kinase C pathway, has begun to receive some attention. An interesting and surprising finding was that the accumulation of inositol phosphates stimulated by muscarinic agonists had a different developmental profile from that of muscarinic receptor binding sites. Although muscarinic receptors in rat cerebral cortex (labeled by [³H]QNB) increase during development, as previously discussed, the phosphoinositide response increases from birth to postnatal days 7–10 and then gradually declines (Fig. 2.1; Balduini *et al.*, 1987, 1990a,b; Heacock *et al.*, 1987; Rooney and Nahorski, 1987). Lee *et al.* (1990) attributed this developmental pattern to a decrease in the incorporation of [³H]inositol into tissues of older rats. However, this explanation is not fully convincing because (1) data were usually corrected to account for differ-

ences in incorporation; (2) incorporation was actually greater on postnatal days 1 and 3, when the effect of muscarinic agonists was less than that observed on days 7 or 10; (3) such a developmental pattern could not be observed for other neurotransmitters. Indeed, the developmental pattern of receptor-stimulated phosphoinositide metabolism has been shown to be neurotransmitter receptor- and brain region-specific (Balduini *et al.*, 1991b).

In addition to rat cortical slices, similar findings have been also reported in a synaptoneurosomal preparation (Dudek *et al.*, 1989). This developmental pattern has also been observed in other tissues, such as the rat cochlea (Bartolami *et al.*, 1990). In brain regions other than the cerebral cortex, the developmental profile of muscarinic receptor-stimulated phosphoinositide metabolism is

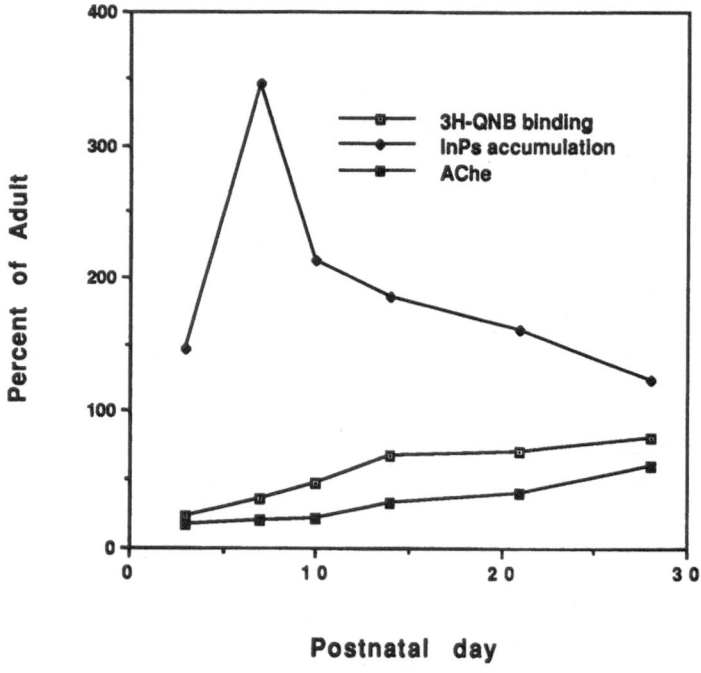

Fig. 2.1. Postnatal development of [³H]QNB binding, carbachol-stimulated inositol phosphates (InsPs) accumulation and acetylcholinesterase (AChE) activity in rat cerebral cortex, expressed as percentages of adult (75-day-old) values. Adult values were: [³H]QNB binding, 1907 ± 128 fmol/mg of protein; carbachol (1 mM)-stimulated InsPs accumulation, 428 ± 24% of basal; AChE activity, 136 ± 21.3 μmol acetylthiocholine hydrolyzed/min/mg of protein. Adapted from Balduini *et al.* (1987).

different. Indeed, while the hippocampus resembles the cerebral cortex, with a maximal stimulation on day 7, the effect of carbachol in brainstem and cerebellum declines from gestational day 19 to adulthood (Balduini *et al.*, 1991b). A similar decrease from day 4 to adulthood has also been observed in the rat retina (Osborne, 1988).

In order to understand the possible mechanism(s) underlying the enhanced stimulation of phosphoinositide metabolism seen in immature rats, relative to adults, a series of pharmacological and biochemical characteristics of the muscarinic receptor-stimulated inositol phosphate metabolism were compared in brain from 7-day-old and adult rats (Balduini *et al.*, 1990a). No significant age differences were found with respect to the effects of agonists, partial agonists and antagonists, calcium and sodium ion dependency, sensitivity to phorbol esters and regional distribution. It was found, however, that a high concentration of potassium ions (at least 12 mM) was capable of potentiating the effect of carbachol on phosphoinositide metabolism in cortices from adult rats, but not in 7-day-old animals (Balduini *et al.*, 1990b). Eva and Costa (1986) suggested that in adult animals K^+ might change the coupling efficiency of the muscarinic recognition site with its transducer. One might speculate that in the 7-day-old animal this 'positive coupler' would already be operating and, therefore, the potentiating effect of K^+ would be minimal. This hypothesis would agree with that of Heacock *et al.* (1987), who suggested the presence of an enhanced receptor–effector coupling as the mechanism underlying the enhanced sensitivity of phosphoinositide metabolism in 7-day-old rats.

In rat cortical membranes, the GTP analog GTP[S] (guanosine 5'-*O*-(3-thiotrisphosphate)) stimulates phosphoinositide metabolism to a higher degree in adult, than in 7-day-old rats (Candura *et al.*, 1992b). This is not surprising since this nucleotide, as well as AlF_4^- which shows a similar pattern of stimulation, is

expected to interact with all G proteins, not solely those coupled to the muscarinic receptors. However, in membranes from neonatal rats, the stimulatory effect of carbachol, which is seen only in the presence of micromolar concentrations of GTP (or GTP[S]) (Wallace and Claro, 1990), is similar to that observed in adult animals (Candura *et al.*, 1992b). Thus, although the enhanced response to muscarinic agonist in the neonatal rat, as observed in slices or synaptoneurosomes, is not seen in this membrane preparation, the effect of muscarinic agonists in 7-day-old animals is still remarkable, since only a fraction (about 30%) of adult muscarinic receptors are present at this developmental stage (Balduini *et al.*, 1987). A somewhat related similar observation was made in electrophysiological recordings from neonatal rat hippocampal slices, which showed that the responses elicited by muscarinic agonists (depolarization of the membrane potential and/or increase in firing frequency) were similar to those observed in mature animals (Reece and Schwartzkroin, 1991).

A few studies have investigated the development of muscarinic receptor-induced inositol phosphate formation in neurons in primary culture. In striatal cultures from mouse brain (prepared from 14–15-day-old fetuses) the effects of carbachol increased up to 14 days *in vitro*, roughly equivalent to day 7–8 *in vivo* (Weiss *et al.*, 1988). Longer times were not investigated. On the other hand, the effect of glutamate peaked earlier and then declined, a finding that led these investigators to suggest that the excitatory amino acid response precedes synaptogenesis, whereas maximal response to carbachol requires complete neuronal differentiation (Weiss *et al.*, 1988). Similar results were obtained in the same system by Schmidt *et al.* (1991), who found a doubling of the carbachol response after 13 days in culture, i.e. after synaptogenesis. Other investigators used primary cultures to study muscarinic receptor-mediated inositol responses (e.g. Ambrosini and Meldolesi, 1989; Akins *et al.*, 1990), but, no time-course studies of developmental maturation were conducted.

One aspect that has also been addressed by studying brain tissue in culture is that of the neuronal and/or glial localization of the muscarinic response. Whereas initial findings suggested that muscarinic receptor-mediated inositide hydrolysis is primarily a neuronal response (Gonzales *et al.*, 1985), subsequent studies showed that a robust phosphoinositide response is also present in astrocytes (Pearce and Murphy, 1988). In slice preparations, the observed response is probably the sum of the effects in both neurons and glial cells, and the contribution of astrocytes to events observed in slices cannot be ignored, especially as these glial cells can outnumber neurons by an order of magnitude (Pearce and Murphy, 1988).

Lee *et al.* (1990) provide the only information on the development of muscarinic receptor-mediated inhibition of cyclic AMP accumulation in rat cerebral cortex. They found that this action of muscarinic agonists does not occur until postnatal week 3, when it is already maximal (about 20% inhibition of forskolin-stimulated cyclic AMP accumulation). This is surprising, since muscarinic receptors of the M_2 subtype are present even in the first two weeks after birth. Since the subunits of G proteins are also present, it has been proposed that during the early stages of development, the interactions between muscarinic receptors and G proteins may be less efficient, thus leading to a lack of muscarinic receptor-mediated inhibition of cyclic AMP accumulation (Lee *et al.*, 1990). The mechanism(s) underlying such diminished coupling remains, however, obscure. Muscarinic receptor-mediated inhibition of cyclic AMP accumulation has also been studied in primary cultures of striatal and brainstem neurons from 14–15-day-old mouse embryos (Ellis *et al.*, 1990). Cells were maintained *in vitro* for seven to eight divisions, up to a developmental stage when there is little synapse formation; no time-course studies on developmental changes

were conducted. No information is available on the development of muscarinic receptor-mediated cyclic GMP accumulation in nerve tissue.

2.5.3   DEVELOPMENT OF OTHER PARAMETERS OF THE CHOLINERGIC SYSTEM

In order to have a better understanding of the possible role(s) of muscarinic receptors in brain development, it is useful to compare their ontogeny with that of other parameters of the cholinergic system. The development of activities of cholineacetyltransferase (ChAT) and acetylcholinesterase, of high affinity sodium-dependent choline uptake, and of levels of choline and acetylcholine in different brain areas, have been studied by several investigators, and the results have been summarized in various reviews (Bajgar *et al.*, 1979; Muller *et al.*, 1985; Hohman and Ebner, 1985; Vaca, 1988); only those aspects which may be relevant to a discussion of the development of muscarinic receptors will be briefly summarized here. Activity of ChAT in rat brain is usually low at birth and increases with age; its development, however, is rather slow. For example, only 19% of adult ChAT activity is present after one week (Fig. 2.2; Ladinsky *et al.*, 1972). Represa *et al.* (1989) similarly reported that ChAT activity in rat striatum was approximately 25% of adult levels at postnatal day 6; in other brain areas, the development was somewhat different. For example, in the septum, a spike in activity was seen on day 9 (up to 40% of adult levels), followed by a decline and by a slow increase (Represa *et al.*, 1989). A study by Large *et al.* (1986) showed that in the olfactory bulb the development of ChAT parallels that of muscarinic receptors.

The development of ChAT and several markers of the cholinergic system was also investigated by Coyle and Yamamura (1976). In whole brain levels of ChAT and high affinity choline uptake were almost undetect-

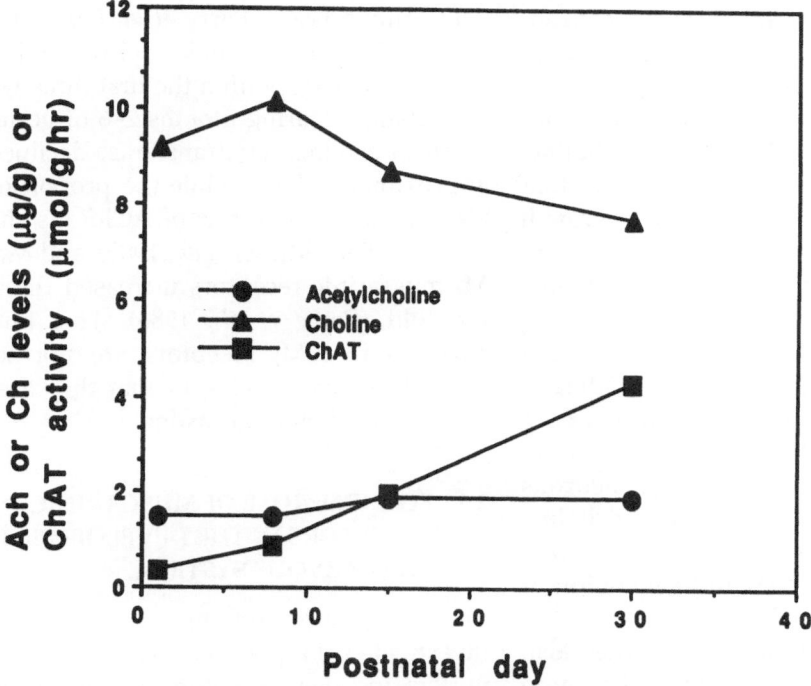

Fig. 2.2. Levels of acetylcholine and choline (μg/g wet wt) and activity of cholineacetyltransferase (ChAT; μmol/g wet wt per h) in brain of developing rat. Data adapted from Ladinsky *et al.* (1972).

able at birth and increased rapidly after the first week. On the other hand, levels of acetylcholine were already 30–40% of adult levels in the prenatal brain and during the first postnatal week. Particularly in the medulla-pons, the levels of acetylcholine in newborn rats were 87% of adult values (Coyle and Yamamura, 1976). Ladinsky *et al.* (1972) similarly reported levels of acetylcholine to be as high as 73% of adult levels by postnatal day 1 (Fig. 2.2). High levels of acetylcholine in the neonatal brain have also been reported by other investigators (Kasa *et al.*, 1982; Dörner *et al.*, 1982).

Choline levels, which are considered to be the rate-limiting factor in the biosynthesis of acetylcholine, are actually higher in the neonatal than in the adult rat brain (Fig. 2.2; Ladinsky *et al.*, 1972). A study by Kotas and Prince (1987) confirmed these findings. Levels of acetylcholine in the hippocampus of the 7-day-old rat were already 40% of adult levels, whereas choline levels were three-fold higher

than in adults. However, a significantly higher percentage of the choline is incorporated into phosphorylcholine at one week of age, than in adult animals. The rate of hemocholinium-3-sensitive uptake increased with age, and reached 37% of adult values at one week of age. High affinity choline uptake sites were also measured by Aubert *et al.* (1990), and found to mature more slowly than M_1 receptors, but similarly to M_2 (presynaptic) receptors. On the other hand, acetylcholinesterase activity is very low at birth (20% of adult values) and increases steadily to reach adult levels at 35 days (Fig. 2.1).

A noteworthy feature of the development of these parameters is that there is a continual apparent lag in the development of ChAT activity, compared to acetylcholine levels. Only by postnatal day 20 does the level of ChAT activity become commensurate with the endogenous levels of acetylcholine (Ladinsky *et al.*, 1972). Two factors may, therefore, play a role in the high concentration of acetylcholine in the neonatal brain: the activity of the metabolizing enzyme, acetylcholinesterase, which is quite low, and the availability of high levels of choline. Studies in *Xenopus* embryos have shown that acetylcholine can be released from the growth cone upon stimulation at the soma (Sun and Pao, 1987). The rate of release of acetylcholine by slices of immature brain, whether resting or electrically stimulated, can greatly exceed those evident in mature synapses (Pedata *et al.*, 1983). Furthermore, the turnover of acetylcholine is high in the adult brain, whereas it is lower in immature neurons (Purpura, 1972). High levels of acetylcholine would, therefore, be available for interaction with muscarinic receptors and generation of second messenger signals.

The regional developmental studies also indicate that the maturation of ChAT is more rapid in the cerebral regions of the brain, probably reflecting the fact that cell division ceases, and synaptogenesis commences in these regions much in advance of the subcortical and cortical areas (Coyle and Yamamura, 1976). Navarro *et al.* (1989) measured ChAT and choline uptake in the developing rat cerebral cortex: both parameters increased with age. However, the uptake/ChAT ratio peaked on postnatal day 9, just at the beginning of a major phase in synaptogenesis, and then declined. Only the cortex displayed a postnatal spike of uptake/ChAT; in the midbrain–brainstem, for example, this ratio decreased monotonically after birth. This developmental pattern resembles those observed for muscarinic receptor-stimulated phosphoinositide metabolism in these two areas (Balduini *et al.*, 1987, 1991b). Although the interrelationship between those observations is still unclear, it appears that in certain brain regions, such as the cerebral cortex, the unique development of cholinergic activity may render this system vulnerable to developmental insult (Hohman *et al.*, 1988; Navarro *et al.*, 1989).

In human brain, Perry *et al.* (1986) found high activity of cholineacetyltransferase in fetal cortex, even within the first three weeks of gestation. During months 2–6 of gestation levels of cholineacetyltransferase declined by approximately 50%, while the proportion of G4/G1 molecular forms of AChE (as shown also by others; Muller *et al.*, 1985), and M_1 and M_2 muscarinic receptors increased three- to five-fold (Perry *et al.*, 1986). This finding suggests that M_2 receptors are not solely located presynaptically, or that they are not confined to cholinergic axons.

2.6 POSSIBLE ROLE OF MUSCARINIC RECEPTORS IN THE DEVELOPMENT OF THE NERVOUS SYSTEM

Though a large number of studies have investigated the possible role of neurotransmitters and their receptors in brain development (see reviews by Lanier *et al.*, 1976; Coyle, 1977; Lauder, 1988; Mattson, 1988; Meier *et al.*, 1991; Mattson and Hauser, 1991), there is only limited information on the cholinergic system, particularly muscarinic receptors. Most findings on the development of various parameters of the cholinergic system (see previous sections) indicate that muscarinic receptor formation precedes the development of the presynaptic cholinergic marker ChAT (Brooksbank *et al.*, 1978; Rotter *et al.*, 1979). Thus, muscarinic receptors may have an effect on synaptogenesis, though, in turn, the latter may influence the final distribution of receptors (Rotter *et al.*, 1979).

Recent findings have suggested that an important role in brain development may be carried out by the second messenger system cascade of events which follows activation of muscarinic receptors. In particular, the peculiar developmental profile of muscarinic receptor-stimulated phosphoinositide metabolism, which peaks between postnatal days 6–8, in correspondence with the brain growth spurt, has led to the hypothesis that this system may have a fundamental role in the regulation of neurocytomorphogenesis and glial cell proliferation (Balduini *et al.*, 1987). Recent observations have lent support to the hypothesis that this effect of acetylcholine may have functional relevance during brain development. Hohman *et al.* (1988) have shown that neonatal lesions of the basal forebrain cholinergic neurons result in abnormal cortical development. The main features of this disruption include a delay in the developmental expression of critical neuronal components in neocortex, such as the large pyramidal cells in layer V, and a more prolonged disruption of the laminar boundaries that characterize the cortex (Hohman *et al.*, 1988). Though the mechanisms that mediate this disruption in cortical morphogenesis linked to transient loss of forebrain cholinergic neurons remain to be defined, it has been suggested that the ontogenetic expression of muscarinic receptor second messenger coupling in neocortex supports a role of this receptor in modulating neocortical differentiation (Hohman *et al.*, 1988).

A second study has shown that activation of muscarinic receptors in nerve growth cones isolated from neonatal rat brain can stimulate the phosphorylation of the nerve-specific protein B-50 (or GAP43), a major substrate for PKC (Van Hoof *et al.*, 1989). Nerve growth cones are the highly motile tips of axons and dendrites that guide the trailing neurite to the correct target (Smalheiser, 1990; Kater and Mills, 1991). Neurotransmitters are thought to modulate neurite outgrowth and maturation of a growth cone into a synaptic terminal (Mattson, 1988) and the phosphoprotein B-50 is presumed to mediate signal transduction in growth cone membranes (Van Hoof *et al.*, 1989). The finding that a muscarinic agonist can stimulate phosphorylation of B-50 in nerve growth cones adds evidence to the hypothesis that it may play a relevant role in synapto-genesis.

Finally, studies by Ashkenazi *et al.* (1989)

have shown that activation of those subtypes of muscarinic receptors which elicit a strong phosphoinositide response, stimulates DNA synthesis in primary astrocytes derived from prenatal rat brain, in an age-dependent fashion. The strong correlation between phosphoinositide response and DNA synthesis and its age-dependence led the authors to conclude that activation of phosphoinositide hydrolysis by acetylcholine is involved in its mitogenic action and is an important factor in astroglial cell growth during brain development (Ashkenazi *et al.*, 1989). Although the same mechanism could also be operating earlier for neurons in their dividing phase, another hypothesis is that for non-dividing cells, neurotransmitter-driven growth processes could involve an increase in cytoplasmatic surface area mediated by an enhancement of process formation and branching (Hanley, 1989). Though higher than ChAT activity, muscarinic receptor density is low in the neonatal brain (Balduini *et al.*, 1987); however, as discussed earlier, acetylcholine is present in significant amounts. It is conceivable, therefore, that acetylcholine could have profound influences on brain development through its 'amplified' phosphoinositide response during the brain growth spurt. An interesting, possibly related, observation (Mattson, 1989) was that in cultured hippocampal neurons acetylcholine potentiates glutamate-induced neurodegeneration, by activating muscarinic receptors. This effect of acetylcholine (which was devoid of neurotoxicity when present alone) may be mediated by increases in intracellular calcium (Mattson, 1989). Clearly, the potential role(s) of muscarinic receptors in brain development remain mostly unexplored. Additional roles of muscarinic receptors in early embryogenesis would also be worth exploring. Filogamo and Marchisio (1971) suggested that acetylcholine may serve a function other than as a neurotransmitter in early embryogenesis and may not be localized exclusively in cholinergic neurons.

Muscarinic receptors have been identified electrophysiologically in human ovarian oocytes, and sperm-carried acetylcholine has been suggested to be involved in activation processes triggered by sperm–egg interaction (Eusebi *et al.*, 1984). Thus, muscarinic receptors may play a significant role at different stages of development.

2.7　MUSCARINIC RECEPTORS AS TARGETS FOR DEVELOPMENTAL NEUROTOXICITY

Drugs, environmental toxicants and disease states have been shown to alter the development of brain muscarinic receptors. Whether these changes are a direct consequence of the action of the exogenous agents, or represent an adaptive response to other lesions, remains, in most cases, to be determined. Neonatal treatment of mice with L-thyroxine or betamethasone caused an accelerated accumulation of muscarinic receptors in the mouse cortex; however, at 30 days after birth, there was a reduction in the quantity of receptors in several brain areas (Ben-Baruch *et al.*, 1981). Undernutrition induced by restriction of the blood supply to the fetus by ligation of the uterine vessels five days before birth (a model known as intrauterine growth retardation or IUGR), has been shown to delay the development of muscarinic receptors in the hippocampus (Represa *et al.*, 1989). Alterations in the development of cerebellar muscarinic receptors have been reported in X-ray irradiated rats, homozygous Gunn rats (jj) and staggerer mice (Soreq *et al.*, 1982).

Exposure to various environmental neurotoxicants (e.g. pesticides, metals) during brain development has been shown to affect the ontogenesis of muscarinic receptors. Two pyrethroid insecticides, bioallethrin and deltamethrin, and the organochlorine DDT, have been shown to alter the development of muscarinic receptors (Eriksson and Nordberg, 1986, 1990). Exposure to the organophosphorus compound, cholinesterase inhibitor, diiso-propylfluorophosphate from postnatal days 8 to 14, resulted in a gradual decrease in the number of muscarinic binding sites in mouse brain (Levy, 1981). However, when the treatment was stopped the density of sites approached the control levels within a week. Similar observations were made by Ben-Barak *et al.* (1981) in Tetram (*O,O*-diethyl *S*-(β-diethyl-amino) ethyl phosphorothiolate)-treated rats. Prenatal exposure to diiso-propylfluorophosphate also caused a delay in the development of brain muscarinic receptors, however, full recovery was observed by day 20 (Michalek *et al.*, 1985). Stamper *et al.* (1988) exposed developing rats to the organophosphorus insecticide parathion from postnatal day 5 to day 20, and observed a decreased receptor density in cerebral cortex on days 21 and 28, which was paralleled by small, but significant, deficits in tests of spatial memory in both the T-maze and the radial arm maze. Developmental exposure of rats to lead (postnatal days 0–21, via the mother's milk) was shown to cause a selective decrease in muscarinic receptor density in the visual cortex (Costa and Fox, 1983), which could be related to observed visual acuity deficits (Fox *et al.*, 1982). Ethanol was reported to decrease muscarinic receptors in the hippocampus of 21-day-old rats following exposure from birth (Serbus and Light, 1990). Similar changes were, however, not observed in the hippocampus of rats exposed to ethanol prenatally (Wigal *et al.*, 1990), or in cerebral cortex of rats exposed between postnatal days 4 and 10 (Balduini and Costa, 1989).

The peculiar developmental profile of the muscarinic receptor-stimulated phospho-inositide metabolism and its possible functional implications (see previous sections) led to the formulation of the hypothesis that this system could represent a target for the developmental neurotoxicity of ethanol. Administration of ethanol to rats during the brain growth spurt (i.e. during the first two postnatal weeks, equivalent to the late second and third trimester of pregnancy in humans) has been shown to cause microencephaly, an effect which is seen in 80% of Fetal Alcohol Syndrome cases (West, 1986). Ethanol (given at the daily dose of 4 g/kg, by gavage, from day 4 to day 9) also caused a significant decrease of muscarinic receptor-stimulated phosphoinositide response in cerebral cortex and hippocampus of 7- and 10-day-old rats, i.e. at the peak of the brain growth spurt, but not at other ages (Balduini and Costa, 1989). The finding that exposure of adult rats to dosages of ethanol resulting in nearly identical blood alcohol levels (approx. 250 mg/dl) had no effects on muscarinic receptor-stimulated phosphoinositide metabolism suggested that this system may represent a molecular target specifically relevant for the developmental neurotoxicity of ethanol. A series of *in vitro* studies (Balduini and Costa, 1990; Balduini *et al.*, 1991a; Candura *et al.*, 1991, 1992a) have addressed this hypothesis. Ethanol was found to inhibit the effect of muscarinic receptor agonists on phosphoinositide metabolism at concentrations equivalent to those present following exposure *in vivo*. The effect was region-dependent, in accordance with the reported differential developmental effects of ethanol in brain areas (West, 1986). For example, in cerebral cortex, hippocampus and cerebellum, the effect of ethanol was more pronounced than in brainstem, which is not affected by ethanol following *in vivo* administration. The effect of ethanol was neurotransmitter-specific: no effect on norepinephrine-, serotonin-, or histamine-stimulated phosphoinositide metabolism was observed. The effect of ethanol was also strongly age-dependent: inhibition of phosphoinositide metabolism was more pronounced in the 7-day-old rat and less so in younger or older animals. Ethanol also inhibited carbachol-stimulated inositol metabolism in cortical membrane preparations from neonatal rats. Inhibition of muscarinic receptor-stimulated phosphoinositide metabolism was also produced by *n*-propanol and *t*-butanol, two aliphatic alcohols which cause

microencephaly when administered during the brain growth spurt in the rat (Grant and Samson, 1982, 1984). Altogether, these data provide strong, though still circumstantial, evidence that the metabolism of phosphoinositides stimulated by activation of muscarinic receptors might represent a relevant and specific target for the developmental neurotoxicity of ethanol. Further studies should investigate the exact molecular mechanism of this effect of ethanol and the consequence of such inhibition on intracellular calcium mobilization and activation of protein kinase C.

2.8 CONCLUSIONS

Over the last several years there has been considerable growth in the knowledge of muscarinic receptors. Five subtypes of muscarinic receptors have been cloned and the role of each structural domain in agonist and antagonist binding and in coupling to second messenger systems is being investigated. While molecular biologists lead the way, research in chemistry and pharmacology lags behind in discovering new chemicals that possess receptor subtype specificity. Particularly for the m_4 and m_5 subtype, no selective agent is yet available. Application of the novel tools for studying muscarinic receptors will find its ways also in investigations on their development in the nervous system. Our knowledge on the development of receptor subtypes and on their second messenger system is still limited. The interaction of acetylcholine with other neurotransmitter systems during brain development is also worthy of further investigation. Progress in developmental neurobiology would certainly be necessary to place all these findings in a functional physiological perspective.

ACKNOWLEDGEMENTS

Work by the author was supported in part by grants from the National Institute of

Environmental Health Sciences (ES-04696), the Alcohol and Drug Abuse Institute, University of Washington, and the Fondazione Clinica del Lavoro, Pavia. Thanks are due to Drs A.F. Castoldi and S.M. Candura for their useful comments and to Ms Chris Sievanen for secretarial assistance.

REFERENCES

Agullo, L. and Garcia, A. (1991) Norepinephrine increases cyclic GMP in astrocytes by a mechanism dependent on nitric oxide synthesis. *Eur. J. Pharmacol.*, **206**, 343–6.

Akins, P.T., Surmeier, D.J. and Kitai, S.T. (1990) M_1 muscarinic acetylcholine receptor in cultured rat neostriatum regulates phosphoinositide hydrolysis. *J. Neurochem.*, **54**, 266–73.

Ambrosini, A. and Meldolesi, J. (1989) Muscarinic and quisqualate receptor-induced phosphoinositide hydrolysis in primary cultures of striatal and hippocampal neurons. Evidence for differential mechanisms of activation. *J. Neurochem.*, **53**, 825–33.

Araujo, D.M., Lapchak, P.A., Meaney, M.J. *et al.* (1990) Effects of aging on nicotinic and muscarinic autoreceptor function in the rat brain: relationship to presynaptic cholinergic markers and binding sites. *J. Neurosci.*, **10**, 3069–78.

Aronstam, R. and Eldefrawi, M.E. (1979) Transition and heavy metal inhibition of ligand binding to muscarinic acetylcholine receptors from rat brain. *Toxicol. Appl. Pharmacol.*, **48**, 489–96.

Aronstam, R.S., Kellog, C. and Abood, L.G. (1979) Development of muscarinic cholinergic receptors in inbred strains of mice: identification of receptor heterogeneity and relation to audiogenic seizure susceptibility. *Brain Res.*, **162**, 231–41.

Ashendel, C.L. (1985) The phorbol ester receptor: a phospholipid-regulated protein kinase. *Biochim. Biophys. Acta*, **822**, 219–42.

Ashkenazi, A., Ramachandran, J. and Capon, D.J. (1989) Acetylcholine analogue stimulates DNA synthesis in brain-derived cells via specific muscarinic receptor subtypes. *Nature*, **340**, 146–50.

Aubert, I., Cecyre, D., Araujo, D.M. *et al.* (1990) Comparative expression of cholinergic markers during brain development. *Soc. Neurosci. Abst.*, **16**, 536.

Bajgar, J., Hrdina, V., Petr, R. and Golda, V. (1979) Development of cholinergic nervous system in the brain of normal and hypertensive rats. *Dev. Neurosci.*, **2**, 94–100.

Balduini, W. and Costa, L.G. (1989) Effects of ethanol on muscarinic receptor-stimulated phosphoinositide metabolism during brain development. *J. Pharmacol. Exp. Ther.*, **250**, 541–7.

Balduini, W. and Costa, L.G. (1990) Developmental neurotoxicity of ethanol: *in vitro* inhibition of muscarinic receptor-stimulated phosphoinositide metabolism in brain from neonatal but not adult rats. *Brain Res.*, **512**, 248–52.

Balduini, W., Murphy, S.D. and Costa, L.G. (1987) Developmental changes in muscarinic receptor-stimulated phosphoinositide metabolism in rat brain. *J. Pharmacol. Exp. Ther.*, **241**, 421–7.

Balduini, W., Murphy, S.D. and Costa, L.G. (1990a) Characterization of cholinergic muscarinic receptor-stimulated phosphoinositide metabolism in brain from immature rats. *J. Pharmacol. Exp. Ther.*, **253**, 573–9

Balduini, W., Murphy, S.D. and Costa, L.G. (1990b) Potassium ions potentiate the muscarinic receptor-stimulated phosphoinositide metabolism in cerebral cortex slices: a comparision of neonatal and adult rats. *Neurochem. Res.*, **15**, 33–9.

Balduini W., Candura, S.M., Manzo, L. *et al.* (1991a) Time-concentration-, and age-dependent inhibition of muscarinic receptor-stimulated phosphoinositide metabolism by ethanol in the developing rat brain. *Neurochem. Res.*, **16**, 1235–40.

Balduini, W., Candura, S.M. and Costa, L.G. (1991b) Regional development of carbachol-, glutamate-, norephinephrine- and serotonin-stimulated phosphoinositide metabolism in rat brain. *Dev. Brain Res.*, **62**, 115–20.

Bartolami, S., Guiramund, J., Lenoir, M. *et al.* (1990) Carbachol-induced inositol phosphate formation during rat cochlea development. *Hear. Res.*, **47**, 229–34.

Ben-Barak, J., Gazit, H., Silman, I. and Dudai, Y. (1981) In vivo modulation of the number of muscarinic receptors in rat brain by cholinergic ligands. *Eur. J. Pharmacol.*, **74**, 73–81.

Ben-Baruch, G., Egozi, Y., Kloog, Y. *et al.* (1981) Altered ontogenesis of muscarinic cholinergic receptor in mouse brain: effect of L-thyroxine and betamethasone. *Endocrinology*, **109**, 235–9.

Berridge, M.J. (1987) Inositol triphosphate and diacylglycerol: two interacting second messengers. *Annu. Rev. Biochem.*, **56**, 159–93.

Boess, F.G., Balasubramian, M.K., Brammer, M.J. and Campbell, I.C. (1990) Stimulation of muscarinic acetylcholine receptors increases synaptosomal free calcium concentration by protein kinase-dependent opening of L-type calcium channels. *J. Neurochem.*, **55**, 230–6.

Bonner, T.I. (1989) The molecular basis of muscarinic receptor diversity. *Trends Neurosci.*, **12**, 148–51.

Brooksbank, B.W.L., Martinez, M., Atkinson, D.J. and Balazs, R., (1978) Biochemical development of the human brain. I. Some parameters of the cholinergic system. *Dev. Neurosci.*, **1**, 267–84.

Brown, J.H. (ed.) (1989) *The Muscarinic Receptors*. Humana Press, Clifton, NJ.

Brown, T.H., Chapman, P.F., Kairiss, E.W. and Keenan, C.L. (1988) Long-term synaptic potentiation. *Science*, **242**, 724–8.

Buckley, N.J., Bonner, T.I. and Brann, M.R. (1988) Localization of a family of muscarinic receptors mRNAs in rat brain. *J. Neurosci.*, **8**, 4646–52.

Candura, S.M., Balduini, W. and Costa, L.G. (1991) Interaction of short chain aliphatic alcohols with muscarinic receptor-stimulated phosphoinositide metabolism in cerebral cortex from neonatal and adult rats. *Neurotoxicology*, **12**, 23–32.

Candura, S.M., Manzo, L. and Costa, L.G. (1992a) Inhibition of muscarinic receptor- and G-protein-dependent phosphoinositide metabolism in cerebrocortical membranes from neonatal rats by ethanol. *Neurotoxicology*, **13**, 281–8.

Candura, S.M., Manzo, L. and Costa, L.G. (1992b) Guanine nucleotide- and muscarinic agonist-dependent phosphoinositide metabolism in synaptoneurosomes from cerebral cortex of immature rats. *Neurochem. Res.*, **17**, 1133–41.

Castoldi, A.F., Manzo, L. and Costa, L.G. (1993) Cyclic GMP formation induced by muscarinic receptors is mediated by nitric oxide synthesis in rat primary cultures. *Brain Res.* (in press).

Chiarugi, V.P., Ruggiero, M. and Corradetti, R. (1989) Oncogenes, protein kinase C, neuronal differentiation and memory. *Neurochem. Int.*, **14**, 1–9.

Chiu, A.S., Li, P.P. and Warsh, J.J. (1988) G-protein involvement in central nervous system muscarinic receptor-coupled polyphosphoinositide hydrolysis. *Biochem. J.*, **256**, 995–9.

Costa, L.G. (1988) Organophosphorus compounds, in *Recent Advances in Nervous System Toxicology* (eds C.L. Galli, L. Manzo and P.S. Spencer), Plenum Press, New York, pp. 203–46.

Costa, L.G. (1990) The phosphoinositide/protein kinase C system as a potential target for neurotoxicity. *Pharmacol. Res.*, **22**, 393–408.

Costa, L.G. and Fox, D.A. (1983) A selective decrease of cholinergic muscarinic receptors in the visual cortex of adult rats following developmental lead exposure. *Brain Res.*, **276**, 259–66.

Coyle, J.T. (1977) Biochemical aspects of neurotransmission in the developing brain. *Int. Rev. Neurobiol.*, **20**, 65–103.

Coyle, J.T. and Yamamura, H.I. (1976) Neurochemical aspects of the ontogenesis of cholinergic neurons in the rat brain. *Brain Res.*, **118**, 429–40.

De Boer, P., Westerink, B.H.C., Rollema, H. *et al.* (1990) An M_3-like muscarinic autoreceptor regulates the *in vivo* release of acetylcholine in rat striatum. *Eur. J. Pharmacol.*, **179**, 167–72.

Diaz-Meco, M.T., Larrodera, P., Lopez-Barahona, M. *et al.* (1989) Phospholipase C-mediated hydrolysis of phosphatidylcholine is activated by muscarinic agonists. *Biochem. J.*, **263**, 115–20.

Dohanick, G.P., Witcher, J.A., Weaver, D.R. and Clemens, L.G. (1982) Alteration of muscarinic binding in specific brain areas following estrogen treatment. *Brain Res.*, **241**, 347–50.

Dörner, G., Bluth, R. and Tönjer, R. (1982) Acetylcholine concentrations in the developing brain appear to affect emotionality and mental capacity in later life. *Acta Biol. Med. Germ.*, **41**, 721–3.

Drummond, A.H., Joels, L.A. and Hughes, P.J. (1987) The interaction of lithium ions with inositol lipid signalling systems. *Biochem. Soc. Trans.*, **15**, 32–5.

Dudai, Y. and Yavin, E. (1978) Ontogenesis of muscarinic receptors and acetylcholinesterase in differentiating rat cerebral cortex in culture. *Brain Res.*, **155**, 368–73.

Dudai, Y., Ben-Barak, J., Silman, I. and Gazik, H. (1980) Ontogenesis and modulation of cholinergic receptors in rat brain, in *Neurotransmitters and Their Receptors* (eds U.Z. Littauer, Y. Dudai, I. Silman *et al.*), Wiley, New York, pp. 217–39.

Dudek, S.M., Bowen, W.D. and Bear, M.F. (1989) Postnatal changes in glutamate-stimulated phosphoinositide turnover in rat neocortical synaptoneurosomes. *Dev. Brain. Res.*, **47**, 123–8.

East, J.M. and Dutton, G.R. (1980) Muscarinic binding sites in developing normal and mutant mouse cerebellum. *J. Neurochem.*, **34**, 657–61.

Egozi, Y., Kloog, Y. and Sokolovsky, M. (1980) Studies on postnatal changes of muscarinic receptors in mouse brain, in *Neurotransmitters and Their Receptors* (eds U.Z. Littauer, Y. Dudai, I. Silman *et al.*), Wiley, New York, pp. 201–15.

Egozi, Y., Sokolovsky, M., Scheijter, E. *et al.* (1986) Divergent regulation of muscarinic binding sites and acetylcholinesterase in discrete regions of the developing human fetal brain. *Cell. Mol. Neurobiol.*, **6**, 55–70.

El-Fakahany, E.E. and Cioffi, C.L. (1990) Molecular mechanisms of regulation of neuronal muscarinic receptor sensitivity. *Membr. Biochem.*, **9**, 9–27.

El-Fakahany, E.E. and Richelson, E. (1983) Effect of some calcium antagonists on muscarinic receptor-mediated cyclic GMP formation. *J. Neurochem.*, **40**, 705–10.

Ellis, J., Huyler, J.H., Kemp, D.E. and Weiss, S. (1990) Muscarinic receptors and second-messenger responses of neurons in primary culture. *Brain Res.*, **511**, 234–40.

Enna, S.J., Yamamura, H.I. and Snyder, S.H. (1976) Development of muscarinic cholinergic and GABA receptor binding in chick embryo brain. *Brain Res.*, **101**, 177–83.

Eriksson, P. and Nordberg, A. (1986) The effects of DDT, DDOH-palmitic acid and a chlorinated paraffin on muscarinic receptors and the sodium-dependent choline uptake in the central nervous system of immature mice. *Toxicol. Appl. Pharmacol.*, **85**, 121–7.

Eriksson, P. and Nordberg, A. (1990) Effects of two pyrethroids, Bioallethrin and Deltamethrin, on subpopulations of muscarinic and nicotinic receptors in the neonatal mouse brain. *Toxicol. Appl. Pharmacol.*, **102**, 456–63.

Eusebi, F., Pasetto, N. and Siracusa, G. (1984) Acetylcholine receptors in human oocytes. *J. Physiol.*, **346**, 321–30.

Eva, C. and Costa, E. (1986) Potassium ions facilitation of phosphoinositide turnover activation by muscarinic receptor agonists in rat brain. *J. Neurochem.*, **46**, 1429–35.

Evans, R.A., Watson, M., Yamamura, H.I. and Roeske, W.R. (1985) Differential ontogeny of putative M_1 and M_2 muscarinic receptor binding sites in the murine cerebral cortex and heart. *J. Pharmacol. Exp. Ther.*, **235**, 612–18.

Fain, J.N. (1990) Regulation of phosphoinositide-specific phospholipase C. *Biochim. Biophys. Acta*, **1053**, 81–8.

Fernandez-Tomè, P. and Segal, M. (1987) Onto-genesis of muscarinic receptors in cultured rat hippocampal cells. *Dev. Brain Res.*, **35**, 158–60.

Filogamo, G. and Marchisio, P.C. (1971) Acetylcholine system and neural development. *Neurosci. Res.*, **4**, 29–64.

Forray, C. and El-Fakahany, E.E. (1990) On the involvement of multiple muscarinic receptor subtypes in the activation of phosphoinositide metabolism in rat cerebral cortex. *Mol. Pharmacol.*, **37**, 893–902.

Fox, D.A., Wright, A.A. and Costa, L.G. (1982) Visual acuity deficits following neonatal lead exposure: cholinergic interactions. *Neurobehav. Toxicol. Teratol.*, **4**, 689–93.

Garthwaite, J., Garthwaite, G., Palmer, R.M.J. and Moncada, S. (1989) NMDA receptor activation induces nitric oxide synthesis from arginine in rat brain slices. *Eur. J. Pharmacol.*, **172**, 413–16.

Gonzales, R.A. and Crews, F.T. (1984) Characterization of the cholinergic stimulation of phosphoinositide hydrolysis in rat brain slices. *J. Neurosci.*, **4**, 3120–7.

Gonzales, R.A., Feldstein, J.B., Crews, F.T. and Raizada, M.K. (1985) Receptor-mediated inositide hydrolysis is a neuronal response: comparison of primary neuronal and glial cultures. *Brain Res.*, **345**, 350–5.

Gonzales, R.A., Greger, P.H., Baker, S.P. *et al.* (1987) Phorbol esters inhibit agonist-stimulated phosphoinositide hydrolysis in neuronal primary cultures. *Dev. Brain Res.*, **37**, 759–66.

Goyal, R.K. (1989) Muscarinic receptor subtypes. Physiology and clinical implications. *N. Engl. J. Med.*, **321**, 1022–9.

Grant, K.A. and Samson, H.H. (1982) Ethanol and tertiary butanol-induced microencephaly in the neonatal rat: comparison of brain growth parameters. *Neurobehav. Toxicol. Teratol.*, **4**, 315–21.

Grant, K.A. and Samson, H.H. (1984) *n*-Propanol-induced microencephaly in the neonatal rat. *Neurobehav. Toxicol. Teratol.*, **6**, 165–9.

Gremo, F., Palomba, M., Marchisio, A.M. *et al.* (1987) Heterogeneity of muscarinic cholinergic receptors in the developing human fetal brain: regional distribution and characterization. *Early Hum. Dev.*, **15**, 165–77.

Haga, K., Haga, T. and Ichiyama, A. (1990) Phosphorylation by protein kinase C of the muscarinic acetylcholine receptor. *J. Neurochem.*, **54**, 1639–44.

Hammer, R., Berrie, C.P., Birdsall, N.J.M. *et al.* (1980) Pirenzepine distinguishes between different subclasses of muscarinic receptors. *Nature*, **283**, 90–2.

Hanley, M.R. (1989) Mitogenic neurotransmitters. *Nature*, **340**, 97.

Harden, T.K. (1989) Muscarinic cholinergic receptor-mediated regulation of cyclic AMP metabolism, in *The Muscarinic Receptors* (ed. J.H. Brown), Humana Press, Clifton, NJ, pp. 221–58.

Harden, T.K. (1990) G Protein-dependent regulation of phospholipase C by cell surface receptors. *Am. Rev. Respir. Dis.*, **141**, S119–22.

Hawkins, P.T., Reynolds, D.J.M., Poyner, D.R. and Hanley, M.R. (1990) Identification of a novel inositol phosphate recognition site: specific ^{3}H-inositol hexakisphosphate binding to brain regions and cerebellar membranes. *Biochem. Biophys. Res. Commun.*, **167**, 819–27.

Heacock, A.M., Fisher, S.K. and Agranoff, B.W. (1987) Enhanced coupling of neonatal muscarinic receptors in rat brain to phosphoinositide turnover. *J. Neurochem.*, **48**, 1904–11.

Hohman, C.F. and Ebner, F.F. (1985) Development of cholinergic markers in mouse forebrain. I. Choline acetyltransferase enzyme activity and acetylcholinesterase histochemistry. *Dev. Brain Res.*, **23**, 225–41.

Hohman, C.F., Pert, C.C. and Ebner, F.F. (1985) Development of cholinergic markers in mouse forebrain. II. Muscarinic receptor binding in cortex. *Dev. Brain Res.*, **23**, 243–53.

Hohman, C.F., Brooks, A.C. and Coyle, J.T. (1988) Neonatal lesions of the basal forebrain cholinergic neurons result in abnormal cortical development. *Dev. Brain Res.*, **42**, 253–64.

Hoover, R.K. and Toews, M.L. (1990) Activation of protein kinase C inhibits internalization and down regulation of muscarinic receptors in 1321N1 human astrocytoma cells. *J. Pharmacol. Exp. Ther.*, **253**, 185–91.

Horwitz, J. (1990) Carbachol and bradykinin increase the production of diacylglycerol from sources other than inositol-containing phospholipids in PC-12 cells. *J. Neurochem.*, **54**, 983–91.

Hulme, E.C., Birdsall, N.J.M. and Buckley, N.J. (1990) Muscarinic receptor subtypes. *Annu. Rev. Pharmacol. Toxicol.*, **30**, 633–73.

Ignarro, L.J. (1991) Signal transduction mechanisms involving nitric oxide. *Biochem. Pharmacol.*, **41**, 485–90.

Irvine, R.F., Moor, R.M., Pollock, W.K. *et al.* (1988) Inositol phosphates: proliferation, metabolism and function. *Philos. Trans. R. Soc. Lond. [Biol.]*, **320**, 281–98.

Jia, W.G., Shaw, C., van Huizen, F. and Cynader, M.S. (1989) Phorbol 12,13-dibutyrate regulates muscarinic receptors in rat cerebral cortical slices by activating protein kinase C. *Mol. Brain Res.*, **5**, 311–15.

Kasa, P., Bansaghy, K., Rakonczay, Z. and Guyla, K. (1982) Postnatal development of the acetylcholine system in different parts of the rat cerebellum. *J. Neurochem.*, **39**, 1726–32.

Kater, S.B. and Mills, L.R. (1991) Regulation of growth cone behavior by calcium. *J. Neurosci.*, **11**, 891–9.

Kendall, D.A. (1986) Cyclic GMP and inositol phosphate accumulation do not share common origins in rat brain slices. *J. Neurochem.*, **47**, 1483–9.

Kikkawa, U., and Nishizuka, Y. (1986) The role of protein kinase C in transmembrane signalling. *Annu. Rev. Cell Biol.*, **2**, 149–78.

Kikkawa, U., Ogita, K., Shearman, M.S. *et al.* (1988) The heterogeneity and differential expression of protein kinase C in nervous tissue. *Philos. Trans. R. Soc. Lond. [Biol.]*, **320**, 313–24.

Kilbinger, H. (1987) Control of acetylcholine release by muscarinic autoreceptors, in *International Symposium on Muscarinic Cholinergic Mechanisms* (eds S. Cohen and M. Sokolovsky), Freund, London, pp. 219–28.

Kotas, A.M. and Prince, A.K. (1987) High-affinity uptake of choline, a marker for cholinergic nerve terminals, is not specific in developing rat brain. *Dev. Brain Res.*, **35**, 175–81.

Kubo, T., Fukuda, R., Mikami, A. *et al.* (1986) Cloning, sequencing and expression of complementary DNA encoding the muscarinic acetylcholine receptor. *Nature*, **323**, 411–16.

Kudo, Y., Ogura, A. and Iijima, T. (1988) Stimulation of muscarinic receptor in hippocampal neuron induces characteristic increase in cytosolic free Ca^{2+} concentration. *Neurosci. Lett.*, **85**, 345–50.

Kuhar, M.J., Birdsall, N.J.M., Burgen, A.S.V. and Hulme, E.C. (1980) Ontogeny of muscarinic receptors in rat brain. *Brain Res.*, **184**, 375–83.

Labarca, R., Janowsky, A., Patel, J. and Paul, S.M. (1984) Phorbol esters inhibit agonist-induced [^{3}H] inositol-1-phosphate accumulation in rat hippocampal slices. *Biochem. Biophys. Res. Commun.*, **123**, 703–9.

Ladinsky, H., Consolo, S., Peri, G. and Garattini, S. (1972) Acetylcholine, choline and choline acetyltransferase activity in the developing brain of normal and hypothyroid rats. *J. Neurochem.*, **19**, 1947–52.

Lai, H., Carino, M.A. and Wen, Y.F. (1989) Repeated noise exposure affects muscarinic cholinergic receptors in the rat brain. *Brain Res.*, **488**, 361–4.

Lai, W.S., Rogers, T.B. and El-Fakahany, E.E. (1990) Protein kinase C is involved in desensitization of muscarinic receptors induced by phorbol esters but not by receptor agonists. *Biochem. J.*, **267**, 23–9.

Lanier, L.P., Dunn, A.J. and van Hartesveldt, C.

(1976) Development of neurotransmitters and their function in brain. *Rev. Neurosci.*, **2**, 195–257.

Large, T.H., Lambert, M.P., Gremillion, M.A. and Klein, W.L. (1986) Parallel postnatal development of choline acetyltransferase activity and muscarinic acetylcholine receptors in the rat olfactory bulb. *J. Neurochem.*, **46**, 671–80.

Lauder, J.M. (1988) Neurotransmitters as morphogens. *Progr. Brain Res.*, **73**, 365–87.

Lee, W., Nichlaus, K.J., Manning, D.C. and Wolfe, B.B. (1990) Ontogeny of cortical muscarinic receptor subtypes and muscarinic receptor-mediated responses in rat. *J. Pharmacol. Exp. Ther.*, **252**, 482–90.

Levine, R.R. and Birdsall, N.J.M. (1989) Nomenclature for muscarinic receptor subtypes recommended by symposium. *Trends Pharmacol. Sci.* Suppl., *Subtypes of Muscarinic Receptors*, p. vii.

Levy, A. (1981) The effect of cholinesterase inhibition on the ontogenesis of central muscarinic receptors. *Life Sci.*, **29**, 1065–70.

Liles, W.C., Hunter, D.D., Meier, K.E. and Nathanson, N.M. (1986) Activation of protein kinase C induces rapid internalization and subsequent degradation of muscarinic acetylcholine receptors in neuroblastoma cells. *J. Biol. Chem.*, **261**, 5307–16.

Llinas, R.R. (1982) Calcium in synaptic transmission. *Sci. Am.*, **247**, 56–65.

Lo, W.W.Y. and Hughes, J. (1987) Receptor-phosphoinositidase C coupling. Multiple G proteins? *FEBS Lett.*, **244**, 1–3.

Mallol, J., Sarraga, M.C., Bartolomè, M. *et al.* (1984) Muscarinic receptor during postnatal development of rat cerebellum: an index of cholinergic synapse formation? *J. Neurochem.*, **42**, 1641–9.

Martinson, E.A., Goldstein, D. and Brown, J.H. (1989) Muscarinic receptor activation of phosphatidylcholine hydrolysis. *J. Biol. Chem.*, **264**, 14748–54.

Mash, D.C., Flynn, D.D. and Potter, L.T. (1985) Loss of M_2 muscarinic receptors in the cerebral cortex in Alzheimer's disease and experimental cholinergic denervation. *Science*, **228**, 1115–17.

Mattson, M.P. (1988) Neurotransmitters in the regulation of neuronal cytoarchitecture. *Brain Res. Rev.*, **13**, 179–212.

Mattson, M.P. (1989) Acetylcholine potentiates glutamate-induced neurodegeneration in cultured hippocampal neurons. *Brain Res.*, **497**, 402–6.

Mattson, M.P. and Hauser, K.F. (1991) Spatial and temporal integration of neurotransmitter signals in the development of neural circuitry. *Neurochem. Int.*, **19**, 17–24.

McKinney, M. and Richelson, E. (1986) Blockade of N1E-115 murine neuroblastoma muscarinic receptor function by agents that affect the metabolism of arachidonic acid. *Biochem. Pharmacol.*, **35**, 2389–97.

McKinney, M. and Richelson, E. (1989) Muscarinic receptor regulation of cyclic GMP and eicosanoid production, in *The Muscarinic Receptors* (ed J.H. Brown), Humana Press, Clifton, NJ, pp. 309–40.

Mei, L., Roeske, W.R. and Yamamura, H.I. (1989) Molecular pharmacology of muscarinic receptor heterogeneity. *Life Sci.*, **45**, 1831–51.

Meier, E., Hertz, L. and Schousboe, A. (1991) Neurotransmitters as developmental signals. *Neurochem. Int.*, **19**, 1–15.

Michalek, H., Pintor, A., Fortuna, S. and Bisso, G.M. (1985) Effects of diisopropylfluorophosphate on brain cholinergic systems of rats at early developmental stages. *Fundam. Appl. Toxicol.*, **5**, S204–12.

Mitchelson, F. (1988) Muscarinic receptor differentiation. *Pharmacol. Ther.*, **37**, 357–423.

Miyoshi, R., Kito, S., Shimizu, M. and Matsubayashi, H. (1987) Ontogeny of muscarinic receptors in the rat brain with emphasis on the differentiation of M_1- and M_2-subtypes – semiquantitative *in vitro* autoradiography. *Brain Res.*, **420**, 302–12.

Muller, F., Dumez, Y. and Massoulie, J. (1985) Molecular forms and solubility of acetylcholinesterase during the embryonic development of rat and human brain. *Brain Res.*, **331**, 295–302.

Nathanson, N.M. (1987) Molecular properties of the muscarinic acetylcholine receptor. *Annu. Rev. Neurosci.*, **10**, 195–236.

Nathanson, N.M. (1989) Regulation and development of muscarinic receptor number and function, in *The Muscarinic Receptors* (ed. J.H. Brown), Humana Press, Clifton, NJ, pp. 419–54.

Navarro, H.A., Seidler, F.J., Eylers, J. *et al.* (1989) Effects of prenatal nicotine exposure on development of central and peripheral cholinergic neurotransmitter systems. Evidence for cholinergic trophic influences in developing brain. *J. Pharmacol. Exp. Ther.*, **251**, 894–900.

Nishizuka, Y. (1988) The molecular heterogeneity of protein kinase C and its implication for cellular regulation. *Nature*, **334**, 661–5.

Nordberg, A. and Winblad, B. (1981) Cholinergic receptors in human hippocampus. Regional distribution and variance with age. *Life Sci.*, **29**, 1937–44.

Nordberg, A., Wahlström, G. and Larsson, C. (1980) Increased number of muscarinic binding sites in brain following chronic barbiturate treatment to rat. *Life Sci.*, **26**, 231–7.

Ohsako, S. and Deguchi, T. (1981) Stimulation by phosphatidic acid of calcium influx and cyclic GMP synthesis in neuroblastoma cells. *J. Biol. Chem.*, **256**, 10945–8.

Olianas, M.C., Onali, P., Neff, N.H. and Costa, E. (1983) Adenylate cyclase activity of synaptic membranes from rat striatum. Inhibition by muscarinic receptor agonists. *Mol. Pharmacol.*, **23**, 393–8.

Osborne, N.N. (1988) Muscarinic stimulation of inositol phosphate formation in rat retina: developmental changes. *Vision Res.*, **29**, 871–81.

Pearce, B. and Murphy, S. (1988) Neurotransmitter receptors coupled to inositol phospholipid turnover and Ca^{2+} flux: consequences for astrocyte function, in *Glial Cell Receptors* (ed. H.K. Kimelberg), Raven Press, New York, pp. 197–221.

Pedata, F., Slavikova, J., Kotas, A. and Pepeu, G. (1983) Acetylcholine release from rat cortical slices during postnatal development and aging. *Neurobiol. Aging*, **4**, 31–5.

Peralta, E.G., Ashkenazi, A., Winslow, J.W. *et al.* (1988) Differential regulation of PI hydrolysis and adenylyl cyclase by muscarinic receptor subtypes. *Nature*, **334**, 434–7.

Peralta, E.G., Ashkenazi, A., Winslow, J.W. *et al.* (1987) Distinct primary structures, ligand-binding properties and tissue-specific expression of four human muscarinic acetylcholine receptors. *EMBO J.*, **6**, 3923–9.

Perry, E.K., Smith, C.J., Atach, J.R. *et al.* (1986) Neocortical cholinergic enzyme and receptor activities in the human fetal brain. *J. Neurochem.*, **47**, 1262–9.

Potter, L.T., Flynn, D.D., Hanchett, H.E. *et al.* (1983) Independent M_1 and M_2 receptors: ligands, autoradiography and functions. *Trends Pharmacol. Sci.* (Suppl), *Subtypes of Muscarinic Receptors*, pp. 22–31.

Purpura, D.P. (1972) Intracellular studies of synaptic organization in the mammalian brain, in *Structure and Function of Synapses* (eds G.D. Pappas and D.P. Purpura), Raven Press, New York, pp. 257–302.

Qian, Z. and Drewes, L.R. (1989) Muscarinic acetylcholine receptor regulates phosphatidyl-choline phospholipase D in canine brain. *J. Biol. Chem.*, **264**, 21720–4.

Rana, R.S. and Hokin, L.E. (1990) Role of phosphoinositides in transmembrane signaling. *Physiol. Rev.*, **70**, 115–64.

Ravikumar, B.V. and Sastry, P.S. (1985) Muscarinic cholinergic receptors in human fetal brain: characterization and ontogeny of ^{3}H-quiniclidinyl benzilate binding sites in frontal cortex. *J. Neurochem.*, **44**, 240–6.

Reece, L.J. and Schwartzkroin, P.A. (1991) Effects of cholinergic agonists on immature rat hippocampal neurons. *Dev. Brain Res.*, **60**, 29–42.

Represa, A., Chanez, C., Flexor, M.A. and Ben-Ari, V. (1989) Development of the cholinergic system in control and intra-uterine growth retarded rat brain. *Dev. Brain Res.*, **47**, 71–9.

Rooney, T.A. and Nahorski, S.R. (1987) Postnatal ontogeny of agonist and depolarization-induced phosphoinositide hydrolysis in rat cerebral cortex. *J. Pharmacol. Exp. Ther.*, **243**, 333–41.

Rotter, A., Field, P.M. and Raisman, G. (1979) Muscarinic receptors in the central nervous system of the rat. III. Postnatal development of binding of ^{3}H-propylbenzilylcholine mustard. *Brain Res. Rev.*, **1**, 185–205.

Sandmann, J. and Wurtman, R.J. (1991) Stimulation of phospholipase D activity in human neuroblastoma (LA-N-2) cells by activation of muscarinic acetylcholine receptors or by phorbol esters: relationship to phosphoinositide turnover. *J. Neurochem.*, **56**, 1312–19.

Schmidt, B.H., Manzoni, O.J.J., Royer, M. *et al.* (1991) Cholinergic inositol phosphate formation in striatal neurons is mediated by distinct mechanisms. *Eur. J. Pharmacol.*, **206**, 87–94.

Schultz, G., Hardman, J.G., Schultz, K. *et al.* (1973) The importance of calcium ions for the regulation of guanosine 3′,5′-monophosphate levels. *Proc. Natl. Acad. Sci. USA*, **70**, 3889–93.

Serbus, D.C. and Light, K.E. (1990) Cholinergic alterations of hippocampus and cerebellum at postnatal days 21 and 25 following twice daily ethanol exposure of rats throughout the suckling period. *Alcoholism*, **14**, 336.

Shapiro, R.A., Scherer, N.M., Habecher, B.A. *et al.* (1988) Isolation, sequence and functional expression of the mouse M_1 muscarinic acetylcholine receptor gene. *J. Biol. Chem.*, **263**, 18397–403.

Singh, M.M., Warburton, D.M. and Lal, H. (eds) (1985) *Central Cholinergic Mechanisms and Adaptive Dysfunctions*. Plenum Press, New York.

Smalheiser, N.R. (1990) Neuronal growth cones: an extended view. *Neuroscience*, **38**, 1–11.

Smrcka, A.V., Hepler, J.R., Brown, K.D. and Sternweis, P.C. (1991). Regulation of phosphoinositide-specific phospholipase C activity by purified Gq. *Science*, **251**, 804–7.

Snider, R.M., McKinney, M., Forray, C. and Richelson, E. (1984) Neurotransmitter receptors mediate cyclic GMP formation by involvement of arachidonic acid and lipoxygenase. *Proc. Natl. Acad. Sci. USA*, **81**, 3905–9.

Snyder, S.H. and Bredt, D.S. (1991) Nitric oxide as a neuronal messenger. *Trends Pharmacol. Sci.*, **12**, 125–7.

Soreq, H., Gurwitz, D., Eliyahu, D. and Sokolovsky, M. (1982) Altered ontogenesis of muscarinic receptors in agranular cerebellar cortex. *J. Neurochem.*, **39**, 756–63.

Stamper, C.R., Balduini, W., Murphy, S.D. and Costa, L.G. (1988) Behavioral and biochemical effects of postnatal parathion exposure in the rat. *Neurotoxicol. Teratol.*, **10**, 261–6.

Stephens, L.R. and Logan, S.D. (1989) Formation of [^{3}H] inositol metabolites in rat hippocampal formation slices prelabeled with [^{3}H] inositol and stimulated with carbachol. *J. Neurochem.*, **52**, 713–21.

Stundermann, K.A., Harris, G.D. and Lovenberg, W. (1988) Characterization of inositol 1,4,5-triphosphate-stimulated calcium release from rat cerebellum microsomal fractions. *Biochem. J.*, **255**, 667–83.

Sun, Y.A. and Pao, M.M. (1987) Evoked release of acetylcholine from the growing embryonic neuron. *Proc. Natl. Acad. Sci. USA*, **84**, 2540–4.

Tietje, K.M., Goldman, P.S. and Nathanson, N.M. (1990) Cloning and functional analysis of a gene encoding a novel muscarinic acetylcholine receptor expressed in chick heart and brain. *J. Biol. Chem.*, **265**, 2828–34.

Vaca, K. (1988) The development of cholinergic neurons. *Brain Res. Rev.*, **13**, 262–86.

Vallejo, M., Jackson, T., Lightman, S. and Hanley, M.R. (1987) Occurrence and extracellular actions of inositol pentakis- and hexakis-phosphate in mammalian brain. *Nature*, **330**, 656–8.

Van Hoof, C.O.M., De Graan, P.N.E., Oestreicher, A.B. and Gispen, W.H. (1989) Muscarinic receptor activation stimulates B50/GAP43 phosphorylation in isolated nerve growth cones. *J. Neurosci.*, **9**, 3753–9.

van Delft, A.M.L., Hagan, J.J. and Tonnaer, J.A.D.M. (1989) Muscarinic receptors in the central nervous system. *Progr. Pharmacol. Clin. Pharmacol.*, **7**, 93–117.

Vicentini, L.M. and Villareal, M.L. (1986) Inositol phosphates turnover, cytosolic Ca^{2+} and pH: putative signals for the control of cell growth. *Life Sci.*, **38**, 2269–76.

Vilarò, M.T., Palacios, J.M. and Mengod, G. (1990) Localization of m_5 muscarinic receptor mRNA in rat brain examined by *in situ* hybridization histochemistry. *Neurosci. Lett.*, **114**, 154–9.

Wallace, M.A. and Claro, E. (1990) Comparison of serotoninergic to muscarinic cholinergic stimulation of phosphoinositide-specific phospholipase C in rat brain cortical membranes. *J. Pharmacol. Exp. Ther.*, **255**, 1296–300.

Wang, S.Z., Hu, J., Long, R.M. *et al.* (1990) Agonist-induced down regulation of m_1 muscarinic receptors and reduction of their mRNA level in a transfected cell line. *FEBS Lett.*, **276**, 185–8.

Weiss, S., Schmidt, B.H., Sebben, M. *et al.* (1988) Neurotransmitter-induced inositol phosphate formation in neurons in primary culture. *J. Neurochem.*, **50**, 1425–33.

Wess, J., Bonner, T.J. and Brann, M.R. (1990) Chimeric m_2/m_3 muscarinic receptors: role of carboxyl terminal receptor domains in selectivity of ligand binding and coupling to phosphoinositide hydrolysis. *Mol. Pharmacol.*, **38**, 872–7.

West, J.R. (1986) *Alcohol and Brain Development*. Oxford University Press, New York.

Wigal, S.B.E., Amsel, A. and Wilcox, R.E. (1990) Cholinergic alterations of hippocampus and cerebellum at postnatal days 21 and 25 following twice daily ethanol exposure of rats throughout the suckling period. *Alcoholism*, **4**, 336.

Worley, P.F., Baraban, J.M. and Snyder, S.H. (1989) Inositol 1,4,5-triphosphate receptor binding: autoradiographic localization in rat brain. *J. Neurosci.*, **9**, 339–46.

Yavin, E. and Harel, S. (1979) Muscarinic binding sites in the developing rabbit brain. *FEBS Lett.*, **97**, 151–4.

3

THE ROLE OF SEROTONIN AND SEROTONIN RECEPTORS IN DEVELOPMENT OF THE MAMMALIAN NERVOUS SYSTEM

Patricia M. Whitaker-Azmitia

The neurons which produce the neuro-transmitter serotonin, make up one of the most widely distributed neuronal systems in the mammalian brain. This neuronal network is also one of the earliest developing systems. The turnover rate of serotonin (i.e. the ratio of metabolite to neurotransmitter) is higher in the immature mammalian brain than at any other time in life (Hamon and Bourgoin, 1981). This distribution, developmental pattern and high level of activity are the key elements in the role which serotonin plays in the immature brain – a role as a growth factor directing both proliferation and maturation (Lauder and Krebs, 1978; Buznikov, 1984; Lauder, 1983; Chubakov *et al.*, 1986; Lauder *et al.*, 1988; Lauder and Zimmerman, 1988; Lauder, 1990).

This chapter will first discuss what is known about the development of the serotonin system itself, since development of this system is so important to overall development of the brain. Then there will be discussion of which serotonin receptors mediate the developmental role of serotonin, since for any neurotransmitter to elicit a response, it must have a receptor.

3.1 DEVELOPMENT OF THE SEROTONIN NEURONAL SYSTEM

The anatomical development of the serotonin neuronal system has been most extensively studied in the rat, although the development has been described in several species, including chick (Wallace, 1985; Wallace *et al.*, 1986), leech (Stuart *et al.*, 1987), sheep (Tillet, 1988), xenopus (Messenger and Warner, 1989), fruitfly (Valles and White, 1988), cat (Gu *et al.*, 1990), mouse (Fujimiya *et al.*, 1986; Ni and Jonakait, 1989) and primate (see references in Jacobs and Azmitia, 1992). However, the following sections refer to the rat, unless noted otherwise.

The cell bodies of serotonin neurons develop as bilateral groups in two distinct regions of the brainstem of rat. First, a rostral collection of cells (caudal to the mesencephalic flexure) on embryonic day (ED) 12 and then in more dorsal regions near the pontine flexure on ED14 (Aitken and Tork, 1988). Through varying degrees of migration, these clusters develop into the adult pattern of nine groups (B 1–9) by ED18 (Lidov and Molliver, 1982a). The rostral group eventually will

Receptors in the Developing Nervous System Vol. 2: Neurotransmitters. Edited by Ian S. Zagon and Patricia J. McLaughlin. Published in 1993 by Chapman & Hall. ISBN 0 412 49400 0. Vols. 1 and 2 (set) ISBN 0 412 54520 9.

become B 4–9 and give rise to ascending fibers. The rostral group develops into B 1–3 and projects mainly into the spinal cord (Wallace and Lauder, 1983; Aitken and Tork, 1988).

As the serotonin cells migrate into nuclei, they continue to divide, with mitosis continuing up to ED18 (Aitken and Tork, 1988). After completion of mitosis, the cells begin to migrate in a new direction – towards the midline. Eventually, six of the nine nuclei become fused nuclei (the midline or raphe nuclei). Fusion takes place in a rostral to caudal direction, beginning at ED18 and ending at postnatal day (PD) 6.

Axonal projections from the nuclei develop very rapidly once the cells are present. In the rostral group, visualized by ED12, fibers are identified within 24 h entering into the diencephalon. In general the fibers appear to migrate along preformed non-serotonergic pathways, although there are some exceptions (Lidov and Molliver, 1982b; Wallace and Lauder, 1983). By ED17, these fibers have reached the frontal neocortical pole (Wallace and Lauder, 1983). Although the nuclei have become fused, they maintain their original projections, that is the projections are mostly ipsilateral (Levitt and Moore, 1978; Goto and Sano, 1984). Complete, adult-like innervation of the cerebral cortex is extremely heterogeneous, continuing up to PD21, since any one particular region is only innervated once it has reached a relatively mature architecture (Lidov and Molliver, 1982b).

Descending pathways also begin to develop almost immediately after the cells are differentiated, with the caudal-most regions of the spinal cord being reached by ED17. However, as with the ascending projections, there is a period of redistribution of terminals, with the final adult pattern being reached at PD21 (Rajaofetra *et al.*, 1989).

Dendritic arborization begins later than axonal growth, with a rapid increase from ED 19 until PD7. At the same time, the serotonin nuclei become less densely packed, as other cell types begin to develop in the vicinity (Lidov and Molliver, 1982b).

As the serotonin neurons are developing, the amount of serotonin produced, released and metabolized also increases rapidly. The enzyme responsible for the synthesis of serotonin, tryptophan hydroxylase, is virtually saturated by the very high levels of tryptophan. Only during this perinatal period are levels of tryptophan so high that this enzyme is saturated. The increased levels of tryptophan are caused by lack of tryptophan binding and an increased carrier system compared to that in the adult (Hamon and Bourgoin, 1979).

Like all neurotransmitter systems, a variety of factors are involved in the regulation of normal serotonin development. If any of these factors are altered, in concentration or time of exposure, the serotonin system may never recover fully. The remainder of this section describes some of these factors, although of course there are likely to be many more yet to be studied. Since serotonin itself is a growth factor, any of the factors which influence serotonin development may then indirectly influence the development of many other systems.

3.1.1. GROWTH AND ATTACHMENT FACTORS

Although many growth factors have been described in brain, few have actually been studied for a direct effect on specific neurotransmitter systems. For the serotonin system, nerve growth factor (NGF), epidermal growth factor (EGF) and insulin have all been shown to be without effect (Azmitia *et al.*, 1990a). However, for some time the presence of a soluble, proteinaceous growth factor for serotonin neurons has been suggested from lesioning studies and transplant studies. Moreover, this factor was shown to be selective for serotonin neurons, having no effect on noradrenergic neurons (Zhou *et al.*, 1987). In our work with astroglial cells, we have identified a substance which may be identical to this factor – the astroglial specific protein, S-100β; this is discussed in more detail later.

In addition to S-100, other astroglial-derived proteins appear to be important. Developing serotonin neurons show selective adhesion to astrocytes and display a greater rate and amount of neurite outgrowth, as compared to their growth on fibroblasts (Lieth *et al.*, 1990). At least some of this attachment and guidance may be due to the presence of laminin (Zhou and Azmitia, 1988).

3.1.2 NEUROTRANSMITTER EFFECTS

The normal course of serotonin development is also influenced by the levels of several neurotransmitters and neuromodulators. This includes an influence of serotonin itself on its own neurons – a circumstance referred to as autoregulation of development. Since this influence takes place through serotonin receptors, this work will be discussed later.

Dopamine is also very important in the regulation of serotonin neuronal development. Since the earliest studies by Breese *et al.* (1984) and more recently by Jackson *et al.* (1988), removal of dopamine by selective lesioning of neonates has been shown to lead to overgrowth of serotonin terminals into the caudate nucleus. Since dopamine had been shown to inhibit neurite outgrowth through a dopamine D_1 receptor (Lankford *et al.*, 1988), we proposed that dopamine may have a tonically inhibitory effect on serotonin development through this receptor. Removal of dopamine would therefore cause overgrowth into regions where D_1 receptors could be localized on serotonin terminals, such as the caudate. To test this hypothesis, animals were treated prenatally with a selective D_1 receptor agonist, SKF 38393. At 90 days of age these animals showed significant loss of serotonin terminals, plus altered behavioral sensitivity to both serotonin and dopamine agonists (Whitaker-Azmitia *et al.*, 1990b).

Substance P also has an influence on developing serotonin neurons, but this influence appears to be important only if the endogenous serotonin influence is lacking (Jonsson and Hallman, 1983). Also, ACTH and related peptides are stimulatory to growth, in the absence of trophic signals from a target region (Azmitia and De Kloet, 1987). Finally, enkephalin is inhibitory to the growth of serotonin neurons (Davila-Garcia and Azmitia, 1989).

3.1.3 EXOGENOUS FACTORS

Changes in serotonin levels and the subsequent development of brain have been shown to be induced by diet and malnutrition (Resnick and Morgane, 1984; Spear and Scalzo, 1985). This is presumably due to the high demand for tryptophan for normal development. Tryptophan, as an essential amino acid, is often lacking in poor diets.

Work by Peters has focused on the role of stress in altering the development of the serotonin system. This work has shown changes in serotonin behaviors and serotonin receptors by stress at various pre- and postnatal timepoints (Peters, 1982, 1988a, b, 1990). Some of these changes may be due to the influence of corticosteroids on the activity of tryptophan hydroxylase.

Finally, drugs of abuse may seriously damage the development of the serotonin system. For example, cocaine, 3,4-methylenedioxy amphetamine (MDMA) (Azmitia *et al.*, 1990b) and alcohol (Goodrich *et al.*, 1986; Schambra *et al.*, 1990) have all been shown to cause permanent changes.

3.2 SEROTONIN RECEPTORS INVOLVED IN DEVELOPMENT

In order for a neurotransmitter to function as a growth factor, there must be receptors present for that neurotransmitter. Characterizing these receptors, in terms of their pharmacology, transduction systems and ontogeny, is thus very important in understanding that neurotransmitter's developmental role.

Our initial studies, begun before all the currently known subtypes of adult serotonin receptors had been described, used 5-methoxytryptamine (5-MT), which is useful in pharmacological studies for its biphasic effects on serotonin receptors. At low doses, this agent stimulates release-regulating autoreceptors, whereas at high doses, it stimulates a postsynaptic receptor.

High affinity serotonin receptors are present prenatally and are, in a sense, functional, as they show appropriate compensatory changes in density in response to chronic stimulation or blockage (Whitaker-Azmitia *et al.*, 1987). To test directly for a role of these receptors in influencing development of serotonin neurons, ED14 rat serotonergic neurons were grown in culture for four days and exposed to varying doses of 5-MT. The growth responses of the serotonin neurons were biphasic, leading to the conclusion that two serotonin receptors were involved in regulation of development. The highest affinity receptor, presumably the presynaptic release-regulating autoreceptor, was inhibitory to growth of terminals. The lower affinity site was stimulatory (Whitaker-Azmitia and Azmitia, 1986a).

In a whole-animal model, similar results were obtained by treating animals prenatally from gestational day 12 until birth, with three different doses of 5-MT. A low dose of 5-MT (1 mg/kg) caused a decrease in terminal density in forebrain, up to PD30. Conversely, 3.0 mg/kg 5-MT caused significant increases up to PD30.

In addition to the neurochemical studies, the animals were tested in several behavioral paradigms – the neonatal serotonin syndrome as described by Spear and Ristene (1982) at PD5, spontaneous alternation and open field activity at PD15, and a passive avoidance paradigm (lick suppression) at PD30. In the neonatal animals, changes in response to quipazine were found in those pups which had been treated prenatally with 1.0 or 3.0 mg/kg, but there was no effect in those animals treated with the lowest dose, 0.1 mg/kg. By PD15, there were still significant behavioral changes in the animals treated prenatally with 1.0 or 3.0 mg/kg, but for the first time there were also significant changes in the lowest dose animals, those treated with 0.1 mg/kg. Finally, by PD30, the animals treated with the highest dose, 3.0 mg/kg were not impaired in the lick suppression paradigm. Interestingly, the animals which were the most impaired were those which had been treated with the lowest dose prenatally, 0.1 mg/kg. In addition to testing unchallenged behaviors, the animals treated prenatally with 1.0 mg/kg 5-MT were also tested at 30 days of age for open field activity, in response to challenges of 5-MT or apomorphine. These animals were found to have a blunted response to both receptor active drugs (Shemer *et al.*, 1991).

The classification of the serotonin receptors involved in development has not been a straightforward task. Although in searching for these receptors it has been assumed that they will be identical to those found in the adult brain, this may not necessarily be the case (Connell and Wallis, 1989). It is possible that the pharmacological profile of the receptor is different, or that the signal transduction mechanism could change. In these cases, strategies other than radioligand binding may be necessary, such as mRNA analysis or the use of specific antibodies. It is also possible that the receptor will be a transiently expressed receptor, which is lacking altogether in the adult brain. With these limitations in mind, we have tested selective drugs, and drawn some conclusions on how they may relate to brain development.

At present, there are thought to be seven different types of serotonin receptor in the adult brain. These are described in the following sections.

3.2.1 5-HT$_{1a}$ RECEPTORS

In adult brain, the 5-HT$_{1a}$ receptor can be both negatively and positively coupled to the production of cAMP. It is often referred to as the limbic serotonin receptor, as it is highly localized in these regions, particularly the hippocampus. It is virtually absent from the basal ganglia but can be found in the raphe nuclei, where it may act as a cell body autoreceptor. Drugs acting at this receptor are clinically useful in the treatment of depression and anxiety. The most commonly used radiolabel for this site is 8-[^{3}H]hydroxy-DPAT. This compound is also commonly used as an antagonist in behavioral and functional studies, as are ipsaperone, buspirone, geperone and tandospirone.

The 5-HT$_{1a}$ receptor is one of a group of neurotransmitter receptors which can be termed a 'transiently expressed receptor'; that is, at specific times in development very high amounts are expressed which then decrease as the animal ages. This peak in receptor number has been shown in rat (Daval *et al.*, 1987) and human fetal tissue (Bar-Peled *et al.*, 1991). A loss of this receptor has been reported in both Down's syndrome and Alzheimer's disease (Middlemiss *et al.*, 1986).

The 5-HT$_{1a}$ receptor appears to be the receptor which is stimulatory to growth of serotonin neurons, as well as to neurons in target regions.

Astroglial cells possess a number of different neurotransmitter receptors, including serotonin receptors, which decrease in number as the astrocyte matures (Whitaker-Azmitia and Azmitia, 1986b). Stimulating this receptor with selective receptor agonists causes production of growth-promoting media, principally by agonists selective for the 5-HT$_{1a}$ receptor (Whitaker-Azmitia and Azmitia, 1989). At the same time as the astrocytes release this growth factor, they attain a mature morphology. To characterize the growth-promoting substance(s), the astroglial-conditioned media (GCM) were preincubated with various antibodies, to see which would eliminate the growth-promoting properties. One such antibody tested was against S-100β. This protein was considered a candidate for several reasons: it has recently been shown to be on chromosome 21 (Allore *et al.*, 1988) (trisomy of which is responsible for Down's syndrome, which has long been considered to have a serotonergic involvement); to be overexpressed in Alzheimer's disease (Griffin *et al.*, 1989); and to be transiently expressed in the fetal rat brain at the time and in the place that serotonin neurons are developing (Van Hartesveldt *et al.*, 1986). Moreover, S-100β is a neurite extension factor (Kligman and Marshak, 1985). These studies indicated that S-100β is indeed at least one of the substances released by stimulation of astroglial 5-HT$_{1a}$ receptors (Whitaker-Azmitia *et al.*, 1990a). This protein can then act as a trophic factor for both serotonin neurons (Azmitia *et al.*, 1990) and cortical neurons (Kligman and Marshak, 1985).

In our most recent studies, we have examined the role of 5-HT$_{1a}$ receptors in the development of whole animals and found that the timing of stimulation of the 5-HT$_{1a}$ receptor is crucial. Selective agonists have limited effect prenatally, but at early postnatal timepoints the effects are profound.

In whole animal studies, pregnant rats were treated from gestational day 12 until birth with 1.0 mg/kg 8-OH-DPAT (a 5-HT$_{1a}$ receptor agonist) or 1.0 mg/kg spiroxatrine (a 5-HT$_{1a}$ receptor antagonist). Both of these drugs were toxic to the fetus, causing a significant amount of fetal reabsorption (based on low litter sizes) as well as high neonatal death rates. However, in the animals that survived, no changes were seen in behavior. In addition, there were no changes in serotonin terminal density, up until 60 days of age, when significant increases were observed in both hippocampus and brainstem.

When given at critical time periods postnatally, either from PD5–9 or PD11–15, 8-OH-DPAT has more significant effects than after prenatal administration. Interestingly, the results on behavioral and neurochemical development of the animals appears to be, in many cases, completely opposite. Treatment with the agonist at the earlier timepoint accelerates development – the animals opened their eyes sooner and gained weight more rapidly. At the later timepoint for treatment, the animals showed an increase in anxiety-related behaviors and development was delayed – the animals gained weight more slowly and showed delayed olfactory responses. Other workers have shown that depletion of serotonin by *p*-chlorophenyl-alanine on PD8–16 but not PD1–7, leads to decreased anxiety and more environmental reactivity in the rats when they reached adulthood (Farabollini *et al.*, 1988).

In tissue culture studies, 8-OH-DPAT induces tryptophan hydroxylase activity in embryonic mouse hypothalamus cells (de Vitry *et al.*, 1986).

3.2.2   5-HT$_{1b/d}$ RECEPTORS

The 5-HT$_{1b}$ receptor occurs in rat brain largely in the basal ganglia, where it functions predominantly as a release-regulating auto-receptor. There are also reports that this receptor is negatively coupled to adenylate cyclase. The receptor has been difficult to study pharmacologically, since the drugs which are potent at this site also affect other serotonin receptors and other neuro-transmitter systems. The most commonly used agonists are mCPP (*m*-chloro-phenylpiperazine) and TFMPP (trifluoro-methphenylpiperazine).

The 5-HT$_{1b}$ receptor has not been characterized for developmental density changes, in terms of a radiolabel binding assay. However, the receptor has been studied for behavioral effects. By four days of age, rat pups respond to mCPP by changes in mouthing and probing behaviors, as well as showing a general behavioral activation (Kirstein and Spear, 1988). There are few data yet to show a role for this receptor in mediating the role of serotonin as a growth factor. The 5-HT$_{1b}$ receptor has been shown to increase the synthesis of DNA in fibroblasts in culture, through inhibition of the production of cAMP (Seuwen *et al.*, 1988).

Drugs selective for the 5-HT$_{1b}$ receptor were found to have no effect on direct application to serotonin neurons growing in culture or on the production of growth factors by astrocytes. In studies using whole animals, rats treated with TFMPP (1 mg/kg) from gestational day 12 until birth had normal physical appearance but their weight gain was significantly greater than controls. In general, the animals were slow, slept excessively and displayed a gait abnormality. However, there were no neurochemical changes at postnatal days 15, 30 or 60. Although this receptor may be the most likely receptor responsible for the developmentally inhibitory effects of 5-MT, since it is a release-regulating autoreceptor, the agonists current-ly available may not be selective enough to adequately test this hypothesis.

3.2.3   5-HT$_{1c}$

The 5-HT$_{1c}$ receptor is highly concentrated in the choroid plexus, although it can be found in other regions at much lower concentra-tions. The receptor is linked to the inositol phosphate second messenger system and is thus in may ways related to the 5-HT$_2$ receptor. In radioligand binding assays, a number of receptor antagonists are useful, including [^{3}H]mesulergine and iodo-[^{3}H]LSD, however, there are no selective agonists. This has limited the ability of researchers to test for a role of this receptor in development. In one report, based on transfection of the gene for this receptor into fibroblasts, there is evidence that the 5-HT$_{1c}$ receptor may act as a protooncogene (Julius *et al.*, 1989). There are no studies on behavioral activation of the 5-

HT_{1c} receptor in either young or adult animals. However, the one study on number of receptors shows a linear increase in binding sites up to the fourth week, when adult levels are reached. Thus, there appears to be no transient peak in these receptors during development (Zilles *et al.*, 1986).

3.2.4 5-HT_2 RECEPTORS

The 5-HT_2 receptor is coupled to the production of inositol phosphates and is widely distributed through the brain, but principally in cortex. The receptor has also been localized to astroglial cells. There are a number of selective agonists, including DOI, as well as antagonists, specifically ketanserin. Both of these drugs are also used in radioligand binding studies.

The 5-HT_2 receptor apparently does not show a transiently increased level of expression during development. Rather, the receptors are not detectable at birth, and increase slowly thereafter to adult levels (Murrin *et al.*, 1985). Interestingly, however, the second messenger system to which this receptor is linked, phosphatidylinositol hydrolysis, does show a developmental peak, with the production of inositol phosphates by serotonin approximately ten-fold greater in the immature brain than in the adult brain (Claustre *et al.*, 1988). Moreover, 3-day-old rat pups respond to DOI by behavioral activation (Kirstein and Spear, 1988) and the effect of LSD on inhibiting firing of dorsal raphe serotonergic neurons is much greater in neonatal animals than in adult animals (Smith and Gallager, 1989). This suggests that the coupling efficiency of the receptor may change with time.

In our studies, using prenatal treatment with DOI, we observed no developmental changes in the offspring, in terms of serotonergic behaviors or neurochemistry.

Although 5-HT_2 receptors may not influence development, they may be what is referred to as programmable receptors, that is

events during development may affect the number, affinity or function of these receptors in the adult brain (Whitaker-Azmitia, 1991). In the case of 5-HT_2 receptors, both prenatal and postnatal stress to the mother significantly increases the number of 5-HT_2 receptors in the offspring, even after they have become adults (Peters, 1988a, b). Since the 5-HT_2 receptor is the predominant serotonin receptor in cortex, postulated to be involved in a variety of animal behaviors and human disease states, it is important that more research is done on examining the factors which program this receptor.

3.2.5 5-HT_3 RECEPTORS

The 5-HT_3 receptor is a ligand-gated ion channel principally transporting potassium. The distribution is quite limited, being concentrated in caudate, spinal cord, area postrema and hippocampus. Selective agonists include phenylbiguanide and 2-methylserotonin. Antagonists include MDL 72222, odansetron and zacopride. 5-HT_3 antagonists have recently become available for use as antiemetics.

In tissue culture studies of 5-HT_3 receptor active drugs, 5-HT_3 agonists are inhibitory to development of ED14 serotonin neurons and inhibit proliferation of the neuronal/glial cell line NG 108-15. Although these tissue culture studies suggest that the 5-HT_3 receptor is inhibitory to the development of serotonin neurons, the results are more divergent in whole animals.

Animals were treated prenatally from gestational day 12 until birth, with 5.0 mg/kg phenylbiguanide, or 5.0 mg/kg MDL 72222 or the equivalent volume of saline. Animals were tested at PD16 for spontaneous alternation (a measure of ascending serotonin terminal function) and at days 10, 18 and 30 for a tail flick response, that is latency to remove their tails from a heated light (a measure of descending serotonin terminal function). Interestingly, the tail flick responses were significantly altered. This is the first case in which we have found serotonergic drugs to have an effect on the development of descending serotonin neurons. No changes were observed in spontaneous alternation, but immunocytochemical analysis of serotonin terminals in hippocampus showed a significant decrease at 30 and 60 days of age. This confirms the tissue culture observations. At the same time, serotonin terminal density in spinal cord was increased (Bell *et al.*, 1991).

The data from these studies, therefore, suggest that the 5-HT$_3$ receptor is not simply related to development in a positive or negative manner. The message may be more complicated, and may result in a positive influence into spinal cord and a negative influence into brain.

3.2.6   5-HT$_4$ RECEPTORS

One of the most intriguing questions regarding the role of serotonin receptors in development has been the observation that serotonin stimulates adenylate cyclase activity in immature brain, but that the activity of this system declines with age (Nelson *et al.*, 1980). Since very little positively coupled cyclase is present in adult brain, it has been difficult pharmacologically to classify this receptor. Recently, however, the 5-HT$_4$ receptor has been described in immature rat colliculi. This receptor has a unique pharmacology from other serotonin receptors and is positively coupled to cyclase. A receptor which stimulates cAMP is of course particularly interesting in development, given the role that this second messenger system plays in cellular differentiation (McMahon, 1974).

3.3   CONCLUSION

Serotonin has clearly been shown to play a very important role in regulating the development of the mammalian brain. In order for serotonin to function in this capacity, there must be specific receptors present in the developing brain, which can respond to serotonin by activating specific transduction mechanisms. There are at least two receptors involved. One receptor is active at low concentrations of 5-MT, is related to the release of serotonin, and is inhibitory to the growth of serotonin neurons. The pharmacology of this receptor may not be identical to any receptor which occurs in the adult brain. The other receptor is stimulatory to growth of serotonin neurons, and is likely to be related to the 5-HT$_{1a}$ receptor. Since this receptor also releases the growth factor S-100β, this may be the mechanism by which serotonin directs development in target tissues. Other receptors have not been as well characterized for a role in development. However, the 5-HT$_3$ receptor and possibly the 5-HT$_4$ receptor may also be involved. Finally, the 5-HT$_2$ receptor, although not involved in regulating development, is greatly influenced by events during development, and can thus be termed the programmable serotonin receptor.

ACKNOWLEDGEMENTS

The author gratefully acknowledges the contributions made by collaborators in both the work described and in the formulation of hypotheses. These collaborators include Efrain Azmitia, Ann Shemer and Jean Lauder.

This work has been supported by grants from the National Institute for Child Health and Human Development and from the National Institute on Neurological Disease and Stroke.

REFERENCES

Aitken, A.R. and Törk, I. (1988) Early development of serotonin-containing neurons and pathways as seen in wholemount preparations of the fetal rat brain. *J. Comp. Neurol.*, **274**, 32–47.

Allore, R., O'Hanlon, D., Price, R. *et al.* (1988) Gene encoding the beta subunit of S-100 protein is on chromosome 21: implications for Down Syndrome. *Science*, **239**, 1311–13.

Azmitia, E.C. and De Kloet, E.R. (1987) ACTH

neuropeptide stimulation of serotonergic neuronal maturation in tissue culture: modulation by hippocampal cells. *Progr. Brain Res.*, **72**, 311–17.

Azmitia, E.C., Dolan, K. and Whitaker-Azmitia, P.M. (1990a) S-100 beta, but not NGF, EGF, insulin or calmodulin is a CNS serotonergic growth factor. *Brain Res.*, **576**, 354–6.

Azmitia, E.C., Murphy, R.B., and Whitaker-Azmitia, P.M. (1990b) MDMA (ecstasy) effects on cultured serotonergic neurons: evidence for Ca2(+)-dependent toxicity linked to release. *Brain Res.*, **510**, 97–103.

Bar-Peled, O., Gross-Isseroff, R., Ben-Hur, H. *et al.* (1991) Fetal human brain exhibits a prenatal peak in the density of serotonin 5HT$_{1A}$ receptors. *Neurosci. Lett.*, **127**, 173–6.

Bell, J., Zhang, X.N. and Whitaker-Azmitia, P.M. (1991) 5-HT$_3$ receptor-active drugs alter development of spinal serotonergic innervation: lack of effect of other serotonergic agents. *Brain Res.*, **571**, 293–7.

Breese, G.R., Baumeister, A.A., McCown, T.J. *et al.* (1984) Behavioral differences between neonatal and adult 6-hydroxydopamine-treated rats to dopamine agonists: relevance to neurological symptoms in clinical syndromes with reduced brain dopamine. *J. Pharmacol. Exp. Ther.*, **231**, 343–54.

Buznikov, G.A. (1984) The action of neurotransmitters and related substances on early embryogenesis. *Pharmacol. Ther.*, **25**, 23–59.

Chubakov, A.R., Gromova, E.A., Konovalov, G.V. *et al.* (1986) The effects of serotonin on the morpho-functional development of rat cerebral neocortex in tissue culture. *Brain Res.*, **369**, 285–97.

Claustre, Y., Rouquier, L. and Scatton, B. (1988) Pharmacological characterization of serotonin-stimulated phosphoinositide turnover in brain regions of the immature rat. *J. Pharmacol. Exp. Ther.*, **244**, 1051–6.

Connell, L.A. and Wallis, D.I. (1989) 5-Hydroxytryptamine depolarizes neonatal rat motorneurones through a receptor unrelated to an identified binding site. *Neuropharmacology*, **28**, 625–34.

Daval, G., Verge, D., Becerril, A. *et al.* (1987) Transient expression of 5-HT$_{1A}$ receptor binding sites in some areas of the rat CNS during postnatal development. *Int. J. Dev. Neurosci.*, **5**, 171–89.

Davila-Garcia, M.I. and Azmitia, E.C. (1989) Effects of acute and chronic administration of leuenkephalin on cultured serotonergic neurons.

Dev. Brain Res., **49**, 97–103.

DeVitry, F., Hamon, M., Catelon, J. *et al.* (1986) Serotonin initiates and autoamplifies its own synthesis during mouse central nervous system development. *Proc. Natl. Acad. Sci. USA*, **83**, 8629–33.

Farabollini, F., Hole, D.R. and Wilson, C.A. (1988) Behavioral effects in adulthood of serotonin depletion by *p*-chlorophenylalanine given neonatally to male rats. *Int. J. Neurosci.*, **41**, 187–99.

Fujimiya, M., Kimura, H. and Maeda, T. (1986) Postnatal development of serotonin nerve fibers in the somatosensory cortex of mice studied by immunohistochemistry. *J. Comp. Neurol.*, **246**, 191–201.

Goodrich, C.A., Baker, P.C. and Bauman, G.P. (1986) Biochemical and functional effects of fenfluramine in maturing mice. *Gen. Pharmacol.*, **17**, 456–60.

Goto, M. and Sano, Y. (1984) Ontogenesis of the central serotonin neuron system of the rat – an immunohistochemical study. *Neurosci. Res.*, **1**, 3–18.

Griffin, W.S.T., Stanley, L.C., Ling, C. *et al.* (1989) Brain interleukin 1 and S-100 immunoreactivity are elevated in Down Syndrome and Alzheimer Disease. *Proc. Natl. Acad. Sci. USA*, **86**, 7611–15.

Gu, Q., Patel, B. and Singer, W. (1990) The laminar distribution and postnatal development of serotonin-immunoreactive axons in the cat primary visual cortex. *Exp. Brain Res.*, **81**, 256–66.

Hamon, M. and Bourgoin, S. (1979) Ontogenesis of tryptophan transport in the rat brain. *J. Neural Transm.*, Suppl., **15**, 93–105.

Hamon, M. and Bourgoin, S. (1981) Possible role of serotonin and other monoamines as growth factors during brain development, in *Physiological and Biochemical Basis for Perinatal Medicine* (eds M. Morset-Couchard and A. Minkowski), Karger, Basel, pp. 286–95.

Jackson, D., Bruno, J.P., Stachowiak, M.K. and Zigmond, M.J. (1988) Inhibition of striatal acetylcholine release by serotonin and dopamine after the intracerebral administration of 6-hydroxydopamine to neonatal rats. *Brain Res.*, **457**, 267–73.

Jacobs, B.L. and Azmitia, E.C. (1992) Structure and function of the brain serotonin system. *Physiol. Rev.*, **72**, 165–229.

Jonsson, G. and Hallman, H. (1983) Effect of substance P on the 5,7-dihydroxytryptamine induced alteration of the postnatal development of central serotonin neurons. *Med. Biol.*, **61**, 105–12.

Julius, D., MacDermott, A.B., Axel, R. and Jessell, T.M. (1989) Molecular characterization of a functional cDNA encoding the serotonin 1c receptor. *Science*, **241**, 558–64.

Kligman, D. and Marshak, D. (1985) Purification and characterization of a neurite extension factor from bovine brain. *Proc. Natl. Acad. Sci. USA.*, **82**, 7136-9.

Kirstein, C.L. and Spear, L.P. (1988) 5-HT$_{1A}$, 5-HT$_2$ receptor agonists induce differential behavioral responses in neonatal rat pups. *Eur. J. Pharmacol.*, **150**, 339–45.

Lankford, K.L., Fernando, F.G. and Klein, W.L. (1988) D$_1$ type dopamine receptors inhibit growth cone motility in cultured retina neurons: evidence that neurotransmitters act as morphogenic growth regulators in the developing central nervous system. *Proc. Natl. Acad. Sci. USA*, **85**, 4567–71.

Lauder, J.M. (1983) Hormonal and humoral influences on brain development. *Psychoneuroendocrinology*, **8**, 121–55.

Lauder, J.M. (1990) Ontogeny of the serotonergic system in the rat: serotonin as a developmental signal. *Ann. NY Acad. Sci.*, **600**, 297–313.

Lauder, J.M. and Krebs, H. (1978) Serotonin as a differentiation signal in early neurogenesis. *Dev. Neurosci.*, **1**, 15–30.

Lauder, J.M. and Zimmerman, E.F. (1988) Sites of serotonin uptake in epithelia of the developing mouse palate, oral cavity, and face: possible role in morphogenesis. *J. Craniofacial Genet. Dev. Biol.*, **8**, 265–76.

Lauder, J.M., Tamir, H. and Sadler, T.W. (1988) Serotonin and morphogenesis. I. Sites of serotonin uptake and binding protein immunoreactivity in the midgestation mouse embryo. *J. Craniofac. Genet. Dev. Biol.*, **102**, 709–20.

Levitt, P. and Moore, R.Y. (1978) Developmental organization of raphe serotonin neuron groups in the rat. *Anat. Embryol. (Berl.)*, **154**, 241–51.

Lidov, H.G. and Molliver, M.E. (1982a) Immunohistochemical study of the development of serotonergic neurons in the rat CNS. *Brain Res. Bull.*, **9**, 559–604.

Lidov, H.G. and Molliver, M.E. (1982b) An immunohistochemical study of serotonin neuron development in the rat: ascending pathways and terminal fields. *Brain Res. Bull.*, **8**, 389–430.

Lieth, E., McClay, D.R. and Lauder, J.M. (1990) Neuronal-glial interactions: complexity of neurite outgrowth correlates with substrate adhesivity of serotonergic neurons. *Glia*, **3**, 169–79.

McMahon, D. (1974) Chemical messengers in development: a hypothesis. *Science*, **185**, 1012–21.

Messenger, N.J. and Warner, A.E. (1989) The appearance of neural and glial cell markers during early development of the nervous system in the amphibian embryo. *Development*, **107**, 43–54.

Middlemiss, D.N., Palmer, A.M., Edel, N. and Bowen, D.M. (1986) Binding of the novel serotonin agonists 8-hydroxy-2-(dipropylamino) tetralin in normal and Alzheimer brain. *J. Neurochem.*, **46**, 993–6.

Murrin, C.L., Gibbens, D.L. and Ferrer, J.R. (1985) Ontogeny of dopamine, serotonin and spirodecanone receptors in rat forebrain – an autoradiographic study. *Dev. Brain Res.*, **23**, 91–109.

Nelson, D.L., Herbert, A., Adrien, J. *et al.* (1980) Serotonin-sensitive adenylate cyclase and ^{3}H-serotonin binding sites in the CNS of the rat – II. Respective regional and subcellular distributions and ontogenic developments. *Biochem. Pharmacol.*, **29**, 2455–63.

Ni, L. and Jonakait, G.M. (1989) Ontogeny of substance P-containing neurons in relation to serotonin-containing neurons in the central nervous system of the mouse. *Neuroscience*, **30**, 257–69.

Peters, D.A. (1982) Prenatal stress: effects on brain biogenic amine and plasma corticosterone levels. *Pharmacol. Biochem. Behav.*, **17**, 721–5.

Peters, D.A. (1988a) Both prenatal and postnatal factors contribute to the effects of maternal stress on offspring behavior and central 5-hydroxytryptamine receptors in the rat. *Pharmacol. Biochem. Behav.*, **30**, 669–73.

Peters, D.A. (1988b) Effects of maternal stress during different gestational periods on the serotonergic system in adult rat offspring. *Pharmacol. Biochem. Behav.*, **31**, 829–43.

Peters, D.A. (1990) Maternal stress increases fetal brain and neonatal cerebral cortex 5-hydroxytryptamine synthesis in rats: a possible mechanism by which stress influences brain development. *Pharmacol. Biochem. Behav.*, **35**, 943–7.

Rajaofetra, N., Sandillon, F., Geffard, M. and Privat, A. (1989) Pre- and postnatal ontogeny of serotonergic projections to the rat spinal cord. *J. Neurosci. Res.*, **22**, 305–21.

Resnick, O. and Morgane, P.J. (1984) Ontogeny of the levels of serotonin in various parts of the brain in severely protein malnourished rats. *Brain Res.*, **303**, 163–70.

Schambra, U.B., Lauder, J.M., Petrusz, P. and Sulik, K.K. (1990) Development of neurotransmitter systems in the mouse embryo following acute ethanol exposure: a histological and immunocytochemical study. *Int. J. Dev. Neurosci.*, **8**, 507–22.

Seuwen, K., Magnaldo, I. and Pouysségur, J. (1988) Serotonin stimulates DNA synthesis in fibroblasts acting through 5-HT$_{1B}$ receptors coupled to a Gi-protein. *Nature*, **335**, 254–6.

Shemer, A.V., Azmitia, E.C. and Whitaker-Azmitia, P.M. (1991) Dose-related effects of prenatal 5-methoxytryptamine (5-MT) on development of serotonin terminal density and behaviour. *Dev. Brain Res.*, **59**, 59–63.

Smith, D.A. and Gallager, D.W. (1989) Electrophysiological and biochemical characterization of the development of alpha 1-adrenergic and 5-HT1 receptors associated with dorsal raphe neurons. *Dev. Brain Res.*, **46**, 173–86.

Spear, L.P. and Ristine, L.A. (1982) Suckling behavior in neonatal rats: psychopharmacological investigations. *J. Comp. Phys. Behav.*, **96**, 244–55.

Spear, L.P. and Scalzo, F.M. (1985) Ontogenic alterations in the effects of food and/or maternal deprivation on 5-HT/5-HIAA and 5-HIAA/5-HT ratios. *Brain Res.*, **350**, 143–57.

Stuart, D.K., Blair, S.S. and Weisblat, D.A. (1987) Cell lineage, cell death, and the developmental origin of identified serotonin- and dopamine-containing neurons in the leech. *J. Neurosci.*, **7**, 1107–22.

Tillet, Y. (1988) Early ontogeny of serotonin-immunoreactivity in the sheep brain. An immunohistochemical study. *Anat. Embryol.*, **178**, 429–40.

Vallés, A.M. and White, K. (1988) Serotonin-containing neurons in *Drosophila melanogaster*: development and distribution. *J. Comp. Neurol.*, **268**, 414–28.

Van Hartesveldt, C., Moore, B. and Hartman, B.K. (1986) Transient midline raphe glial structure in the developing rat. *J. Comp. Neurol.*, **253**, 175–84.

Wallace, J.A. (1985) An immunocytochemical study of the development of central serotoninergic neurons in the chicken embryo. *J. Comp. Neurol.*, **236**, 443–53.

Wallace, J.A. and Lauder, J.M. (1983) Development of the serotonergic system in the rat embryo: an immunocytochemical study. *Brain Res. Bull.*, **10**, 459–79.

Wallace, J.A., Allgood, P.C., Hoffman, T.J. *et al.* (1986) Analysis of the change in number of sero-

tonergic neurons in the chick spinal cord during embryonic development. *Brain Res. Bull.*, **17**, 297–305.

Whitaker-Azmitia, P.M. (1991) Role of serotonin and other neurotransmitter receptors in brain development: basis for developmental pharmacology. *Pharmacol. Rev.*, **43**, 553–61.

Whitaker-Azmitia, P.M. and Azmitia, E. (1986a) Autoregulation of fetal serotonergic neuronal development: role of high affinity serotonin receptors. *Neurosci. Lett.*, **67**, 307–12.

Whitaker-Azmitia, P.M. and Azmitia, E.C. (1986b) ^{3}H-5-Hydroxytryptamine binding to brain astroglial cells: differences between intact and homogenized preparations and mature and immature cultures. *J. Neurochem.*, **46**, 1186–90.

Whitaker-Azmitia, P.M. and Azmitia, E.C. (1989) Stimulation of astroglial serotonin receptors produces media which regulates development of serotonergic neurons. *Brain Res.*, **497**, 80–5.

Whitaker-Azmitia, P.M., Lauder, J.M., Shemer, A. and Azmitia, E.C. (1987) Prenatal plasticity of high affinity serotonin receptors: further evidence for a role of receptors in development. *Dev. Brain Res.*, **33**, 285–90.

Whitaker-Azmitia, P.M., Murphy, R. and Azmitia, E.C. (1990a) Stimulation of astroglial 5-HT$_{1a}$ receptors releases the serotonergic growth factor, protein S-100, and alters astroglial morphology. *Brain Res.*, **528**, 155–8.

Whitaker-Azmitia, P.M., Quartermain, D. and Shemer, A.V. (1990b) Prenatal treatment with SKF 38393, a selective D-1 receptor agonist: longterm consequences on ^{3}H-paroxetine binding and on dopamine and serotonin receptor sensitivity. *Dev. Brain Res.*, **57**, 181–5.

Zhou, F.C. and Azmitia, E.C. (1988) Laminin facilitates and guides fiber growth of transplanted neurons in adult brain. *J. Chem. Neuroanat.*, **1**, 133–46.

Zhou, F.C., Auerbach, S. and Azmitia, E. (1987) Denervation of serotonergic fibers in the hippocampus induced a trophic factor which enhances the maturation of transplanted serotonergic neurons but not norepinephrine neurons. *J. Neurosci. Res.*, **17**, 235–46.

Zilles, K., Rath, M., Schleicher, A. *et al.* (1986) Ontogenesis of serotonin (5-HT) binding sites in the choroid plexus of the rat brain. *Brain Res.*, **380**, 201–3.

NEUROKININ AND SUBSTANCE P RECEPTORS IN THE DEVELOPING RAT CENTRAL NERVOUS SYSTEM

4

Than-Vinh Dam, Gail E. Handelmann and Rémi Quirion

4.1 INTRODUCTION

It is now established that mammalian neurokinins (NK) are derived from two different but related genes namely preprotachykinin (PPT) I and II (Nawa *et al.*, 1983, 1984; Kawaguchi *et al.*, 1986; Kotani *et al.*, 1986; Krause *et al.*, 1987, 1989; Nakanishi, 1987). The RNA transcribed from the PPT-I gene is alternatively spliced to yield different mRNAs encoding the α-, β- and γ-preprotachykinin (α-, β- or γ-PPT, respectively) precursors. Among the various NKs, it has been shown that substance P (SP) may arise from the post-translational processing of α-, β- or γ-PPT whereas neurokinin A (NKA) is generated by the processing of β- or γ-PPT (Nawa *et al.*, 1984; Kawaguchi *et al.*, 1986; Kotani *et al.*, 1986; Krause *et al.*, 1987, 1989; Nakanishi, 1987). Neurokinin B (NKB) is exclusively derived from the PPT-II gene (Kotani *et al.*, 1986; Nakanishi, 1987; Krause *et al.*, 1989).

The ratio between these various NKs varies according to the differential splicing of the PPT in RNAs in brain and peripheral tissues (Nawa *et al.*, 1984; Krause *et al.*, 1987). For example, in the adult rat brain, γ-PPT mRNA represents up to 80% of total PPT-I derived mRNAs whereas α-PPT mRNA is almost absent (Krause *et al.*, 1987, 1988). Moreover, it was recently shown that the three PPT-I mRNAs can potentially generate multiple biologically active peptides, in addition to the three better known NKs (Dam *et al.*, 1990c; Helke *et al.*, 1990). Extended forms of NKA such as neuropeptide K and neuropeptide γ (γ-PTT-(72–92)-peptide amide), as well as NKA(3–10) are known to exist in various brain regions and peripheral tissues (Tatemoto *et al.*, 1985; MacDonald *et al.*, 1988). It is likely that similar extended forms of either SP and NKB could exist in some tissues (MacDonald *et al.*, 1988).

The diversity of the biological effects induced by these various NKs (subsequent sections and Watson *et al.*, 1983; Buck *et al.*, 1984, 1986; Quirion, 1985; Maggio, 1988; Quirion and Dam, 1988; Regoli *et al.*, 1988, 1989; Krause *et al.*, 1989; Rovero *et al.*, 1989; Takeda and Krause, 1989b; Helke *et al.*, 1990) provides indirect evidence for the existence of multiple classes of NK receptors. This hypothesis has been supported by receptor binding (Cascieri and Liang, 1983; Viger *et al.*, 1983; Buck and Burcher, 1986; Lee *et al.*, 1986; Quirion and Dam, 1986; Iversen *et al.*, 1987; Maggio, 1988) and autoradiographic data (Mantyh *et al.*, 1984a; Quirion and Dam, 1985, 1988). Thus far, three major classes of NK receptors (NK-1, NK-2 and NK-3) have been found and have been recently cloned (Masu *et al.*, 1987; Yokota *et al.*, 1989; Hershey and Krause, 1990; Shigemoto *et al.*, 1990). All three

Receptors in the Developing Nervous System Vol. 2: Neurotransmitters. Edited by Ian S. Zagon and Patricia J. McLaughlin. Published in 1993 by Chapman & Hall. ISBN 0 412 49400 0. Vols. 1 and 2 (set) ISBN 0 412 54520 9.

show important sequence homologies and are members of the seven transmembrane domain G-protein-coupled receptor family, which also includes catecholamine, dopamine and serotonin receptors, the *mas* oncogene, the rhodopsin receptors and several receptors of undefined specificity (O'Dowd *et al.*, 1989). SP, NKA (and its extended forms) and NKB preferentially, but not exclusively, bind to the NK-1, NK-2 and NK-3 receptor class, respectively (Quirion and Dam, 1988; Dam *et al.*, 1990c; Helke *et al.*, 1990).

We have previously reported on the differential distribution of these various classes of NK receptors in mammalian brain (Quirion *et al.*, 1983; Quirion and Dam, 1985, 1988; Dam and Quirion, 1986; Dam *et al.*, 1988) and on the ontogenic profile of NK-1 (Quirion and Dam, 1986) and other NK receptors, especially in the cortical areas (Dam *et al.*, 1988). These results revealed, for example, that the density of NK-1 binding sites in the brainstem is extremely high during the first postnatal week but is very low in the same area in the adult. Moreover, the respective distribution of NK-2 and NK-3 receptor binding sites undergoes modifications during postnatal brain maturation. This suggests the possible involvement of NKs and NK receptors in the developmental functional organization of the CNS. This chapter reviews some of the possible functional roles of NKs in the CNS and focuses on the ontogenic profile of NKs and NK receptor subtypes in the rat brain. The possible regulation of NKs with regard to brain maturation is also discussed.

4.2 BIOLOGICAL FUNCTIONS OF NEUROKININS

4.2.1 NOCICEPTION AND OTHER SENSORY PROCESSES

NKs play an important role in the transmission of sensory information from peripheral receptors to the central nervous system. In the peripheral nervous system, SP and NKA are contained in primary sensory gan-

glion neurons which give rise to the small calibre, unmyelinated C fibers (Holzer *et al.*, 1982; Ogawa *et al.*, 1985; Takano *et al.*, 1986). Primary sensory neurons are contained in the dorsal root ganglia, which project to the spinal cord. The peripheral terminals of the dorsal root ganglia are distributed in the skin and other organs and are found in close apposition to blood vessels, sweat glands, hair follicles, and in autonomic ganglia such as the colonic and hypogastric ganglia. These neurons are activated by painful stimulation, particularly that caused by heat or chemical irritants. Activation of the dorsal root ganglion cells induces the release of SP in the superficial laminae (I and II) of the dorsal horn of the spinal cord (Otsuka and Konishi, 1976). NK-1, NK-2 and NK-3 receptors are concentrated in the dorsal horn, particularly in laminae I and II, where the C fibers terminate (Charlton and Helke, 1985; Buck *et al.*, 1986; Yashpal *et al.*, 1990, 1991a). Several lines of evidence support the role of SP in nociception. SP depolarizes those dorsal horn neurons which are known to be activated by painful stimuli (Randic and Miletic, 1977; Sastry, 1979). Intrathecal administration of SP causes a specific scratching syndrome, as though the animals were responding to a perceived noxious stimulus (Lembeck and Gamse, 1982). Intrathecal SP also temporarily decreases the latency to respond to noxious cutaneous stimulation, suggesting sensitization to this stimulation (Lembeck and Donnerer, 1981). Additional support comes from experiments using the neurotoxin capsaicin. Capsaicin administered to neonatal animals selectively destroys SP-containing neurons in sensory ganglia, but has no effect on SP-containing neurons in the CNS (Gamse *et al.*, 1980). Capsaicin administered to adults does not destroy sensory neurons, but temporarily depletes their supply of SP. Treatment with capsaicin in both neonates and adults causes a dramatic decrease in sensitivity to painful stimulation of the skin, particularly when the pain is induced by heat or chemical irritants. Finally, SP immunoreactivity is absent in the dorsal horn of patients having diminished sensitivity to pain as a result of familial dysautonomia (Pearson *et al.*, 1982). Together, these data indicate a role for NKs in nociception at the level of the spinal cord.

NKs may also play a role in sensory information processing in higher central nervous system structures. SP-containing neurons are found in the trigeminal ganglia and visceral sensory ganglia, which send primary sensory afferent fibers directly to the brain (Cuello *et al.*, 1982). The peripheral terminals of the trigeminal ganglia are distributed in various tissues of the head, including glands, dental pulp, the nasal mucosa and the tongue. The dental C fibers, which contain SP are especially reactive to heat stimuli (Byers, 1984). At the level of the brainstem, there is abundant SP innervation of the general somatic sensory nuclei, such as the substantia gelatinosa rolandi, the nucleus of the tractus spinalis trigemini and the nucleus principalis trigemini. The general visceral sensory nuclei of the brainstem are also innervated by SP fibers (Cuello and Kanazawa, 1978). The origin of these fibers is most likely the respective sensory ganglia outside the brain. Sensory fibers from other organ systems, such as the cardiovascular, gastrointestinal and respiratory systems, terminate in the nucleus of the tractus solitarius, where integration of central peripheral afferent information occurs. This nucleus contains a high density of NK receptors and SP is excitatory to neurons in the nucleus (Morin-Surun *et al.*, 1984). SP binding sites are also found in sensory brain regions such as the central gray, superior and inferior colliculi and cerebral cortex. In cortical areas, SP is excitatory to neurons in the somatosensory cortex, particularly in laminae Vb and VIb (Lamour *et al.*, 1983). Although all three NK receptors are present in cortex, NK-3 receptors are especially dense in these deep cortical layers (Dam *et al.*, 1988, 1990a, b, c; Mantyh *et al.*, 1989a; Stoessl and Hill, 1990).

In addition to their role in the conduction

of pain sensation, NKs may also play a role in the processing of olfactory and visual information. NK-1 receptors are contained in a number of components of the olfactory system. NK-1 and NK-2 receptors are present in high concentration in the olfactory bulb, olfactory tubercule, primary olfactory cortex and septum (Shults *et al.*, 1984; Dam *et al.*, 1990b, c). Surprisingly, only a few scattered SP fibers are present in these structures suggesting that another NK may be the endogenous ligand for the receptors present in the olfactory system. SP sensory fibers are also contained in the nasal mucosa, but their function there appears to be related to the regulation of local blood supply (Holzer *et al.*, 1982; St-Jarne *et al.*, 1989) as will be discussed below.

SP is also highly represented in visual pathways. SP is contained in most layers of the eye and in the optic nerve and SP-containing terminals are present in the superior colliculi and lateral geniculate (Cuello and Kanazawa, 1978). In most parts of the eye, the SP fibers are of sensory origin (Miller *et al.*, 1981). However, sensory denervation does not reduce SP contents in the retina (Tervo *et al.*, 1982). The rat retina contains both SP immunoreactivity, primarily in amacrine cells, and also NK-1 receptors suggesting a local action of SP within the retina as well as a role in the transmission of information to visual centers in the brain. Neonatal treatment with monosodium glutamate, which destroys retinal ganglion cells and the inner nuclear layers of the retina, severely reduced NK-1 receptor content in the retina. This suggests that NK-1 are located on the ganglion cells and/or on cells of the inner nuclear layer of the retina (Lee and Cheng, 1988). SP applied locally enhances light-evoked excitation of retinal ganglion cells, at least in non-mammalian species (Dick *et al.*, 1980; Glickman *et al.*, 1980). This suggests a possible role for SP as a positive modulator of the transmission of visual information.

The possibility that SP might also be involved in the conduction of gustatory information is suggested by the finding that SP-containing nerve terminals are present in the taste buds of the tongue (Lundberg *et al.*, 1980; Nishimoto *et al.*, 1982), presumably derived from the glossopharyngeal nerve (Pernow, 1983). The SP terminals, however, do not make synaptic contact with underlying cells in the taste buds (Yamasaki *et al.*, 1985). Therefore, although SP may be released in the vicinity of the taste buds, the SP fibers are probably not involved in gustatory neurotransmission.

4.2.2   REGULATION OF CARDIOVASCULAR ACTIVITY

Another well-characterized activity of NKs is in the regulation of blood pressure and local blood supply. A unique feature of the primary sensory neurons is their symmetrical nature: both their central and peripheral branches contain and release SP. SP is supplied to sensory nerve branches by fast axonal transport and much of the SP produced in sensory ganglion cells is directed toward the periphery. SP nerve networks derived from these sensory neurons are present in vascular beds in virtually every tissue. Peripheral administration of SP causes vasodilation leading to hypotension (Hassessian *et al.*, 1988; Helke *et al.*, 1990). In addition, SP acts centrally to alter blood pressure. In contrast to its action in the periphery, central administration of SP or NK-3 agonists causes hypertension (Unger *et al.*, 1981; Nagashima *et al.*, 1989), apparently through the activation of the sympathetic nervous system (Unger *et al.*, 1981).

SP released from peripheral sensory nerve endings regulates local blood flow (Lembeck and Holzer, 1979). Antidromic stimulation of sensory C fibers, which contain and release SP, causes vasodilation (Hinsey and Gasser, 1930). Treatment with capsaicin to deplete SP in C fibers abolishes the antidromic vasodilation (Lembeck and Holzer, 1979). Local administration of SP mimics the effects of antidromic nerve stimulation on vasodilation in dental pulp and the eye (Olgart *et al.*, 1977;

Bill *et al.*, 1979). SP is active only in tissues with intact endothelium. The smooth muscle relaxation appears to be due to the stimulation of an endothelial NK-1 receptor, whose activation brings about the release of a vasodilator factor (D'Orléans-Juste *et al.*, 1985). This endothelium-derived relaxing factor has recently been identified as nitric oxide (Schini *et al.*, 1990). NK-2 receptors also appear to be localized in smooth muscle membrane (D'Orléans-Juste *et al.*, 1986; Mastrangelo *et al.*, 1986) and may be similarly involved in vasodilation. In general, however, NK-1 agonists are more potent in their effects on blood pressure than are NK-2 and NK-3 agonists (Hassessian *et al.*, 1988).

SP is present in cerebrovascular nerves and can be depleted by capsaicin, suggesting that primary sensory afferents innervate cerebral blood vessels (Duckles and Buck, 1982). Bundles of SP fibers run in the pia matter of the spinal cord; they appear to arrive there via both the dorsal and ventral roots, again suggesting that they are of sensory origin (Dalsgaard *et al.*, 1982). SP fibers have also been demonstrated around pial arteries in the medulla oblongata (Edvinsson *et al.*, 1981; Edvinsson and Uddman, 1982). As in other tissues, the release of SP at these nerve terminals causes vasodilation and thereby increases sensory nerves; on the other hand, they may represent a pathway for vascular headache (Mayberg *et al.*, 1981).

NKs administered centrally affect blood pressure. NK receptors are located in several brain areas implicated in central control of blood pressure, including the nucleus of the tractus solitarius. Increased numbers of SP receptors were found in several brain regions in spontaneously hypertensive rats (Shigematsu *et al.*, 1987), which are supersensitive in terms of their cardiovascular response to centrally administered SP (Unger *et al.*, 1981).

In the spinal cord, NK-1 binding sites are located postsynaptically on preganglionic sympathetic neurons in the intermediolateral cell column. This location indicates a role for SP in regulating sympathetic outflow from these neurons which influence the cardio-vascular system. SP terminals in this region are derived from projections of the ventral medulla (Helke *et al.*, 1982) and SP is excitatory to the preganglionic neurons (Gilbey *et al.*, 1983; Backman and Henry, 1984). Intrathecal administration of a stable SP analog increased blood pressure and heart rate, and these effects were blocked by a sympathetic ganglionic blocker (Keeler *et al.*, 1985). Therefore, SP neurons in the ventral medulla, which send axons to the intermedio-lateral cell column, may be capable of elevat-ing sympathetic activity and thereby play a role in the maintenance of vasomotor tone.

4.2.3 NON-VASCULAR SMOOTH MUSCLE MOTILITY

In addition to its effects on vascular smooth muscle, SP also influences motility of non-vascular smooth muscle (Lundberg *et al.*, 1980). A major site of this activity is the gastrointestinal tract. In the gastrointestinal tract, the majority of SP is contained in intrinsic neurons (Costa *et al.*, 1980; Holzer *et al.*, 1980). The intrinsic neurons, including those which contain SP, are embedded in the intestinal wall in small ganglia connected by small nerve bundles to form two ganglionated plexuses, the myenteric and submucosal plexuses. From these plexuses, nerve bundles emerge to supply the muscle layers, blood vessels, glands and villi. Although SP is found in only a small proportion of myenteric and submucous neurons, these neurons produce a profuse innervation of muscle, mucosa and the submucous ganglia (Costa *et al.*, 1980). SP is also contained in some extrinsic sensory nerves which supply the submucosal blood vessels and the submucous ganglia (Furness *et al.*, 1982). SP induces contraction of intestinal muscle and it is more potent than acetylcholine in this regard. Both longitudinal and circular muscle layers are contracted directly by SP

(Bury and Mashford, 1977). In addition, SP has been shown, *in vitro*, to stimulate cholinergic neurons, which in turn contract intestinal muscle (Holzer and Lembeck, 1980).

Distinct NK-1 and NK-3 receptors have been characterized in guinea-pig ileum. The NK-1 receptors are present on smooth muscle while the NK-3 receptors are located on cholinergic neurons (Laufer *et al.*, 1985; Cascieri *et al.*, 1986). Despite the presence of NK-3 receptors, NKB is not detectable by radioimmunoassay in the guinea-pig ileum (Too *et al.*, 1989) suggesting that other NKs may act as endogenous ligands at this site. NKA, on the other hand, is found in high concentrations in the ileum (Takano *et al.*, 1986), although NK-2 receptors are not detectable. In general, NK-2 and NK-3 ago-nists are more potent than NK-1 agonists in stimulating smooth muscle contraction (Erspamer *et al.*, 1980; Regoli *et al.*, 1989; Jacques and Couture, 1990).

4.2.4 SALIVATION

SP innervation of the salivary glands is derived from the otic and trigeminal ganglia parasympathetic nerves (Sharkey and Templeton, 1984) and SP stimulates salivation by a direct action on secretory cells in the salivary glands. The effect is dose-dependent and mediated by NK-1 receptors (Buck and Burcher, 1985; Murray *et al.*, 1988). NKA, neuropeptide K and neuropeptide γ are also present in salivary glands (Takano *et al.*, 1986; Takeda *et al.*, 1990). NKA stimulates saliva-tion although less potently than SP (Murray *et al.*, 1988). Neuropeptide K, on the other hand, is even more potent than SP in produc-ing salivation; it also potentiates the actions of SP on the salivary glands (Takeda and Krause, 1989a). Neuropeptide γ can potentiate the effect of SP on salivation although it has little sialogogic activity alone (Takeda and Krause, 1989b). These results may be consis-tent with the presence of more than one NK receptor functioning in the salivary glands.

4.2.5   PROINFLAMMATORY ACTIONS AND WOUND HEALING

In addition to vasodilation, release of SP from peripheral terminals of sensory nerves produces several proinflammatory responses. One of these is plasma extravasation or edema (Lembeck and Holzer, 1979). In the rat, cutaneous plasma extravasation can be induced by sensory nerve stimulation or by injection of SP either into an artery or into the spinal cord (Jacques and Couture, 1990). In the rat paw, plasma extravasation, induced either by sensory nerve stimulation or by intra-arterial SP, is blocked by histamine blockers (Lembeck and Holzer, 1979). Direct release of histamine from mast cells by SP has been demonstrated (Erjavec et al., 1981). Therefore, SP appears to mediate edema by activation of mast cells. The action of SP on mast cells is not mediated by NK-2 or NK-3 receptors, because NKA and NKB also stimulate local edema but have no effect on mast cells (Brain and Williams, 1989). No selective membrane receptors for SP have been identified in those preparations and recent evidence suggests that SP may act on mast cells via a receptor-independent mechanism. Mousli et al. (1990a, b) have found that SP directly activates G proteins *in vitro*. They propose that SP crosses the mast cell membrane to interact directly with the intracellular G protein subunits. This implies a specific interaction of SP with the G protein that other NKs do not share (by virtue of basic charged N-terminal amino acid residues).

Release of SP in the joints also appears to induce proinflammatory activity. The joints are innervated by small unmyelinated sensory nerve fibers containing SP (Wyke, 1981; Pernow, 1983) and SP has been detected in synovial fluids (Inman et al., 1986). SP stimulates rheumatoid synoviocytes to produce prostaglandin E_2 and collagenase and also causes them to proliferate (Lotz et al., 1987). These actions may explain the ability of SP to increase the severity of adjuvant-induced arthritis in rats (Levine et al., 1984).

Studies of various populations of immune cells indicate that NKs induce a variety of responses in these cells. NKA and physalaemin stimulate mouse thymocyte proliferation in tissue culture, although SP had no effect (Soder and Hellstrom, 1989). SP also stimulates proliferation of human T-lymphocytes and enhances the mitogenic response of these cells to phytohemaglutinin (Payan et al., 1983), and there are specific SP receptors on human IM-9 lymphoblasts (Payan et al., 1984). SP also stimulates phagocytosis by macrophages and polymorphonuclear leukocytes (Bar-Shavit et al., 1980). SP stimulates IgA synthesis by organs of the secretory immune system of the mucosal tissues and IgM synthesis by the spleen (Stanisz et al., 1986). SP promotes chemotaxis of monocytes and neutrophils (Ruff et al., 1985), although the chemotaxis may be mediated by a receptor-independent mechanism (Kroegel et al., 1990). SP stimulates lysosomal enzyme release from human, rat and rabbit neutrophils (Serra et al., 1988). Finally, SP induces generation of oxygen radicals and thromboxane A_2 in macrophages and neutrophils (Hartung et al., 1986). Local release of NKs may therefore act as immunoregulatory factors through a variety of mechanisms. In turn, immune system cell products may regulate SP synthesis. The cytokine interleukin 1β increases SP contents of superior cervical ganglion explants (Freidin and Kessler, 1990). In dissociated cultures, this action required the presence of non-neuronal cells, suggesting an indirect effect on the neurons via glia or fibroblasts. The immunosuppressant dexamethasone prevented the interleukin effect. This suggests that an increase in SP synthesis is a response to local injury or inflammation.

Within the central nervous system, certain immune functions are carried out by glial cells. Recent evidence suggests that NK receptors are expressed by glial cells. *In vitro* studies indicate that cortical astrocytes from the mouse express SP receptors in culture (Torrens et al., 1986). NK-1 receptors were also found on cells of a human astrocytoma cell line (Lee *et al.*, 1989). Binding of SP to these cells was inhibited by guanyl-5'-imidodiphosphate, which is a characteristic of G-protein-coupled receptors (Snyder, 1979), indicating that the glial receptors share some of the characteristics of the neuronal NK-1 receptors. SP added to these cultures increased RNA synthesis. *In vivo*, NK-1 receptors are also present on Schwann cells of the giant squid. SP causes a long-lasting hyperpolarization of these Schwann cell membranes (Evans *et al.*, 1990). In a mammalian system, NK-1 receptors were found to be expressed by reactive astrocytes forming a glial scar following section of the optic nerve in adult rabbits (Mantyh *et al.*, 1989b).

SP release in response to irritation or tissue damage can therefore regulate inflammatory and immune reactions. In addition, SP has actions on cutaneous tissue which would promote wound healing. For example, NKs are mitogens for several cell types. SP and NKA stimulate DNA synthesis in cultured arterial smooth muscle cells and human skin fibroblasts. This effect is inhibited by the SP antagonist spantide (Nilsson *et al.*, 1985). However, because NKA was more potent than SP, the effect is probably mediated via NK-2 or NK-3 receptors. SP but not NKA stimulated DNA synthesis in cultured mouse epidermal cells, although the effect required the presence of serum, indicating the requirement for a synergistic growth factor (Tanaka *et al.*, 1988). SP also stimulates neovascularization in the rabbit cornea (Ziche *et al.*, 1990), which may result from the proliferation of endothelial cells, as SP stimulates endothelial cell proliferation in tissue culture (Dalsgaard *et al.*, 1989; Ziche *et al.*, 1990). The effect of SP on cultured endothelial cells appears to be mediated via NK-1 receptors, as NK-2 or NK-3 agonists had no effect. SP also stimulates endothelial cell migration via an NK-1 receptor-mediated mechanism, which also plays a major role in the process of neovascularization (Ziche *et al.*, 1991). These results suggest that the SP innervation of blood vessels has two functions, vasodilation and stimulation of growth. Therefore NK released during inflammation may participate in wound healing processes.

4.2.6 GROWTH AND TROPHIC EFFECTS

In addition to its possible functions in promoting cell proliferation in the immune systems and during tissue repair, SP may have other roles as a trophic factor or growth-promoting factor in various tissues. For example, the atrophy of the salivary glands which occurs after denervation or feeding a liquid diet can be prevented by daily infusions of SP (Mansson *et al.*, 1990). In addition, administration of physalaemin but not eledoisin increases salivary gland weight in intact rats (Bertaccini *et al.*, 1966; Cantalamessa *et al.*, 1975). Therefore, SP may have a trophic effect on salivary glands, probably via NK-1 receptors. The fact that other parasympathetic agonists did not have this effect on salivary glands (Mansson *et al.*, 1990) suggests that this is a function which is distinct from increasing nerve reflex activity or nervous tone.

High doses of SP also stimulate neurite outgrowth in embryonic dorsal root ganglia in tissue culture (Narumi and Fujita, 1978) and in cultured neuroblastoma cells (Narumi and Maki, 1978). The mechanism appears to involve the production of cAMP. The doses of SP required to produce this effect are six orders of magnitude greater than those required for nerve growth factor. Although this comparison would suggest that SP is not a particularly potent stimulator of neurite outgrowth, experiments *in vivo* suggest that SP may stimulate regrowth of neurites following damage. SP counteracts the toxic effects of the specific neurotoxin 6-hydroxydopamine in the neonate (Jonsson and Hallman, 1982b). Treatment of newborn rats with 6-hydroxydopamine causes permanent degeneration of the distant nerve terminals of noradrenergic neurons and excessive growth of those nerve terminals which innervate nearby targets. SP injected intracisternally blocked these effects. SP may act by stimulating neurite regrowth following the toxic damage. In kittens, SP infusion into the fourth ventricle increased the rate of reinnervation of a cortical area denervated by local application of 6-hydroxydopamine (Nakai and Kasamatsu, 1984). Because SP did not have this effect when infused at the site of the lesion, SP is probably acting on the noradrenergic cell bodies in the locus coeruleus. However, the effect on reinnervation is probably not simply due to excitation of the noradrenergic neurons because other compounds excitatory to these cells, such as bethanechol chloride, did not stimulate reinnervation.

Wall *et al.* (1982) have suggested that SP may play a role in the maintenance of synaptic connections between sensory afferents and central neurons. They found that neonatal capsaicin treatment of mice resulted in a loss of specificity of the connections between individual whiskers and their receptive cells in the somatosensory cortex. In addition, they found that capsaicin treatment of sensory nerves in the adult expanded the receptive fields of dorsal horn neurons (Wall *et al.*, 1982). These data suggest that disorganization of somatotrophic maps occurs in the absence of SP. Alternatively, however, such disorganization may result from sprouting from intact nerve terminals in the vicinity of the denervated neurons (Shortland *et al.*, 1990).

Activity of NKs in stimulating growth is also suggested by their second messenger. All three NK receptors have been shown to be associated with inositol phospholipid hydrolysis in different tissues (Mantyh *et al.*, 1984b; Dam *et al.*, 1986; Bristow *et al.*, 1987; Nakanishi, 1987; Guard *et al.*, 1988). Receptors acting on the inositol phospholipid pathway may ultimately stimulate cell division. The agonist-induced hydrolysis of inositol phospholipid into inositol triphosphate and diacylglycerol is an early cellular response to mitogenic stimulation. Inositol triphosphate and diacylglycerol are believed to function as second messengers through their ability to mobilize calcium from intracellular stores. The ability of SP to stimulate DNA synthesis, at least in a lymphoblast cell line, is closely related to its ability to mobilize intracellular calcium (Payan *et al.*, 1986). Finally, the NK-2 receptor shares sequence homology to the receptor-like protein coded by the *mas* oncogene (44% homology). The *mas* oncogene has been predicted to encode a membrane receptor which activates a critical component in a growth regulatory pathway (Young *et al.*, 1986). Therefore, the NK-2 receptor and *mas* could define a subfamily of receptors with growth-control activities (Hanley and Jackson, 1987). Additionally, the fact that the distribution of NK-1 and NK-2 receptors undergoes major reorganization during brain ontogeny, suggests neurotrophic roles for NKs in the CNS (Quirion and Dam, 1986; Dam *et al.*, 1988).

4.2.7   MOTOR FUNCTION

A role for NKs in motor function is suggested by the location of NKs in various motor systems of the brain and spinal cord. SP and its receptors are found in many of the structures comprising the extrapyramidal motor system. Some of these structures are the striatum, substantia nigra, globus pallidus, inferior olivary complex and cerebellum. In fact, one of the heaviest SP innervations in the midbrain is found in the substantia nigra (Cuello *et al.*, 1982). The SP network of the substantia nigra originates in the striatum and SP terminals have been seen in association with dendrites of the substantia nigra (Somogyi *et al.*, 1982). The substantia nigra apparently does not contain NK-1 receptors which is surprising in light of its dense SP innervation. Low levels of NK-2 and NK-3 receptors, however, are found in both the substantia nigra and striatum (Saffroy *et al.*, 1988; Mantyh *et al.*, 1989a; Dam *et al.*, 1990a,c). SP- and dopamine-containing neurons in these structures appear to participate in a mutual regulation which is important in the control of movement (Bannon *et al.*, 1987) and sensory motor integration (Iversen, 1982). SP increases the firing rate of nigral dopamine neurons (Davies and Dray, 1976) and infusion of SP into the substantia nigra produces a grooming response in rats which is characteristic of activation of the striatonigral system (Iversen, 1982). The potential clinical importance of SP in the substantia nigra is underscored by the fact that injection of a SP antagonist into this brain region reduced muscle tone in spastic rats (Turski *et al.*, 1990).

In the ventral horn of the spinal cord, SP terminals of ventral medulla projections are located in close apposition to motor neurons (Ljungdahl *et al.*, 1978; Helke *et al.*, 1982). NK-1 receptors are present on motor neurons (Helke *et al.*, 1985, 1986; Yashpal *et al.*, 1990), which are depolarized by SP (Otsuka and Yanagisawa, 1980). SP fibers and NK-1 receptors are also present in the phrenic motor nucleus of the spinal cord, which controls the diaphragm (Charlton and Helke, 1985). These findings indicate a direct influence of SP on spinal motor mechanisms. NK-2 and NK-3 receptor sites are more discretely distributed in the cord, mostly present in the superficial laminae of the dorsal horn (Yashpal *et al.*, 1990, 1991a).

SP may also influence motor behavior through the mesolimbic dopamine system, serotonergic pathways and the corticospinal tract. When infused into the A10 dopamine cell body region, SP stimulated locomotor and exploratory behavior (Kelly *et al.*, 1979) which suggest that SP increases the functional activity of these dopaminergic neurons. Evidence suggests that NK-3 receptor activation may influence serotonergic neurons. In support of this hypothesis, the NK-3 agonist, senktide, has been shown to elicit 5-HT-mediated motor behaviors following intracisternal or subcutaneous administration in the mouse and rat (Stoessl *et al.*, 1988). In the corticospinal tract, SP is excitatory to the giant pyramidal cells of Betz in motor cortex

which give rise to the descending fibers of the tract (Phillis and Limacher, 1974).

4.2.8. MEMORY

In the limbic system, SP cell bodies are found in the hippocampus, septum and amygdala suggesting that NKs might influence learning and memory. In fact, the heaviest SP fiber network in the forebrain is the medial amygdaloid nucleus (Cuello and Kanazawa, 1978) suggesting that SP may play an important role in the function of this structure. In addition, high concentrations of NK-1 receptors are found in the hippocampus and amygdala (Saffroy *et al.*, 1988). Several reports indicate that SP, NKA and neuropeptide K can each modulate memory retention after central or peripheral administration (Houston and Staubli, 1979; Hecht *et al.*, 1979; Staubli and Houston, 1980; Kafetzopoulos *et al.*, 1986; Flood *et al.*, 1990). Whether this modulation occurs through direct effects on mechanisms of memory storage or through indirect effects such as increased sensitivity to negative reinforcement, remains to be determined.

4.2.9 ENDOCRINE AND EXOCRINE REGULATION

The presence of NKs and their receptors in the hypothalamus, pituitary and adrenal glands suggests a possible role for these peptides in endocrine regulation. Numerous reports have indicated that SP alters the release of many hormones, including gonadotropins, prolactin, growth hormone and TRH (for references see O'Donohue *et al.*, 1990). In contrast, SP inhibits insulin release from the pancreas (Brown and Vale, 1976). Moreover, SP stimulates exocrine pancreatic secretion *in vitro* (Konturek *et al.*, 1981) and inhibits hepatic bile output (Holm *et al.*, 1978). Finally, the central inhibition of gastric acid output appears to be mediated entirely by NK-2 and NK-3 receptors. Central administration of NK-2 and NK-3 agonists caused this inhibition, while selective NK-1 agonists had no effect (Improta and Broccarde, 1990).

4.2.10 ONTOGENY OF NK-LIKE IMMUNOREACTIVITY IN THE BRAIN

Both NKs and their receptors (see below) are present early in fetal development. SP immunoreactivity is detectable in the mouse brain at embryonic day (ED) 12 (Ni and Jonakait, 1981), in the rat brain at ED14 (Johansson *et al.*, 1981; Inagaki *et al.*, 1982a) and in the human brain after week 11 of gestation (Yew *et al.*, 1990). SP appears in the spinal cord and dorsal root ganglia soon after (Gilbert and Emson, 1979; Senba *et al.*, 1982; Charnay *et al.*, 1983). SP innervation of some brainstem nuclei and peripheral structures, such as the taste buds of the tongue, occurs postnatally (Sakanaka *et al.*, 1982; Yamasaki *et al.*, 1985). In the cerebellum, there is a transient SP innervation in the newborn, which can be traced to the lower brainstem, but which disappears within two weeks (Inagaki *et al.*, 1982b). In general, the distribution of SP and PPT mRNA in the nervous system is comparable to that of the adult at an early stage of development, but the concentrations of SP and the prohormone mRNA increase steadily after birth (Brene *et al.*, 1990; Walker *et al.*, 1991). Adult concentrations are reached between 7 and 60 days after birth, depending on the tissue. NKA, NKB and neuropeptide K are also detectable in the rat brain at birth. Their levels increase during the second week of birth, then decline to adult levels after postnatal day (PD) 15 (Diez-Guerra *et al.*, 1989).

4.3 ONTOGENIC PROFILE OF NK RECEPTOR SUBTYPES

4.3.1 ONTOGENY OF NK-1 RECEPTORS

The distribution of NK-1 binding sites undergoes major modifications during ontogeny (Figs 4.1–4.5) of the rat brain. For example,

although high densities of NK-1 binding sites are present very early in most brainstem nuclei, very low concentrations of sites are seen in the same region at PD21 as in the adult CNS (Quirion and Dam, 1988; Dam *et al.*, 1988). A detailed examination of the profile is presented in the following sections.

PD1

At one day postnatally, NK-1 binding sites are concentrated in most nuclei of the brainstem (Fig. 4.1E). High densities of NK-1 binding sites are also present in the striatum (Fig. 4.1A–C), olfactory tubercle (Fig. 4.1A,B), dentate gyrus of the hippocampus (Fig. 4.1C,D), various hypothalamic and amygdaloid nuclei (Fig. 4.1C,D), the habenula (Fig. 4.1D), amygdalohippocampal area (Fig. 4.1D), the colliculi (Fig. 4.1E) and the entorhinal cortex (Fig. 4.1E). Moderate densities of sites are seen in the lateral septum (Fig. 4.1A,B) and various thalamus nuclei (Fig. 4.1C). Low densities are found in most cortical areas (Fig. 4.1). The corpus callosum is devoid of specific labeling (Fig. 4.1).

PD4

The distribution of NK-1 binding sites is relatively similar at one and four days after birth (Figs 4.1 and 4.2). Very high densities of sites are found in various brainstem nuclei (Fig. 4.2E), the locus coeruleus (not shown) and inferior olive (not shown). High densities of sites are present in the striatum (Fig. 4.2A, B), habenula (Fig. 4.2C), dentate gyrus (Fig. 4.2C,D), certain thalamic nuclei (Fig. 4.2C) and the superior colliculus (Fig. 4.2D). Moderate densities are found in the lateral septum (Fig. 4.2A,B) and various hypothalamic nuclei (Fig. 4.2C). Low to moderate densities are present in most cortical areas (Fig. 4.2). Very low densities of sites are present in the substantia nigra throughout postnatal ontogeny (Figs 4.1–4.5).

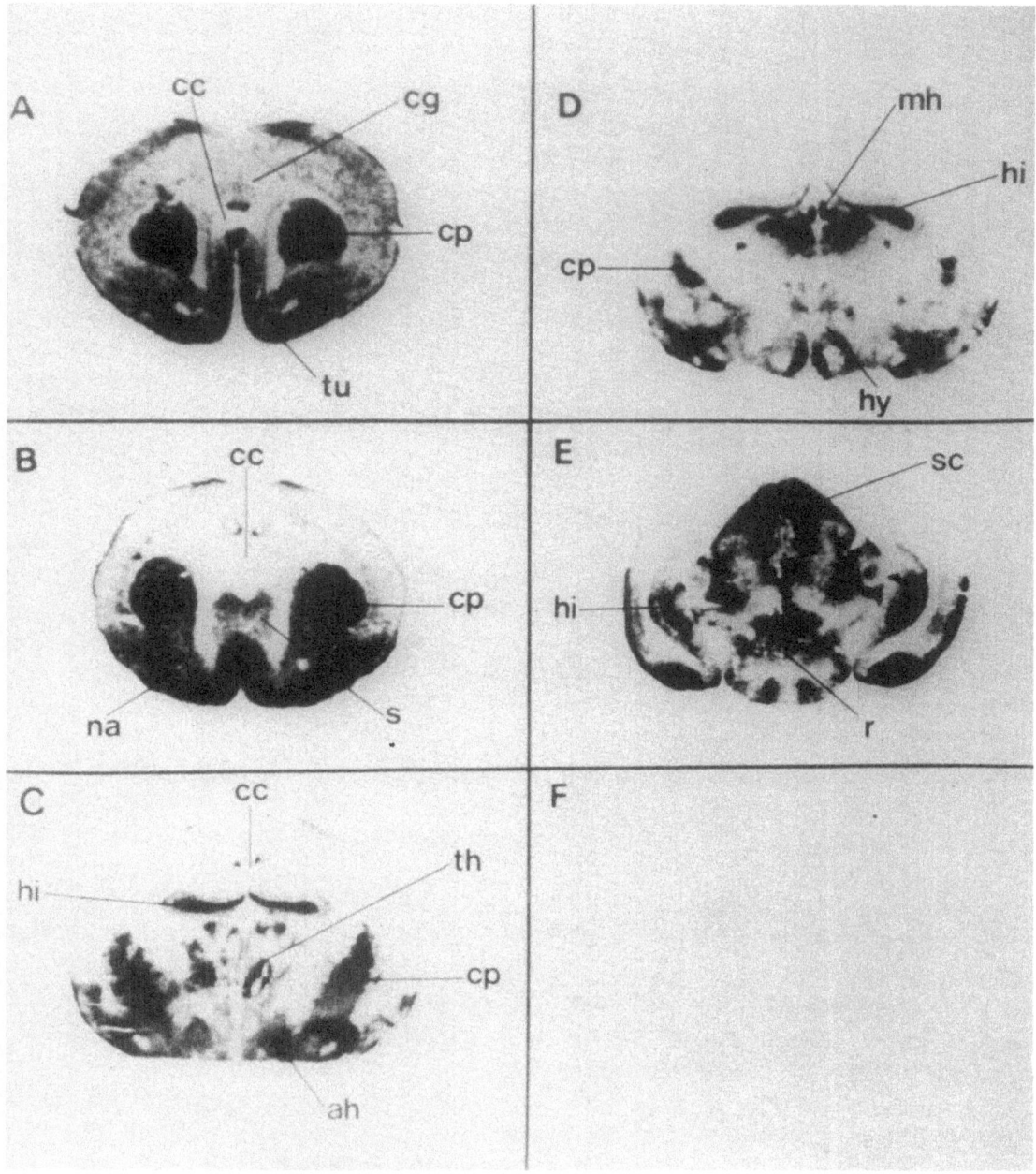

Fig. 4.1. Photomicrographs of the distribution of NK-1 binding sites in coronal brain sections from one day old rats (PD1). High densities of sites are present in the striatum (A,B), olfactory tubercle (A,B), dentate gyrus (C,D), amygdala (C,D) and colliculus (E). Moderate densities are found in the lateral septum (A,B) and hypothalamus (C,D). Section incubated in presence of 1.0 μM SP (F). Abbreviations: ah, anterior hypothalamus; cc, corpus callosum; cg, cingulate cortex; cp, caudate putamen; hi, hippocampus; hy, hypothalamus; mh, medial habenula; na, nucleus accumbens; r, raphe; s, septum; sc, superior colliculus; th, thalamus; tu, olfactory tubercle.

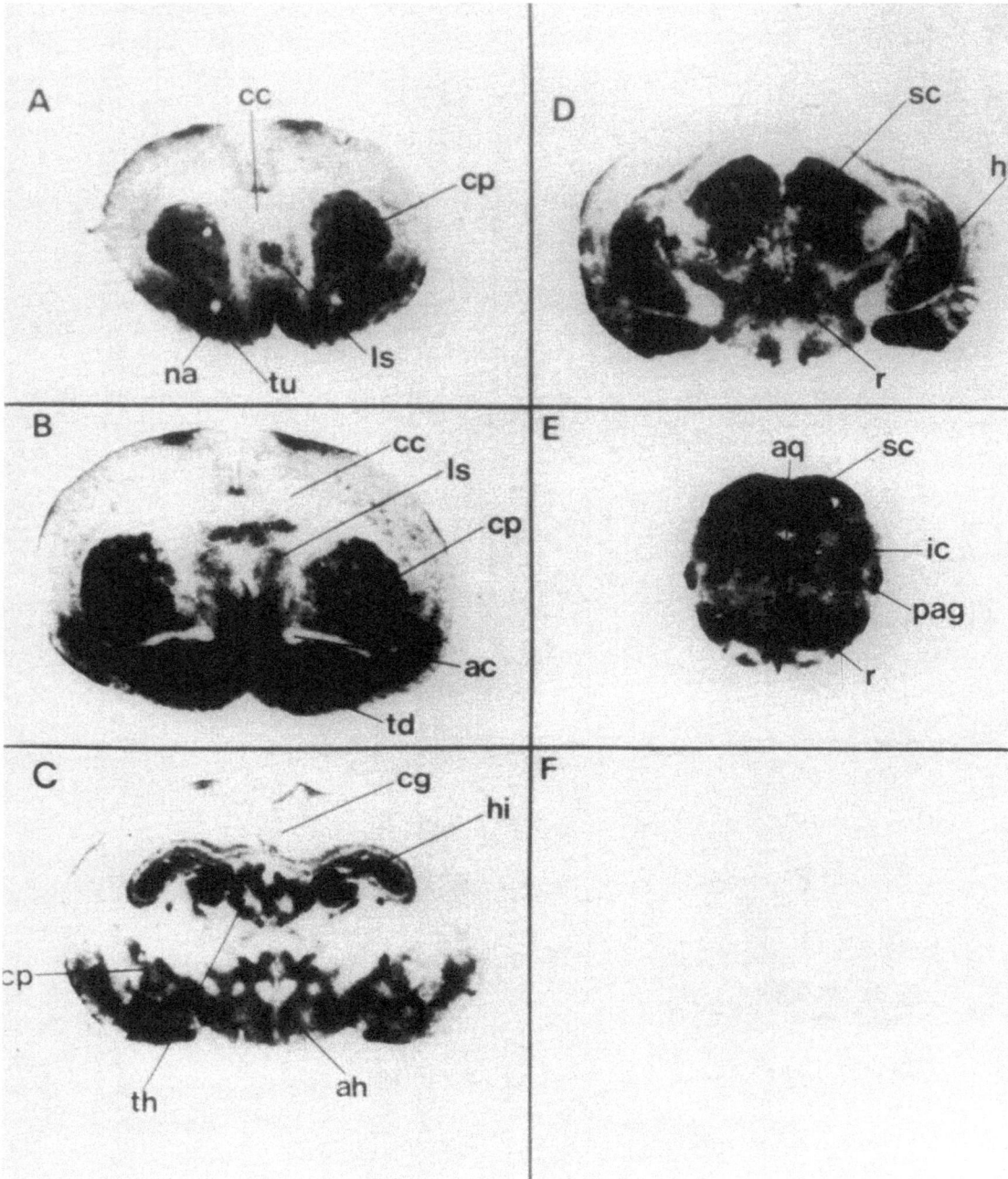

Fig. 4.2. Photomicrographs of the distribution of NK-1 binding sites in coronal brain sections from four-day-old rats (PD4). High densities of sites are present in the striatum (A,B), olfactory tubercle (A,B), dentate gyrus (C,D), certain thalamic nuclei (C), superior colliculus (D,E) and most brainstem nuclei (E). Moderate densities of sites are found in the lateral septum (A,B) and hypothalamus (C). Section incubated in the presence of 1.0 μM SP (F). Abbreviations: ac, anterior commissura; ah, anterior hypothalamus; aq, cerebral aqueduct; cc, corpus callosum; cg, cingulate cortex; cp, caudate putamen; hi, hippocampus; ic, inferior colliculus; ls, lateral septum; na, nucleus accumbens; pag, periaqueductal gray matter (central gray); r, raphe; sc, superior colliculus; td, tractus diagonalis; th, thalamus; tu, olfactory tubercle.

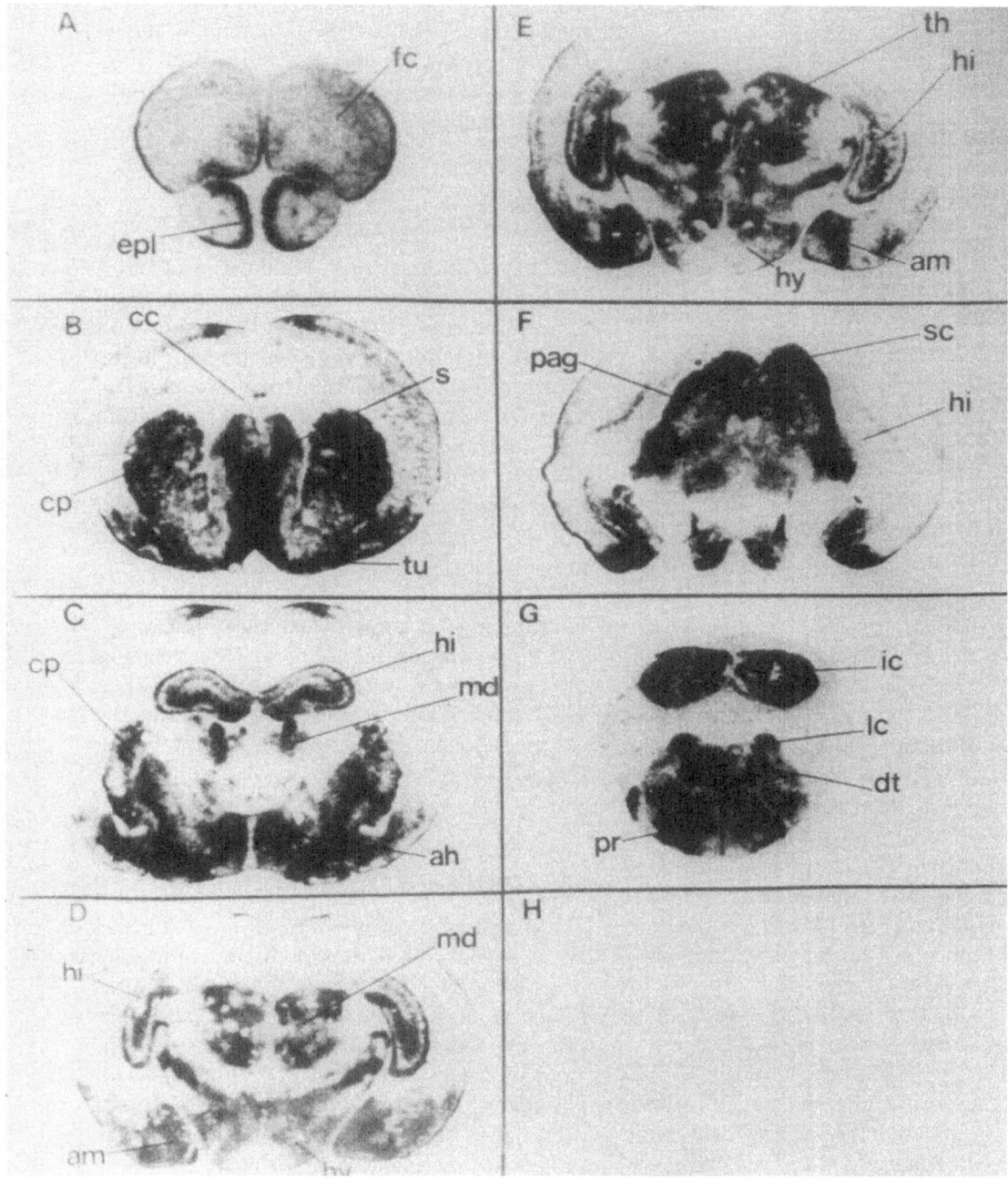

Fig. 4.3. Photomicrographs of the distribution of NK-1 binding sites in coronal brain sections from seven-day-old rats (PD7). High densities of sites are seen in the external plexiform layer of the olfactory bulb (A), striatum (B,C), septum (B), dentate gyrus (C–E), certain thalamic nuclei (C–E), superior colliculus (F), medial geniculate nuclei (F), inferior colliculus (G) and most brainstem nuclei (G) present at this level. Section incubated in the presence of 1.0 μM SP (H). Abbreviations: ah, anterior hypothalamus; am, amygdala; cc, corpus callosum; cp, caudate putamen; dt, dorsal tegmentum; epl, external plexiform layer of the olfactory bulb; fc, frontal cortex; hi, hippocampus; hy, hypothalamus; ic, inferior colliculus; lc, locus coeruleus; md, mediodorsal thalamic nuclei; s, septum; sc, superior colliculus; th, thalamus; tu, olfactory tubercle.

PD7

Very high densities of NK-1 sites are still present over most upper and lower brainstem nuclei (Fig. 4.3). High densities of SP binding sites are found in the external plexiform layer of the olfactory bulb (Fig. 4.3A), striatum (Fig. 4.3B,C), lateral septum (Fig. 4.3B), olfactory tubercle (Fig. 4.3B), dentate gyrus of the hippocampus (Fig. 4.3C,D), amygdala (Fig. 4.3C,D) habenula (Fig. 4.3C), anterior hypothalamic nucleus (Fig. 4.3C), certain thalamic nuclei (Fig. 4.3C,D), the colliculi (Fig. 4.3E,F,G), central gray matter (Fig. 4.3F), entorhinal cortex (Fig. 4.3E,F), medial geniculate nuclei (Fig. 4.3F), locus coeruleus (Fig. 4.3G), olive nuclei and nucleus of the trigeminal nerve. Low densities of sites are present in cortex (Fig. 4.3), and only background levels are seen in the cerebellum (Fig. 4.3G). White matter areas are devoid of NK-1 binding sites (Fig. 4.3).

PD14

At 14 days after birth, the autoradiographic distribution of NK-1 binding sites in rat brain is relatively similar to that seen in adult brain (Quirion *et al.*, 1983; Shults *et al.*, 1984), except for the higher densities in the brainstem (Fig. 4.4F,H) and the somewhat higher density in the hypothalamus (Fig. 4.4D,E). In the forebrain, NK-1 sites are found in various areas including the external plexiform layer of the olfactory bulb (Fig. 4.4A), striatum (Fig. 4.4B–D), olfactory tubercle (Fig. 4.4B,C), lateral septum (Fig. 4.4C), habenula (Fig. 4.4D,E), certain thalamic nuclei (Fig. 4.4D,E), amygdalohippocampal area (Fig. 4.4D,E), amygdala (Fig. 4.4D,E), zona incerta (Fig. 4.4E) and dentate gyrus (Fig. 4.4D,E). At this age, the laminar distribution of NK-1 binding sites in the hippocampus is highly apparent (Fig. 4.4D–F). Low to moderate densities of sites are also present, in a laminated fashion, in various cortical areas (Fig. 4.4). More caudally, high densities of NK-1 sites are found in the superior colliculus (Fig. 4.4F,G), central gray matter (Fig. 4.4F,G), medial geniculate nuclei (Fig. 4.4F,G), pre- and parasubiculum (Fig. 4.4G), locus coeruleus (Fig. 4.4H) and certain brainstem nuclei (Fig. 4.4H).

PD21

At 21 days after birth, low densities of NK-1 sites are seen in most brainstem nuclei (Fig. 4.5) as in the adult rat. However, as shown in Fig. 4.5, high densities of sites are seen in the striatum (Fig. 4.5B–F), olfactory tubercle (Fig. 4.5B–D), septum (Fig. 4.5C,D), amygdala (Fig. 4.5F,G), amygdalohippocampal area (Fig. 4.5G,H), habenula (Fig. 4.5F,G), anterior hypothalamus (Fig. 4.5E,F), dentate gyrus (Fig. 4.5F–H), superior colliculus (Fig. 4.5I,J), central gray matter (Fig. 4.5J) and locus coeruleus (Fig. 4.5K). Moreover, NK-1 binding sites are distributed in a laminar fashion in the cortex and hippocampus (Fig. 4.5). Very low densities of sites are seen in the substantia nigra (Fig. 4.5I,J) and white matter areas such as the corpus callosum are devoid of specific labeling (Fig. 4.5).

4.3.2   ONTOGENY OF NK-2 RECEPTORS

The distribution of NK-2 binding sites in rat brain is markedly altered during ontogeny (Figs 4.6–4.11). As in the case of NK-1 sites (Quirion and Dam, 1986), NK-2 sites are present in high amounts in different brainstem nuclei from PD1 to PD7 (Figs 4.6–4.8). However, the density of these receptors diminishes thereafter to reach adult levels rather rapidly (Figs 4.9–4.10).

PD1

One day after birth, very high densities of NK-2 sites are observed in the amygdalohippocampal area (Fig. 4.6C), inferior colliculus (Fig.

4.6E), locus coeruleus (Fig. 4.6E), raphe nucleus (Fig. 4.6D–G) and periaqueductal gray matter (Fig. 4.6F). Regions moderately enriched with NK-2 sites include the striatum (Fig. 4.6A,B), layers III and IV of the prefrontal cortex (Fig. 4.6A,B), the lateral septum (Fig. 4.6A), certain thalamic nuclei (Fig. 4.6B,C), diagonal band of Broca (Fig. 4.6B), hippocampus (Fig. 4.6B,C), medial habenula (Fig. 4.6C) and zona incerta (Fig. 4.6C).

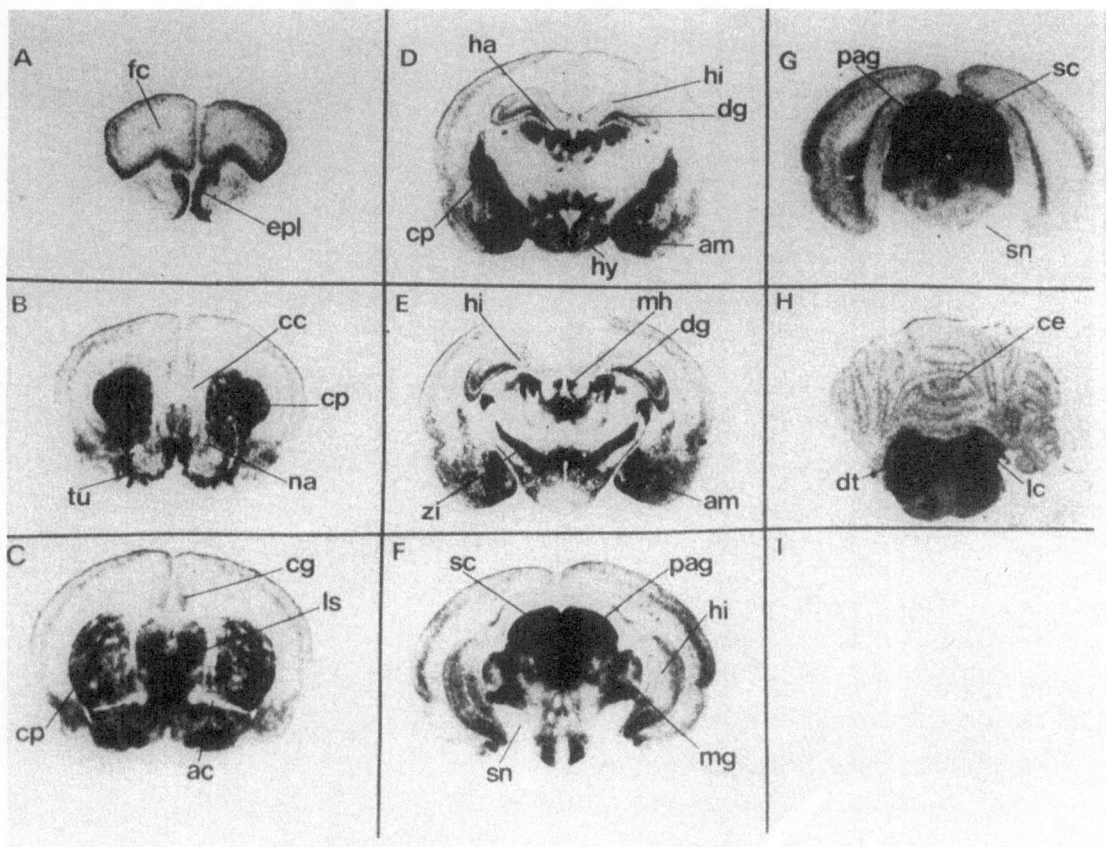

Fig. 4.4. Photomicrographs of the distribution of NK-1 binding sites in coronal brain sections from 14-day-old rats (PD14). High densities of sites are found in the external plexiform layer of the olfactory bulb (A), olfactory tubercle (B,C), striatum (B–D), septum (C), habenula nuclei (D,E), dentate gyrus (D,E), amygdalohippocampal area (D,E), certain hypothalamic nuclei (D,E), superior colliculus (F,G), central gray matter (F,G) and most brainstem nuclei including the locus coeruleus (H). Moderate densities are seen in the hippocampus (D–F) and pre- and parasubiculum (G). Very low densities are present in the substantia nigra (F,G). Section incubated in the presence of 1.0 μM SP (I). Abbreviations: ac, anterior commissura; am, amygdala; cc, corpus callosum; ce, cerebellum; cg, cingulate cortex; cp, caudate putamen; dg, dentate gyrus; dt, dorsal tegmentum; epl, external plexiform layer of the olfactory bulb; fc, frontal cortex; ha, habenula; hi, hippocampus; hy, hypothalamus; lc, locus coeruleus; ls, lateral septum; mh, medial habenula; mg, medial geniculate nucleus; na, nucleus accumbens; pag, periaqueductal gray matter (central gray); sc, superior colliculus; sn, substantia nigra; tu, olfactory tubercle; zi, zona incerta.

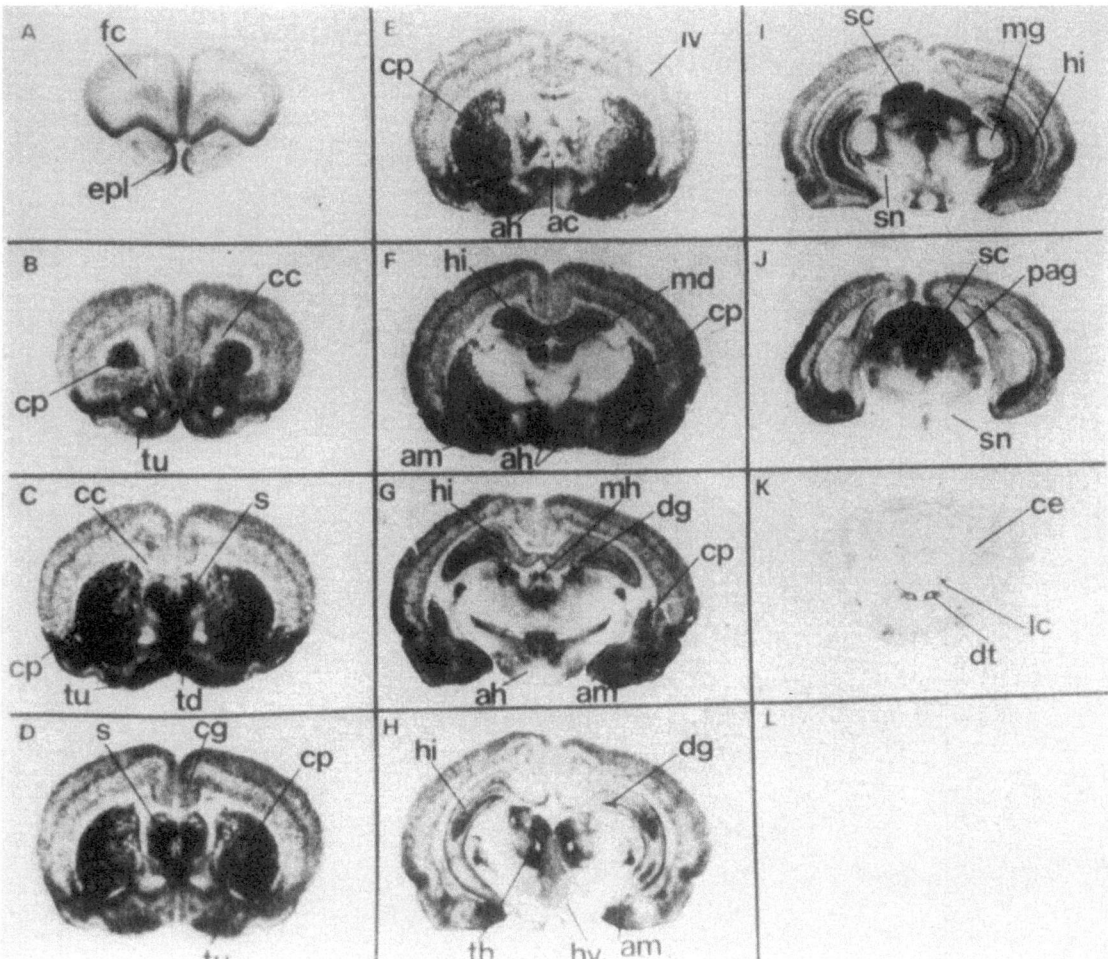

Fig. 4.5.   Photomicrographs of the distribution of NK-1 binding sites in coronal brain sections from 21-day-old rats (PD21). High densities of sites are found in various regions including the external plexiform layer of the olfactory bulb (A), striatum (B,F), lateral septum (C,D), olfactory tubercle (C,D), habenula (F,G), hypothalamus (E–G), dentate gyrus (F,I), certain thalamic nuclei (G,H), amygdalohippocampal area (G,H), superior colliculus (I,J), central gray matter (J) and locus coeruleus (K). At this age, very few brainstem nuclei contain high densities of SP binding sites (K). The substantia nigra is virtually devoid of NK-1 binding sites (I,J). Section incubated in the presence of 1.0 μM SP (L). Abbreviations: ac, anterior commissura; ah, anterior hypothalamus; am, amygdala; cc, corpus callosum; ce, cerebellum; cg, cingulate cortex; cp, caudate putamen; dg, dentate gyrus; dt, dorsal tegmentum; epl, external plexiform layer of the olfactory bulb; fc, frontal cortex; hi, hippocampus; hy, hypothalamus; lc, locus coeruleus; md, medial thalamic nuclei; mg, medial geniculate nucleus; mh, medial habenula; pag, periaqueductal gray matter (central gray); s, septum; sc, superior colliculus; sn, substantia nigra; td, tractus diagonalis; th, thalamus; tu, olfactory tubercle; IV, fourth layer of the cortex.

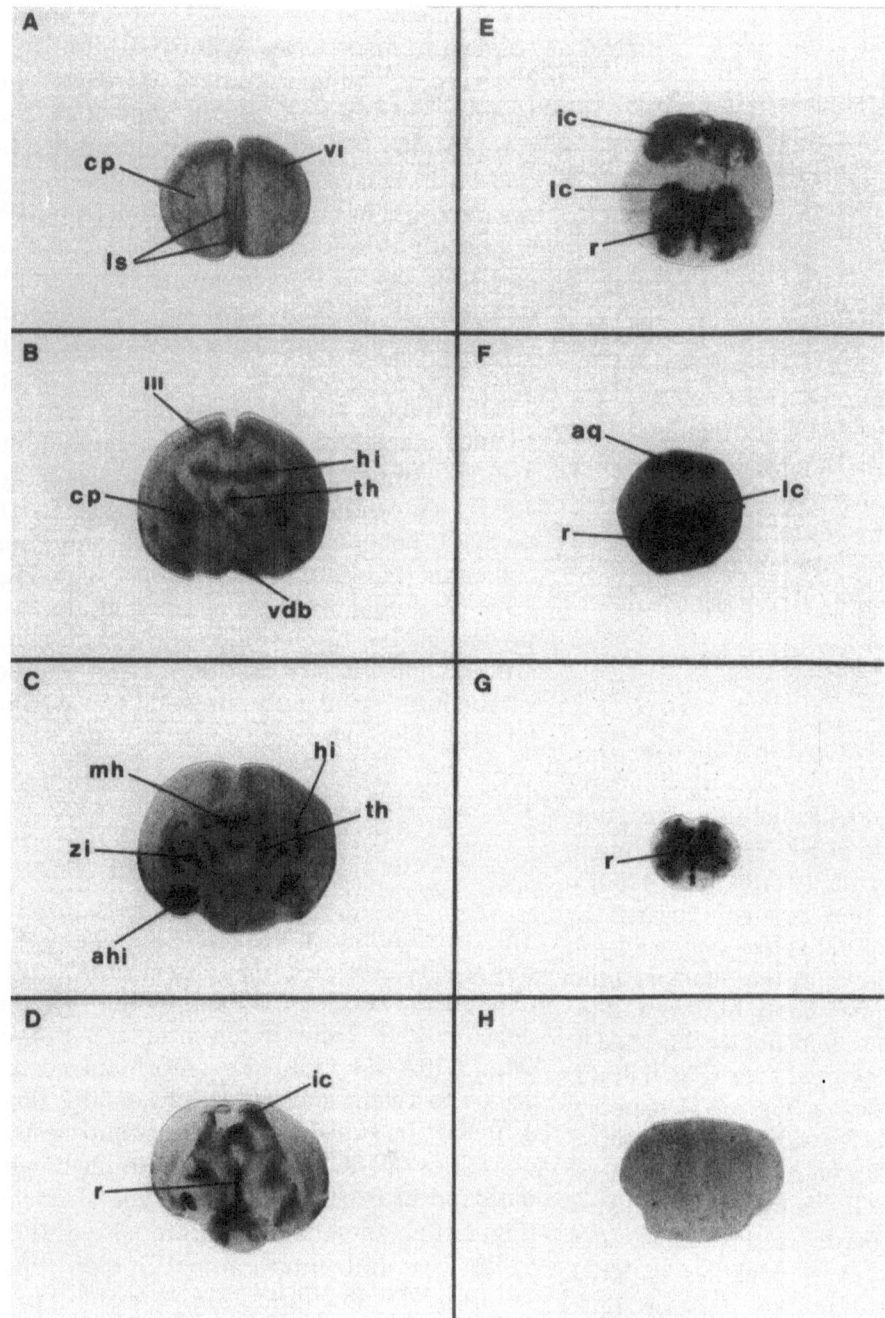

Fig. 4.6. Photomicrographs of the distribution of NK-2 receptor binding sites in coronal brain sections from 1-day-old rats (PD1). Sections were incubated with 50 pM of ^{125}I-labeled NKA. Non-specific (NS) labeling seen in the presence of 1.0 μM unlabeled NKA is shown in H. Abbreviations: ahi, amygdalo-hippocampal area; aq, cerebral aqueduct; cp, caudate putamen; hi, hippocampus; ic, inferior colliculus; lc, locus coeruleus; ls, lateral septum; mh, medial habenula; r, raphe; th, thalamus; vdb, diagonal band of Broca; zi, zona incerta; III, VI, layer III, VI of the frontal cortex.

PD4

Four days after birth, the density of NK-2 sites in the cortex (laminae III and IV) is markedly increased. However, its distribution remains similar to that seen at PD1 (Figs 4.6 and 4.7). Very high densities of NK-2 sites are seen in the frontal cortex (Fig. 4.7A–C) and the external plexiform layer of the olfactory bulb (Fig. 4.7A). Other regions such as the nucleus of diagonal band of Broca (Fig. 4.7B), amygdalohippocampal area (Fig. 4.7C–E), mediodorsal thalamic nucleus (Fig. 4.7D), superior and inferior colliculi (Fig. 4.7E–G), raphe nucleus (Fig. 4.7F,G) and medial habenula (Fig. 4.7D) are also enriched with NK-2 sites at this age. Lower but still significant quantities of sites are found in other regions including striatum, hippocampus (Fig. 4.7B–D) and periaqueductal gray area (Fig. 4.7C,D).

PD7

Seven days after birth, the distribution of NK-2 sites is rather similar to that observed at early ages (Figs 4.6, 4.7 and 4.8) with the exception of a certain scattering of NK-2 sites in laminae III and IV of the occipital cortex (Fig. 4.8E,F). High densities of sites are still located in laminae VI of the frontal cortex (Fig. 4.8B–D), external plexiform layer of the olfactory bulb (Fig. 4.8A,B), diagonal band of Broca (Fig. 4.8C), various amygdaloid nuclei (Fig. 4.8D), the amygdalohippocampal area (Fig. 4.8E,F), medial geniculate nucleus (Fig. 4.8G), superior and inferior colliculi (Fig. 4.8G–I), mediodorsal thalamic nucleus (Fig. 4.8D,E), periaqueductal gray matter (Fig. 4.8G), ventral tegmental area (Fig. 4.8I), locus coeruleus (Fig. 4.8H) and raphe (Fig. 4.8H). Lower densities of NK-2 sites are also found in the striatum (Fig. 4.8B–D), hippocampal formation (Fig. 4.8D–F) and pontine nucleus (Fig. 4.8G).

PD21

Major modifications are observed in the distribution of NK-2 sites in 21-day-old rats (Fig. 4.9). For example, the density of specific NK-2 labeling in the cerebral cortex is markedly diminished as compared to that observed in younger animals. Moreover, the laminar distribution of NK-2 sites in the cortex is altered, being mostly concentrated in mid-layers (Fig. 4.9D–F). Marked differences are also seen in most brainstem nuclei which are mostly devoid of specific labeling at this age (Fig. 4.9H,I). However, very high densities of NK-2 sites are still observed in the external plexiform layer of olfactory bulb (Fig. 4.9A–C), diagonal band of Broca (Fig. 4.9D), preoptic, suprachiasmatic and periventricular nuclei of the hypothalamus (Fig. 4.9E,F), amygdalohippocampal area (Fig. 4.9H), entorhinal cortex (Fig. 4.9E,G,H), medial habenula (Fig. 4.9H), inferior colliculus (Fig. 4.9I) and locus coeruleus (Fig. 4.9I). Moderate amounts of labeling are also present in the hippocampal formation (Fig. 4.9G,H) and the striatum (Fig. 4.9D–G). The cerebellum is usually devoid of specific labeling (Fig. 4.9I).

PD35

The distribution of NK-2 sites in 35-day-old rats is very similar to that observed in adults (Fig. 4.10) (Dam *et al.*, 1990c). By this age, the density of NK-2 sites in the cortex is very low (Fig. 4.10A–E). Only the entorhinal cortex seems to retain specific NK-2 labeling (Fig. 4.10F,G). In subcortical regions, substantial quantities of NK-2 sites are seen in the diagonal band of Broca (Fig. 4.10A), septal nuclei (Fig. 4.10B), amygdalohippocampal area (Fig. 4.10D), ventral hippocampal region (Fig. 4.10E), parasubiculum (Fig. 4.10F,G), presubiculum (Fig. 4.10F,G), superior colliculus (Fig. 4.10E,F) and periaqueductal gray (Fig. 4.10G). Moderate densities of labeling are also detected in the striatum (Fig. 4.10A–C), certain thalamic nuclei (Fig. 4.10C), medial habenula (Fig. 4.10D) and zona incerta (Fig. 4.10C). Other regions are apparently devoid of specific NK-2 sites.

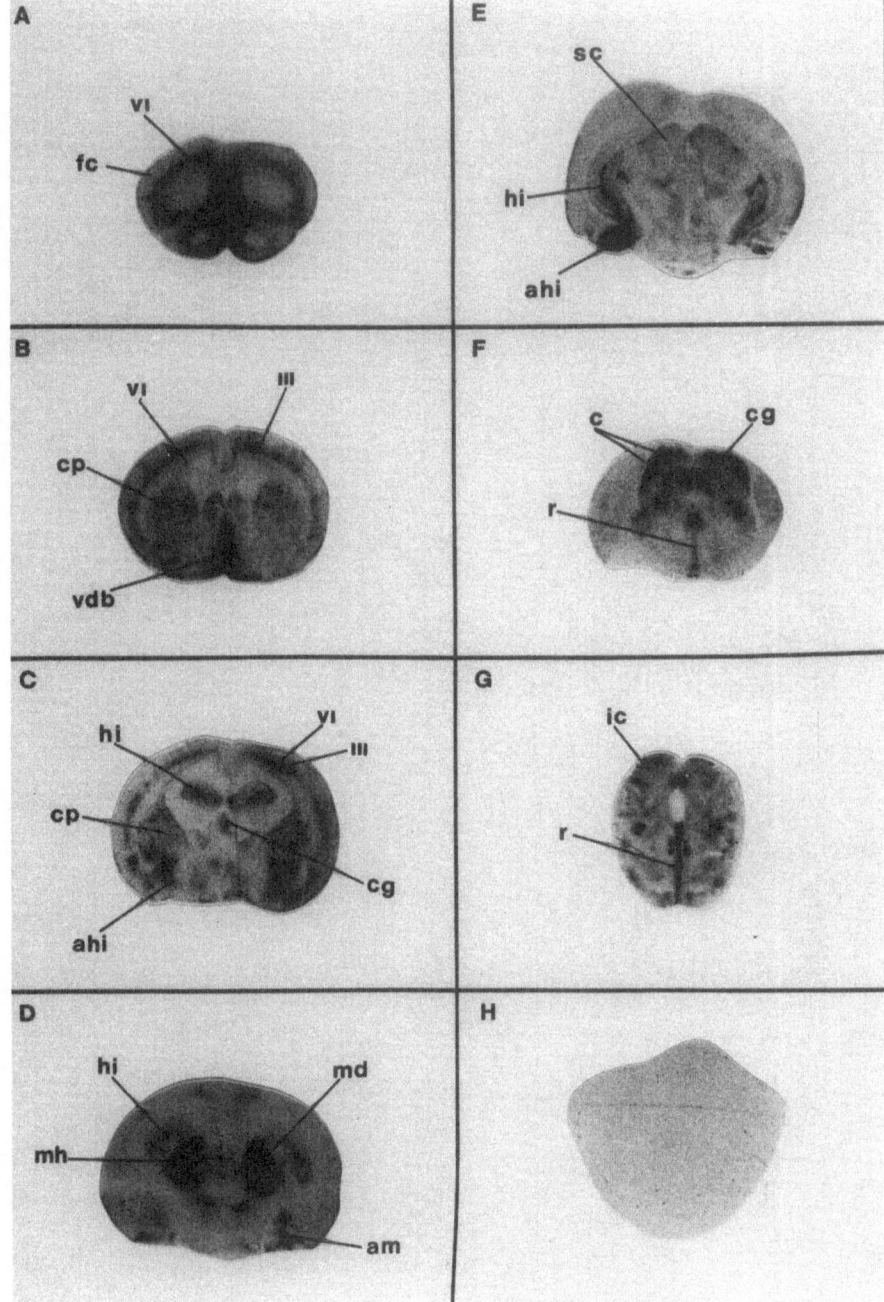

Fig. 4.7. Photomicrographs of the distribution of NK-2 receptor binding sites in coronal brain sections from 4-day-old rats (PD4). Sections were incubated with 50 pM ^{125}I-labeled NKA. Non-specific (NS) labeling seen in presence of 1.0 µM unlabeled NKA is shown in section H. Abbreviations: ahi, amygdalohippocampal area; am, amygdala; c, colliculus; cg, central gray; cp, caudate putamen; fc, frontal cortex; hi, hippocampus; ic, inferior colliculus; md, mediodorsal thalamic nuclei; mh, medial habenula; r, raphe; sc, superior colliculus; vdb, diagonal band of Broca; III, VI, layer III, VI of the frontal cortex.

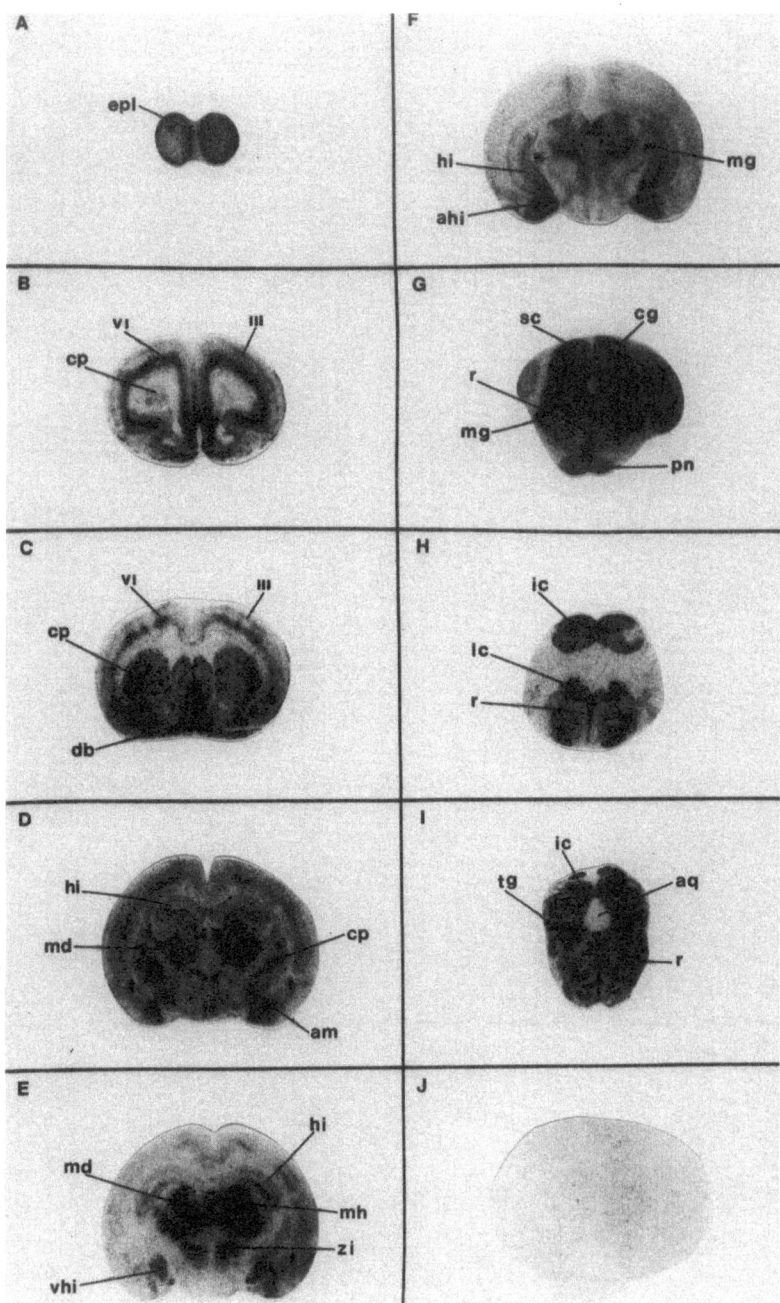

Fig. 4.8.   Photomicrographs of the distribution of NK-2 receptor binding sites in coronal brain sections from 7-day-old rats (PD7). Sections were incubated with 50 pM of ^{125}I-labeled NKA. Non-specific (NS) labeling seen in the presence of 1.0 μM unlabeled NKA is shown in section J. Abbreviations: ahi, amygdalohippocampal area; am, amygdala; aq, cerebral aqueduct; cg, central gray; cp, caudate putamen; db, diagonal band of Broca; epl, external plexiform layer of the olfactory bulb; hi, hippocampus; ic, inferior colliculus; lc, locus coeruleus; md, mediodorsal thalamic nuclei; mg, medial geniculate nucleus; mh, medial habenula; pn, pontine nuclei; r, raphe; sc, superior colliculus; tg, tegmental nuclei; vhi, ventral hippocampal area; zi, zona incerta; III, VI, layer III, VI of the frontal cortex.

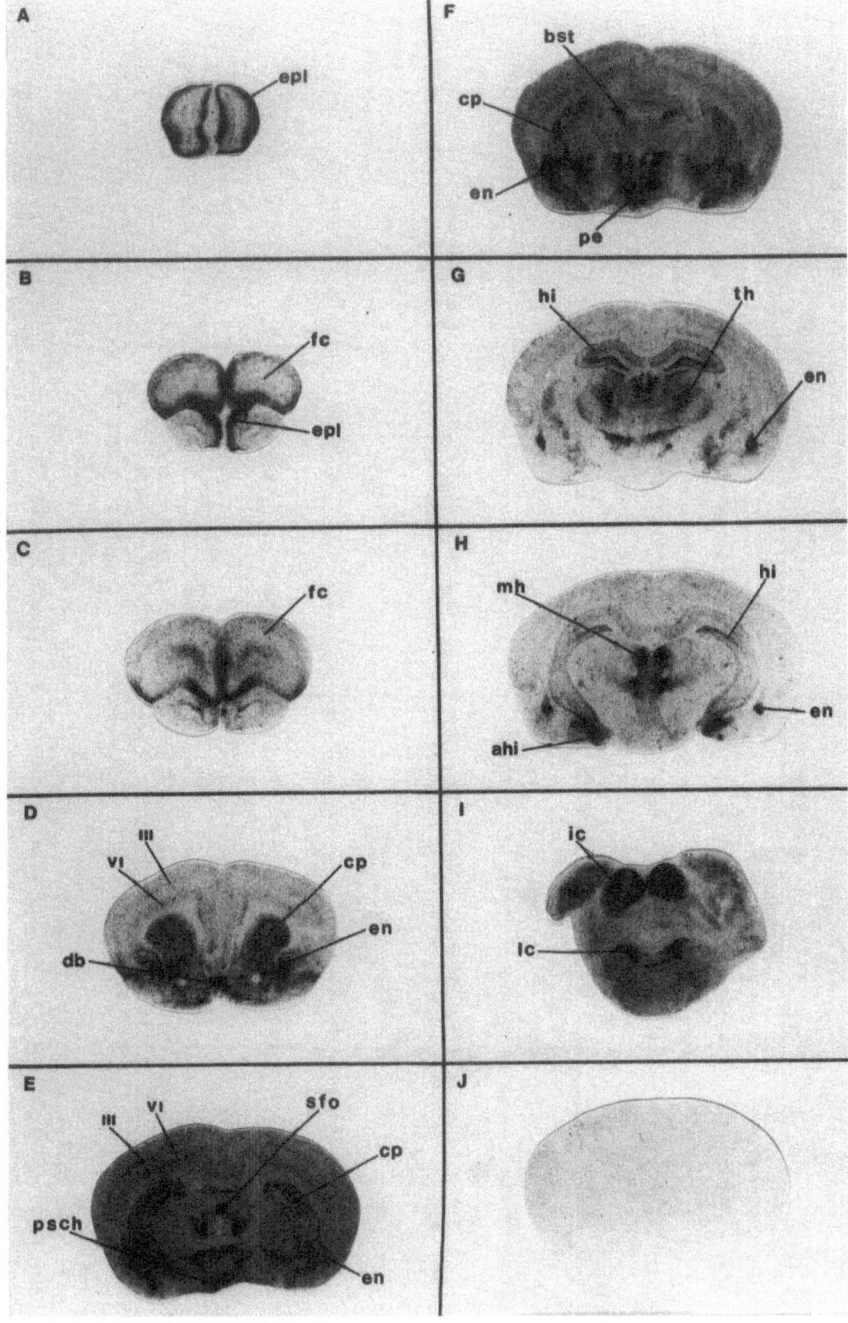

Fig. 4.9. Photomicrographs of the distribution of NK-2 receptor binding sites in coronal brain sections from 21-day-old rats (PD21). Sections were incubated with 50 pM ^{125}I-labeled NKA. Non-specific (NS) labeling seen in the presence of 1.0 μM unlabeled NKA is shown in section J. Abbreviations: ahi, amygdalohippocampal area; bst, stria terminal bed nuclei; cp, caudate putamen; db, diagonal band of Broca; en, entorhinal cortex; epl, external plexiform layer of the olfactory bulb; fc, frontal cortex; hi, hippocampus; ic, inferior colliculus; lc, locus coeruleus; mh, medial habenula; pe, periventricular hypothalamic nuclei; psch, preoptic suprachiasmatic nuclei; sfo, subfornical organ; th, thalamus; III,VI, layer III, VI of the frontal cortex.

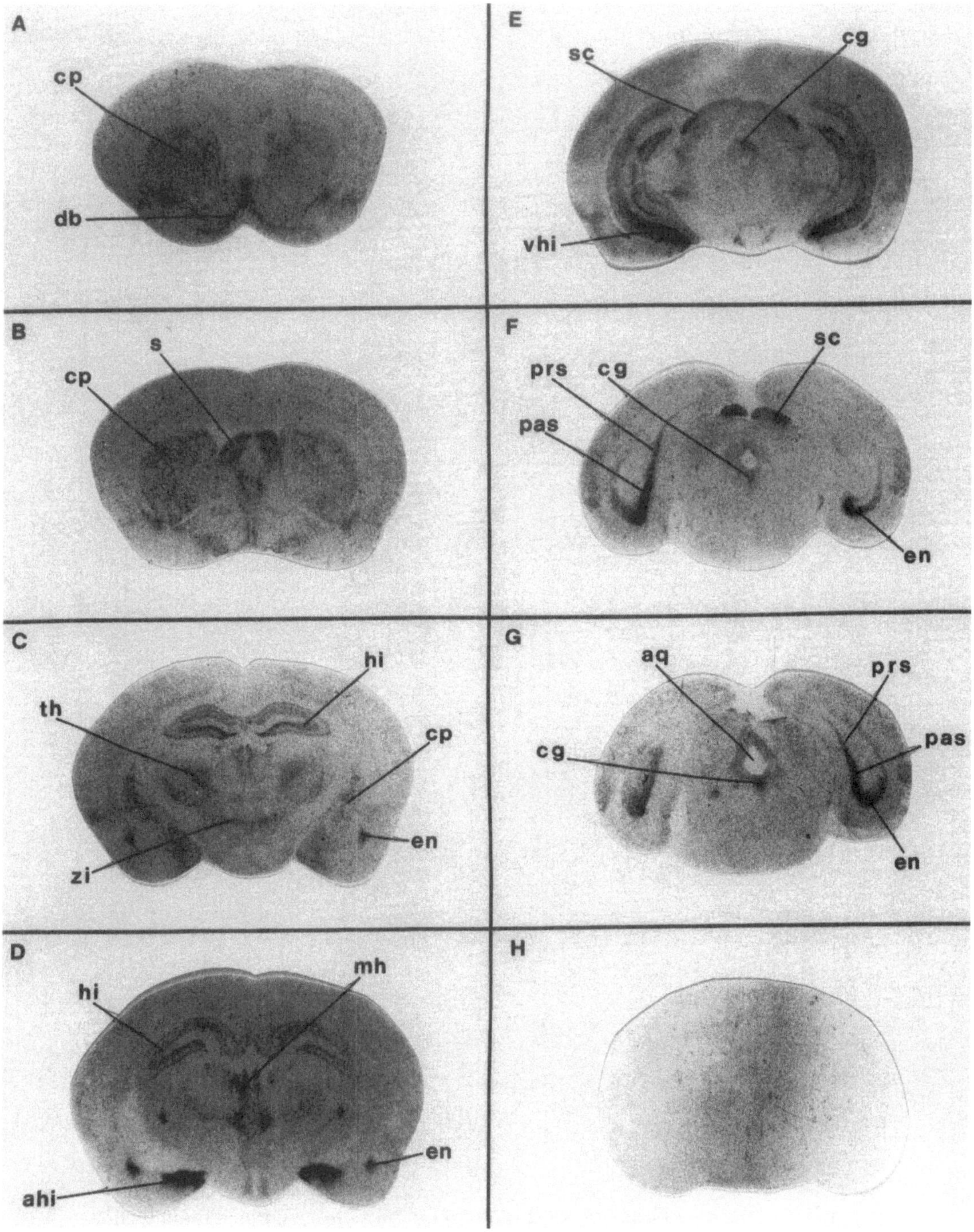

Fig. 4.10.   Photomicrographs of the distribution of NK-2 receptor binding sites in coronal brain sections from 35-day-old rats (PD35). Sections were incubated with 50 pM ^{125}I-labeled NKA. Non-specific (NS) labeling seen in the presence of 1.0 μM unlabeled NKA is shown in section H. Abbreviations: ahi, amygdalohippocampal area; aq, cerebral aqueduct; cg, central gray; cp, caudate putamen; db, diagonal band of Broca; en, entorhinal cortex; hi, hippocampus; mh, medial habenula; pas, parasubiculum; prs, presubiculum; s, septum; sc, superior colliculus; th, thalamus; vhi, ventral hippocampus; zi, zona incerta.

4.3.2 ONTOGENY OF NK-3 RECEPTORS

The distribution of NK-3 binding sites in the rat brain does not undergo major modification during the first postnatal week (Figs 4.11–4.13). However, minor modifications are observed thereafter (Figs 4.14–4.16).

ED20

One day before birth, high densities of NK-3 sites are present in the external plexiform layer of the olfactory bulb (Fig. 4.11A), striatum (Fig. 4.11B), septal area (Fig. 4.11B), hippocampal formation (Fig. 4.11D) as well as various thalamic, hypothalamic and amygdaloid nuclei (Fig. 4.11C,D). At this age, a low density of labeling is seen in most cortical layers (Fig. 4.11A–D).

PD1

One day after birth, NK-3 sites are concentrated in the external plexiform layer of the olfactory bulb (Fig. 4.12A), lateral septum (Fig. 4.12B), CA$_2$ and CA$_3$ subfields of the hippocampus (Fig. 4.12C,D), various nuclei of the thalamus, hypothalamus and amygdala (Fig. 4.12C, D) and the zona incerta (Fig. 4.12C). As at ED20, low densities of NK-3 binding sites are present in cortical areas.

PD6

As shown in Fig. 4.13, the distribution of NK-3 sites at this age is rather similar to that observed at PD1. High densities of NK-3 binding sites are found in the external plexiform layer of the olfactory bulb (Fig. 4.13A), striatum (Fig. 4.13B,C), diagonal band of Broca (Fig. 4.13B), CA$_2$ and CA$_3$ subfields of

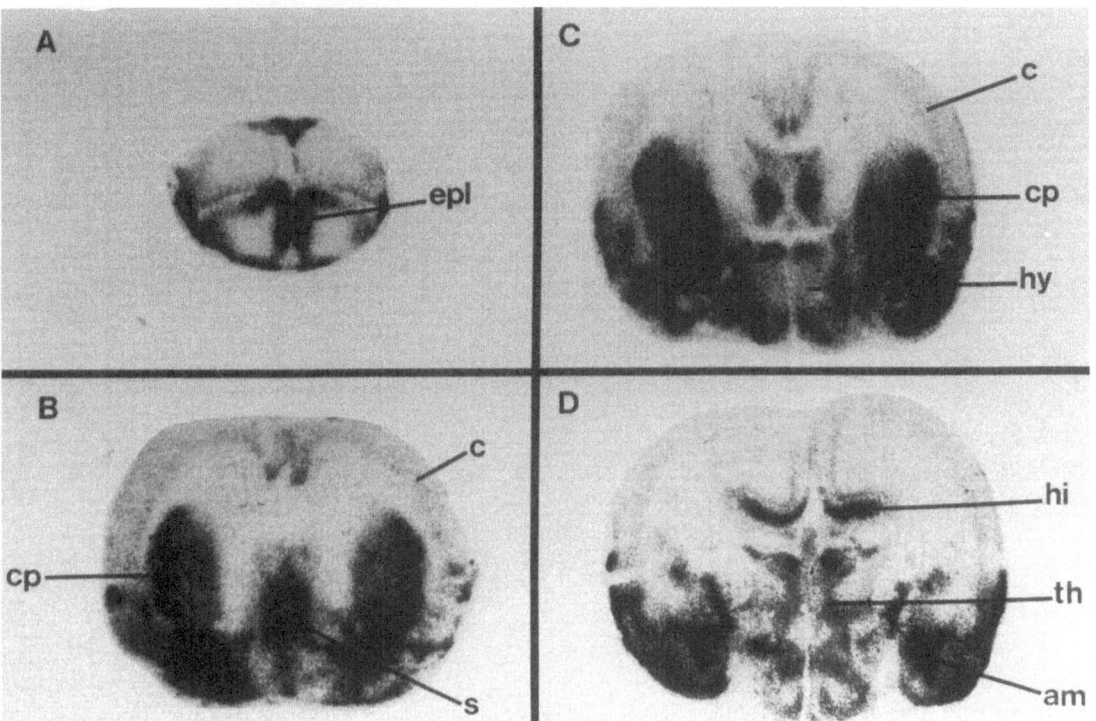

Fig. 4.11. Photomicrographs of the distribution of NK-3 receptor binding sites in coronal brain sections from embryonic 20-day-old rats (ED20). Abbreviations: am, amygdala; c, colliculus; cp, caudate putamen; epl, external plexiform layer of the olfactory bulb; hi, hippocampus; hy, hypothalamus; s, septum; th, thalamus.

the hippocampus (Fig. 4.13C,D), zona incerta (Fig. 4.13D), various thalamic, hypothalamic and amygdaloid nuclei (Fig. 4.13C,D), medial habenula (Fig. 4.13D), superior colliculus and central gray (Fig. 4.13E). However, it is interesting to note that in the frontal cortex, moderate densities of sites are now detected in laminae III and IV (Fig. 4.13B) as well as in the retrosplenial cortex (Fig. 4.13B).

PD14

As observed at younger ages, at PD14, high densities of NK-3 sites are seen in the external and internal plexiform layers of the olfactory bulb (Fig. 4.14A), nucleus of the horizontal limb of the diagonal band of Broca (Fig. 4.14C,D), olfactory tubercle (Fig. 4.14C,D), medial habenula (Fig. 4.14E), ventral hippo-

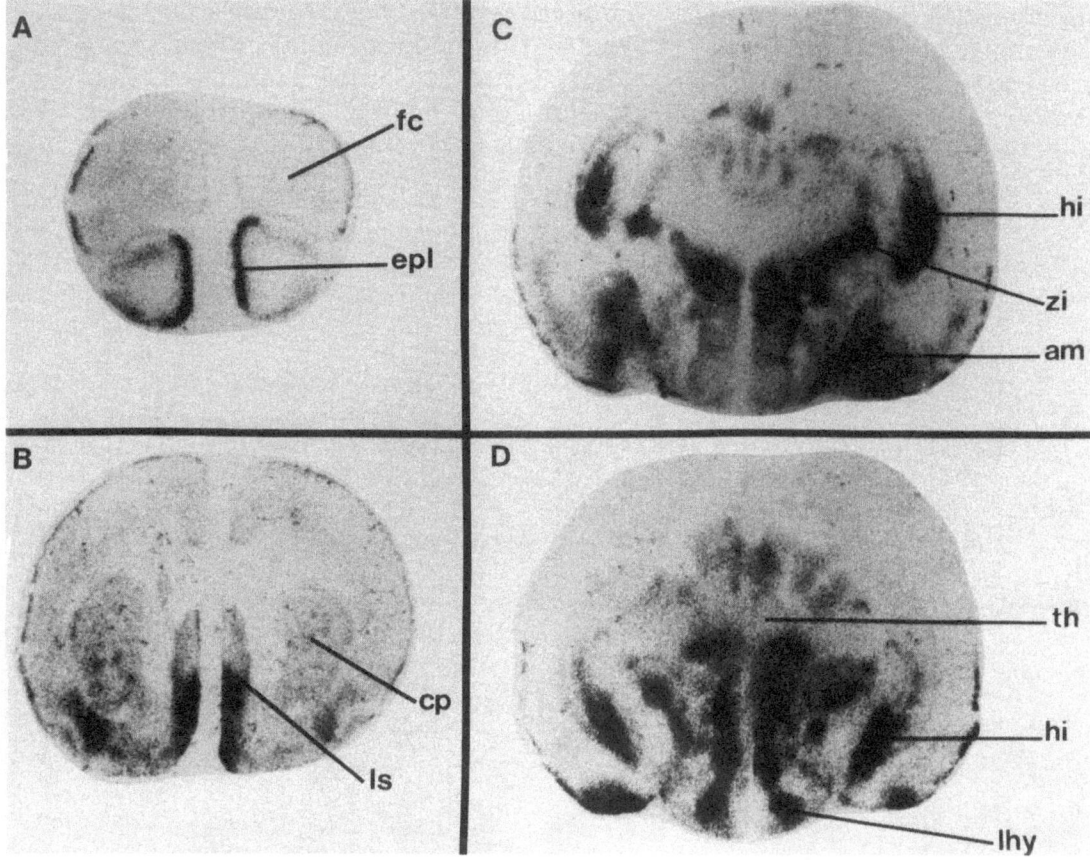

Fig. 4.12.   Photomicrographs of the distribution of NK-3 receptor binding sites in coronal brain sections from 1-day-old rats (PD1). Abbreviations: am, amygdala; cp, caudate putamen; epl, external plexiform layer of the olfactory bulb; fc, frontal cortex; hi, hippocampus; lhy, lateral hypothalamic area; ls, lateral septum; th, thalamus; zi, zona incerta.

campal commissura (Fig. 4.14F), supraoptic and paraventricular hypothalamic nuclei (Fig. 4.14E), certain nuclei of the amygdalohippocampal area (Fig. 4.14E,F), superior and inferior colliculi (Fig. 4.14G,H), central gray (Fig. 4.14G), ventral tegmental area (Fig. 4.14G) and the interpeduncularis nucleus (Fig. 4.14G). However, in the frontal cortex, very high densities of NK-3 binding sites are now clearly evident in laminae IV, even

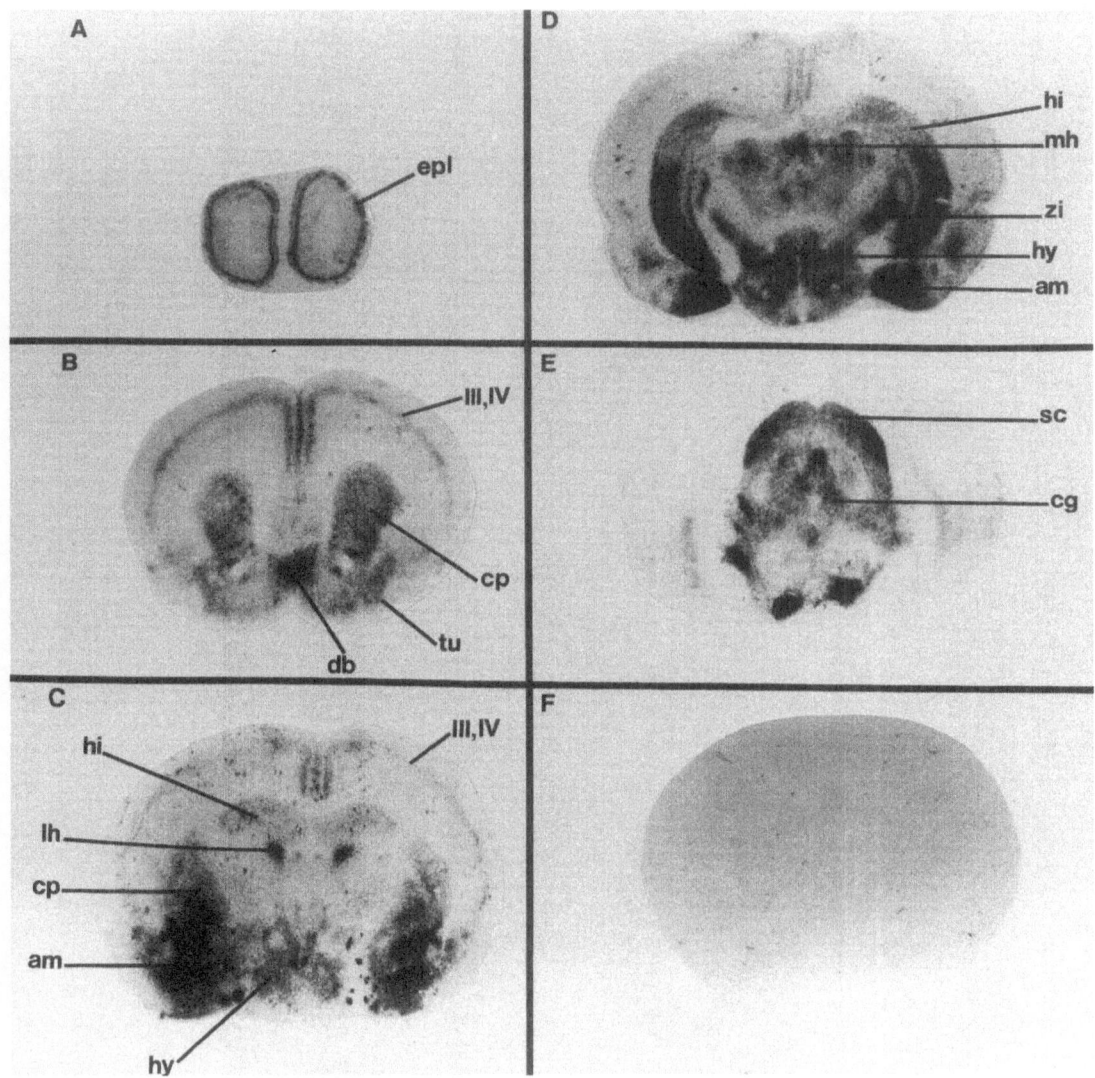

Fig. 4.13. Photomicrographs of the distribution of NK-3 receptor binding sites in coronal brain sections from 6-day-old rats (PD6). Sections were incubated with 50 pM ^{125}I-labeled BH-eledoisin. Non-specific (NS) labeling seen in the presence of 1.0 μM unlabeled eledoisin is shown in section F. Abbreviations: am, amygdala; cg, central gray; cp, caudate putamen; db, diagonal band of Broca; epl, external plexiform layer of the olfactory bulb; hi, hippocampus; hy, hypothalamus; lh, lateral hypothalamic area; mh, medial habenula; sc, superior colliculus; tu, olfactory tubercle; zi, zona incerta; III, IV, layer III, IV of the frontal cortex.

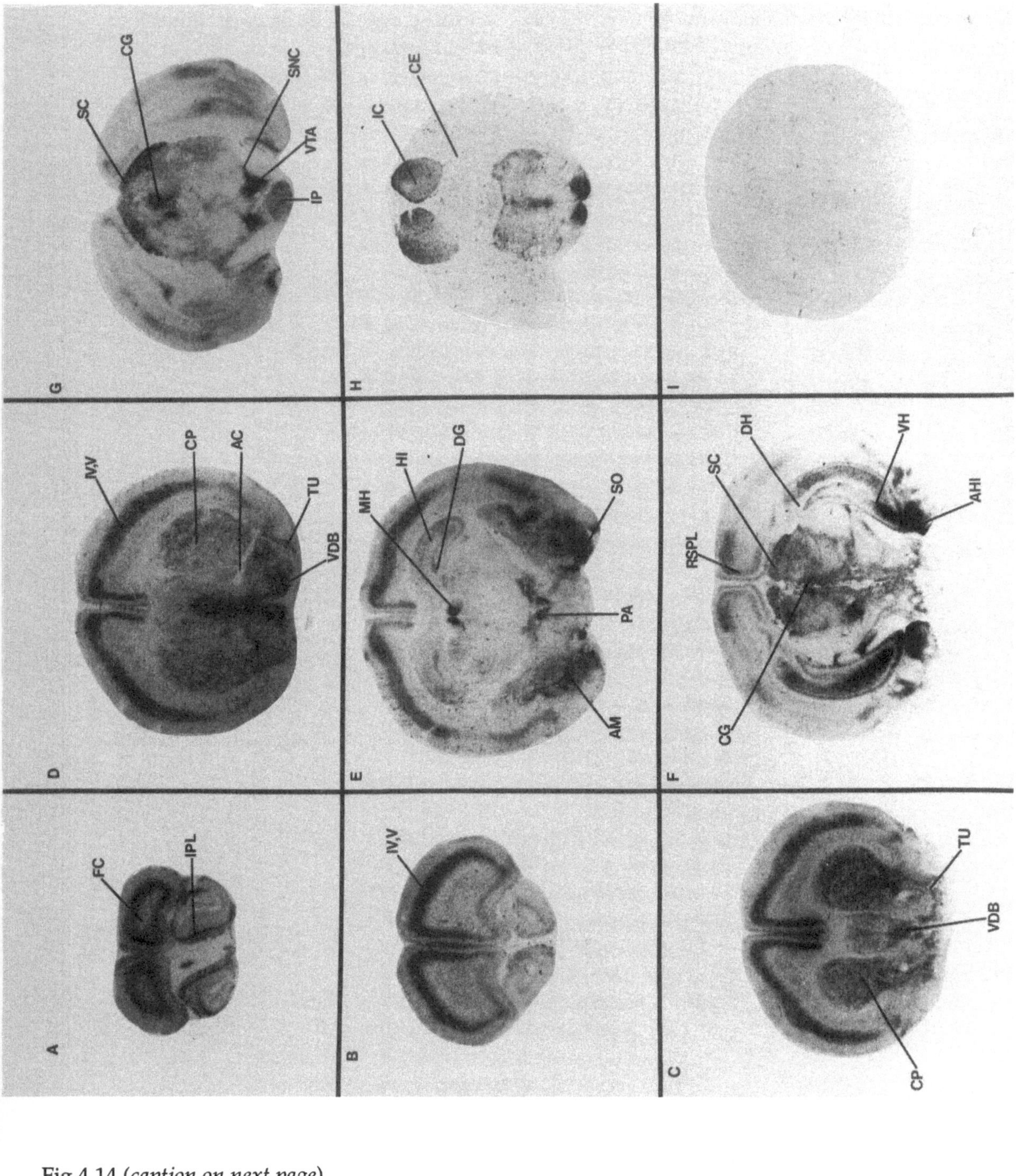

Fig 4.14 (*caption on next page*)

expanding to laminae V (Fig. 4.14A–E). In the occipital cortex, only the retrosplenial portion is enriched with NK-3 sites (Fig. 4.14F,G). The nucleus accumbens and the cerebellum are clearly devoid of NK-3 binding sites (Fig. 4.14D,H).

PD28

At this age (Fig. 4.15), the distribution of NK-3 sites is rather similar to that observed at PD14 (Fig. 4.14) except for the superficial layers of the cortex which are now relatively enriched with NK-3 sites (Fig. 4.15) in contrast to the apparent absence of labeling detected at earlier ages (Fig. 4.12–4.14). Lobules 9 and 10 of the cerebellum also contain NK-3 sites at PD28 (Fig. 4.15G).

PD35

The CNS distribution of NK-3 binding sites of 35-day-old rats is shown in Fig. 4.16. As at PD28 high densities of sites are seen in laminae IV and V of the cortex (Fig. 4.16A–H), in the supraoptic nucleus (Fig. 4.16B), amygdalohippocampal area (Fig. 4.16C,D), zona incerta (Fig. 4.16D), interpeduncular nucleus (Fig. 4.16G), superior colliculus (Fig. 4.16H), locus coeruleus (Fig. 4.16I) and nucleus tractus solitarius (Fig. 4.16J,K). Moderate densities are found in the striatum (Fig. 4.16A–D), septum (Fig. 4.16A,B), dorsal hippocampus (Fig. 4.16D,E), substantia nigra pars compacta (Fig. 4.16F), inferior colliculus (Fig. 4.16I) as well as lobules 9 and

10 of the cerebellum (Fig. 4.16J). Other regions of the cerebellum are devoid of specific labeling NK-3 (Fig. 4.16J). Thus, NK-3 sites are similarly distributed at PD28 and PD35 in the rat brain suggesting their complete profile of maturation during the second and third postnatal week. This is further supported by the discrete localization of NK-3 sites observed in adult (3-month-old) brain tissue.

4.4 POSSIBLE SIGNIFICANCE OF ONTOGENIC RECEPTOR MODIFICATIONS

4.4.1 NK-1 RECEPTORS

NK-1 binding sites appear very early during embryonic development since highly significant amounts of NK-1 labeling are present three days before birth in the rat brain (Quirion and Dam, 1986). Moreover, the density of NK-1 sites increases markedly 1 day before birth suggesting that SP might play a very important role in the early maturation process and organization of the CNS (Quirion and Dam, 1986).

It is evident that the distribution of NK-1 sites undergoes major modifications during postnatal ontogeny. This is especially striking in the brainstem since very high densities of sites are seen in this region up to 14 days after birth but not thereafter. This suggests that SP, by activating NK-1 receptors, is associated with the ontogenic development and maturation of this brain region. This is of special interest since it has been shown that SP can act as a trophic factor for brainstem

Fig. 4.14. Photomicrographs of the distribution of NK-3 receptor binding sites in coronal brain sections from 14-day-old rats (PD14). Sections were incubated with 50 pM ^{125}I-labeled BH-eledoisin. Non-specific (NS) labeling seen in the presence of 1.0 μM unlabeled eledoisin is shown in section I. Abbreviations: AC, anterior commissura; AHI, amygdalohippocampal area; AM, amygdala; CE, cerebellum; CG, central gray; CP, caudate putamen; DG, dentate gyrus; DH, dorsal hippocampus; FC, frontal cortex; HI, hippocampus; IC, inferior colliculus; IP, interpeduncular nuclei; IPL, internal plexiform layer of the olfactory bulb; MH, medial habenula; PA, paraventricular hypothalamic nuclei; RSPL, retrosplenial cortex; SC, superior colliculus; SNC, substantia nigra pars compacta; SO, supraoptic nuclei; TU, olfactory tubercle; VDB, diagonal band of Broca; VH, ventral hippocampus; VTA, ventral tegmental area; IV, V, layer IV, V of the frontal cortex.

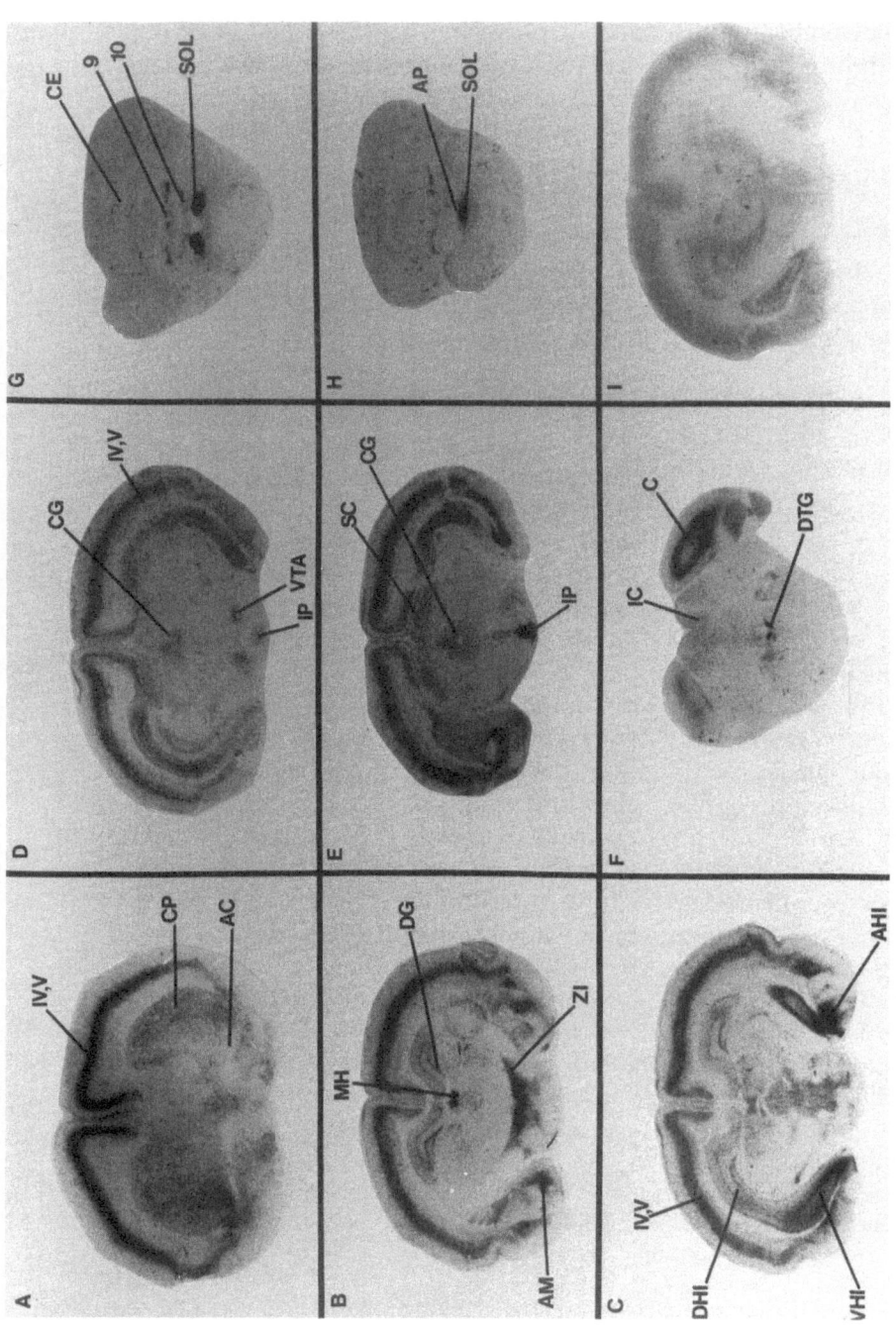

Fig. 4.15. Photomicrographs of the distribution of NK-3 receptor binding sites in coronal brain sections from 28-day-old rats (PD28). Sections were incubated with 50 pM ^{125}I-labeled BH-eledoisin. Non-specific (NS) labeling seen in the presence of 1.0 μM unlabeled eledoisin is shown in section I. Abbreviations: AC, anterior commissura; AHI, amygdalohippocampal area; AM, amygdala; C, cortex; CE, cerebellum; CG, central gray; CP, caudate putamen; DG, dentate gyrus; DHI, dorsal hippocampus; DTG, dorsal tegmentum; IC, inferior colliculus; IP, interpeduncular nuclei; MH, medial habenula; SC, superior colliculus; SOL, tractus solitarius nuclei; VHI, ventral hippocampus; VTA, ventral tegmental area; ZI, zona incerta; IV, V, layer IV, V of the frontal cortex; 9, 10, lobules 9 and 10 of the cerebellum.

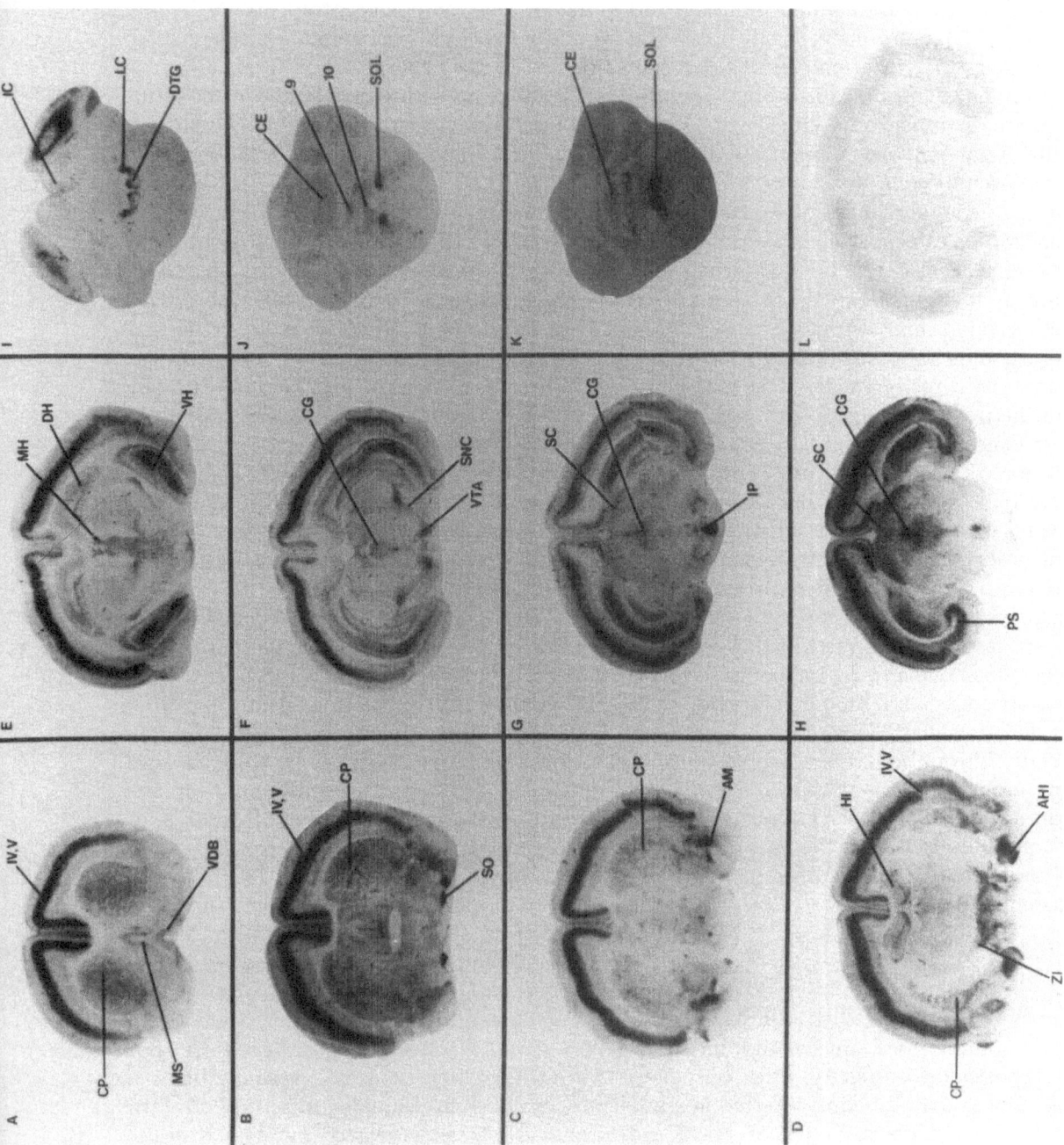

Fig. 4.16. Photomicrographs of the distribution of NK-3 receptor binding sites in coronal brain sections from 35-day-old rats (PD35). Sections were incubated with 50 pM ^{125}I-labeled BH-eledoisin. Non-specific (NS) labeling seen in the presence of 1.0 μM unlabeled eledoisin is shown in section L. Abbreviations: AHI, amygdalohippocampal area; AM, amygdala; CE, cerebellum; CG, central gray; CP, caudate putamen; DH, dorsal hippocampus; DTG, dentate gyrus; HI, hippocampus; IC, inferior colliculus; IP, interpeduncular nuclei; LC, locus coeruleus; MS, median septum; MH, medial habenula; PS, presubiculum; SC, superior colliculus; SNC, substantia nigra pars compacta; SO, supraoptic nuclei; SOL, tractus solitarius nuclei; VDB, diagonal band of Broca; VH, ventral hippocampus; VTA, ventral tegmental area; ZI, zona incerta; IV, V, layer IV, V of the frontal cortex; 9, 10, lobule 9, 10 of the cerebellum.

serotonergic neurons during ontogeny following neurotoxin treatment (Jonsson and Hallman, 1983), can prevent the degeneration of damaged noradrenergic neurons (Jonsson and Hallman, 1982a,b), can stimulate neurite outgrowth in embryonic chick dorsal root ganglia (Narumi and Fujita, 1978) and neuroblastoma cells (Naruma and Maki, 1978) and can accelerate the growth and maturation of catecholaminergic cells of the locus coeruleus (Nakai and Kasamatsu, 1984). Moreover, it has recently been shown that growth cones are enriched with NK-1 receptors suggesting that SP and NK-1 receptors may be directly involved in growth of certain neuronal propagation by modulating growth cones (Lockerbie *et al.*, 1988).

SP could also be an important factor involved in the maturation of other brain pathways on the basis of its early appearance during embryonic development. For example, it has been shown that SP is present in rat brain at gestation day 14, reaching adult level between days 5 and 15 after birth (Inagaki *et al.*, 1982a; Sakanaka *et al.*, 1982). Similar data have been obtained in human fetal brain with high densities of SP-like fibers identified, especially throughout the lower brainstem during embryogenesis (Namura *et al.*, 1982; Charnay *et al.*, 1983; Del Fiacco *et al.*, 1984). This correlates well with the high density of NK-1 binding sites visualized here in the brainstem of neonatal rats.

4.4.2 NK-2 SITES

As for NK-1 sites, the distribution of NK-2 receptors undergoes major modifications during postnatal ontogeny, with only low densities of specific labeling detected in adult animals. Very high densities of NK-2 are present very early (PD1) in the superior and inferior areas of brainstem. Subsequently, the concentrations of NK-2 sites in these regions gradually decrease to reach the adult level by PD21. This suggests that, in addition to SP, NKA could play an important role during the

development and organization of this brain region. The ontogeny profile of NKA-like immunoreactivity remains to be established. However, it appears that SP and NKA are distributed in a similar manner in the adult rat brain (Maggio and Hunter, 1984; Deutch *et al.*, 1985; Dalsgaard *et al.*, 1985; Linderfors *et al.*, 1985; Lee *et al.*, 1986) and are derived from the same precursor (Nakanishi, 1987; Krause *et al.*, 1989). Since SP appears very early during brain embryogenesis (Inagaki *et al.*, 1982a,b; Sakanaka *et al.*, 1982; Del Fiacco *et al.*, 1984; Paulin *et al.*, 1986), it is probable that NKs are also present during the early phase of CNS development. It is also noteworthy that NKA is able to stimulate cell growth in connective tissues (Nilsson *et al.*, 1985) and that the NK-2 receptor is related to the *mas* oncogene (Young *et al.*, 1986). It is likely that NKs, by activating NK-2 receptors, would modulate the growth and maturation of certain brain pathways.

In the basal ganglia, moderate levels of NK-2 sites are present during the various phases of brain development. Moreover, their distribution remains homogeneous and does not reflect the high amount of NKA present in this region in adult rat brain (Minamino *et al.*, 1984). This is rather different from the heterogeneous redistribution of NK-1 sites observed in the striatum during brain maturation (Quirion and Dam, 1986) demonstrating that NK-2 receptors are distinct from NK-1 sites.

This is supported further by the redistribution of NK-1 and NK-2 sites observed in cortical areas during postnatal ontogeny (Dam *et al.*, 1988). NK-2 sites appear very early after birth (PD1); their densities markedly increasing by PD4 in frontal cortex and PD7 in remaining areas. However, by the end of the second week, the density of cortical NK-2 sites begins to decrease and totally disappears by PD35. This is clearly different from NK-1 receptors which are present in various cortical areas in neonatal and adult brain (Quirion and Dam, 1986; Dam *et al.*, 1988).

4.4.3 NK-3 SITES

As for other NK receptors, NK-3 sites appear very early during embryonic development. For example, very high densities of NK-3 sites are observed in the external plexiform layer of the olfactory bulb and in the hippocampus at ED20. Thereafter, the amounts of NK-3 sites remain relatively constant during the first postnatal week. However, the distribution of NK-3 sites undergoes major modification during the second week of ontogenic development to reach the adult level by the end of the third week. As for NK-1 (Quirion and Dam, 1986) and NK-2 sites, various brainstem nuclei are enriched with NK-3 sites early postnatally but not during adulthood. Overall, the distribution of NK-3 sites is not as modified as that of NK-1 and NK-2 receptors during ontogeny suggesting that NK-3 receptors are less likely to mediate trophic effects of NKs.

The ontogenic profile of cortical NK-3 sites is unique. They appear later (PD6) than NK-1 and NK-2 sites and concentrate mostly in the deeper laminae (IV and V) of the frontal cortex. Moreover, NK-3 sites present in the cortex do not disappear (as for NK-2 sites) suggesting that they could be involved in the maintenance of normal cortical functions. It is thus of interest that Lamour *et al.* (1983) have shown that the excitatory responses elicited by NKs in deeper cortical laminae could be related to the activation of NK-3 sites. The distribution of NKB, the putative endogenous ligand of the NK-3 subtype, remains to be fully established either in adult tissues or during brain development.

4.5 ONTOGENIC PROFILE OF NK FUNCTIONS

The development of NK-mediated functions has not been explored as extensively as the development of the NK immunoreactivity and receptors. It appears, however, at least from studies in the spinal cord, that the presence of NK fibers and receptors alone does not predict the onset of function. SP is present in dorsal root ganglia, sensory nerves, spinal cord and skin at birth in the rat (Fitzgerald and Gibson, 1984) indicating that the distribution of SP in the primary sensory afferent system occurs early in development. SP-containing C fibers enter the dorsal horn on ED19–20 in the rat (Senba *et al.*, 1982; Fitzgerald, 1987). On PD 1, SP fibers are concentrated in laminae 1 but on following days they become concentrated in laminae 2. By PD8, the staining pattern is comparable to that in the adult, but the adult density is reached on PD15. Foot withdrawal responses to pinching and heating of the hindfoot skin are present at birth and are exaggerated in amplitude and duration compared to the adult (Fitzgerald and Gibson, 1984). It is possible that the high density of NK-1 receptors in the spinal cord at birth may play a role in this exaggerated response (Charlton and Helke, 1986; Yashpal *et al.*, 1991b). Withdrawal responses to irritant chemicals and neurogenic extravasation, however, did not occur until PD 10. The reason for this developmental difference in functions of the C fibers is not apparent, although it suggests that there may be different synaptic pathways for the responses to these different stimuli and that certain elements in these different pathways may develop at different times.

In comparison, in the peripheral nervous system, the responses to SP in the porcine trachea and rabbit bladder are in place at birth (Haxhiu *et al.*, 1990; Zderic *et al.*, 1990). Similar to the spinal cord, the response of the bladder to SP was greater in the neonate than in the adult. This may also suggest a greater receptor number in the developing animal.

4.6 FACTORS WHICH CAN INFLUENCE THE DEVELOPMENT OF SP AND NK-1 RECEPTORS

4.6.1 SP CONTENT AND APPEARANCE

Several factors have been identified which influence the development of SP-containing

neurons. For example, postnatal administration of nerve growth factor increases the
content of SP in the rat dorsal root ganglia
(Kessler and Black, 1980). Rats exposed
during development to anti-nerve growth
factor antibodies had a significant decrease in
SP levels in the dorsal root ganglia and spinal
cord and also showed increased sensitivity to
pain (Otten *et al.*, 1982).

Recent evidence suggests that the presence
of dopamine may regulate the postnatal
development of the SP content of the striatum. As described previously, there is a close
interrelationship between dopamine- and SP -
containing neurons in the adult striatum.
Depletion of striatal dopamine by 6-hydroxy-
dopamine administration on PD3 depressed
the expression of PPT mRNA and SP content
in the striatum at least through PD35 (Sivam
et al., 1991). The distribution of SP neurons in
the striatum was unaffected, however, by
neonatal dopamine depletion (Snyder-Keller,
1991), suggesting that early developmental
abnormalities in the dopamine system could
have a long-term influence on the SP system
as well.

The development of SP neurons in the
dorsal horn of the spinal cord appears to be
influenced by thyroid hormone and serotonin. Rats made hypothyroid on the first
day after birth had twice the content of SP in
the dorsal horn at maturity than normal rats.
Similarly, inhibition of serotonin synthesis by
PCPA treatment during the neonatal period
increased SP in the spinal cord (Savard *et al.*,
1983).

Contact with the appropriate target organ
may also play a role in regulating SP in
developing neurons. In dissociated culture
of superior cervical ganglion neurons, the
content of SP increased dramatically when
the cells were grown on pineal or salivary
gland cells (normal targets) compared to
when they were grown on non-target cells
such as heart or intestine (Kessler *et al.*,
1984). This effect was not mimicked by nerve
growth factor.

4.6.2 SP RECEPTOR SITES

The presence of SP itself may be an important
factor in regulating the development of SP
receptors. SP receptors were up-regulated in
adult rats which were administered SP (1
µg/day) during the first week after birth. The
increased receptor number was apparent in
the salivary gland and various brain regions,
including the hypoglossal nucleus, dorsal
tegmental nucleus, locus coeruleus, septo-
fimbrial nucleus, dorsal raphe and central
gray (Handelmann *et al.*, 1987). SP concentrations in microdissected brain regions were
not affected by the neonatal treatment. The
increase in receptor expression was sufficient
to alter some SP-mediated behaviors in the
adult rats. For example, the sensitivity of the
rat to painful stimulation of the paws was
increased. In addition, the rats salivated in
response to lower doses of SP injected
intravenously and they produced a greater
volume of saliva (Handelmann *et al.*, 1984).
There were no changes, however, in cardiovascular responses to SP. These data indicate
that neonatal exposure to increased levels of
SP increased the expression of SP receptors in
several tissues and also increased the sensitivity of the tissues to SP. In contrast, adult
rats which had been treated with a specific
substance P antiserum on PD2 had an
attenuated cardiovascular response to SP (De
Felipe *et al.*, 1989). Treated animals also failed
to demonstrate an increased nociceptive
threshold in response to SP administered
intracerebroventricularly, in contrast to
control rats. This suggests that neonatal
exposure to a SP antiserum, which might be
expected to decrease the amount of SP available synaptically, caused decreased sensitivity to SP in the adults. These investigators,
however, failed to observe differences in SP
binding to membranes derived from the
spinal cord and the central gray matter. This
may suggest alterations in transduction
mechanisms instead of receptor density and/
or affinity.

The amount of SP present during development can therefore have a long-lasting impact on SP receptors. SP can also regulate SP receptors in the adult but the change in receptor number is in the opposite direction. In the adult, the number of postsynaptic SP receptors increases following removal of the SP innervation (Helke *et al.*, 1986). The mechanism of action by which SP influences the development of its receptors is unknown. The presence of SP may up-regulate the density of SP receptors expressed on a given cell. Alternatively, given the possibility that SP has trophic influences on cells during development, an excess of SP could promote the survival of cells containing SP receptors. In either event, it would be expected that factors or events such as those described above could have long-lasting impact on behaviors mediated by SP.

Although there is little other information concerning the developmental regulation of SP receptors, it has been shown that SP receptor gene transcription is blocked by glucocorticoids in adult brain (Ihara and Nakanishi, 1990). If a similar action occurs during development, the ontogeny of SP receptors could be influenced by increased circulating glucocorticoids resulting from maternal or fetal stress.

4.7 SUMMARY

NKs are neuropeptides with a very broad distribution in the central and peripheral nervous systems. Thus, factors which influence the development of NK pathways would also affect the functioning of numerous organ systems.

The early appearance of NKs and their receptors in the fetal nervous system suggests that the NKs themselves might play an important role in the maturation of nervous and other tissues. In fact, a number of the functions of the NKs already characterized would be important in development. For example, the effects of these peptides on cell proliferation and migration would be of obvious importance during ontogeny. The appearance of SP receptors on growth cones (Lockerbie *et al.*, 1988) also suggests a role in synaptic development of the nervous system and it has been proposed that SP is important in maintaining synaptic connections once they are formed (Wall *et al.*, 1982). The regulation of local blood flow by NKs could also be important for the supply of nutrients and blood-borne trophic factors necessary for cell growth and survival. Finally, the regulation of hormones such as growth hormone and insulin by NKs would have a broad influence on developmental processes. Although the significance of the NKs in neural development is still only poorly understood, these neuropeptides are likely to have a far-reaching impact on processes of growth and development throughout the body, as their receptors are differentially expressed by many cell types during various developmental periods.

ACKNOWLEDGEMENTS

This work was supported by grants from the Canadian Parkinson Foundation and the Scottish Rite Foundation for Schizophrenia. TVD is a holder of a studentship from the Fonds de Chercheur et d'aide à la recherche, RQ is a Chercheur-Boursier of the Fonds de la Recherche en Santé du Québec.

REFERENCES

Backman, S.B. and Henry, J.L. (1984) Effects of substance P and thyrotropin-releasing hormone on sympathetic preganglionic neurones in the upper thoracic intermediolateral nucleus of the cat. *Can. J. Physiol. Pharmacol.*, **62**, 248–51.

Bannon, M.J., Freeman, A.S., Chiodo, L.A. *et al.* (1987). The electrophysiological and biochemical pharmacology of the mesolimbic and mesocortical dopamine neurons, in *Handbook of Psychopharmacology* (eds L.L. Iversen, S.D. Iversen and S.H. Snyder), Plenum, New York, Vol. 19, pp 329–74.

Bar-Shavit, Z., Goldman, R., Stabinsky, Y. *et al.* (1980) Enhancement of phagocytosis: a newly found activity for substance P residing in its N-terminal tetrapeptide sequence. *Biochem. Biophys. Res. Commun.*, **94**, 1445–51.

Bertaccini, G., De Caro, G. and Cheli, R. (1966) Enlargement of salivary glands in rats after chronic administration of physalaemin or isoprenaline. *J. Pharm. Pharmacol.*, **18**, 312–16.

Bill, A., Stjernschantz, J., Mandahl, A. *et al.* (1979) Substance P: release on trigeminal nerve stimulation, effects in the eye. *Acta Physiol. Scand.*, **106**, 371–3.

Brain, S.D. and Williams, T.J. (1989) Interactions between the tachykinins and calcitonin gene-related peptide lead to the modulation of oedema formation and blood flow in rat skin. *Br. J. Pharmacol.*, **97**, 77–82.

Brene, S., Linderfors, N., Friedman, W.J. and Persson, H. (1990) Preprotachykinin A mRNA expression in the rat brain during development. *Dev. Brain Res.*, **57**, 151–62.

Bristow, D.R., Curtis, N.R., Suman-Chauhan, N. *et al.* (1987) Effects of tachykinins on inositol phospholipid hydrolysis in slices of hamster urinary bladder. *Br. J. Pharmacol.*, **90**, 211–17.

Brown, M. and Vale, W. (1976) Effect of neurotensin and substance P on plasma insulin, glucagon and glucose levels. *Endocrinology*, **98**, 819–22.

Buck, S.H. and Burcher, E. (1985) The rat submaxillary gland contains predominantly P-type tachykinin binding sites. *Peptides*, **6**, 1079–84.

Buck, S.H. and Burcher, E. (1986) The tachykinins: a family of peptides with a brood of receptors. *Trends Pharmacol. Sci.*, **7**, 65–8.

Buck, S.H., Burcher, E., Shults, C.W. *et al.* (1984) Novel pharmacology of substance K-binding sites: a third type of tachykinin receptor. *Science*, **226**, 987–9.

Buck, S.H., Helke, C.J., Burcher, E. *et al.* (1986) Pharmacologic characterization and autoradiographic distribution of binding sites for iodinated tachykinins in the rat central nervous system. *Peptides*, **7**, 1109–20.

Bury, R.W. and Mashford, M.L. (1977) A pharmacological investigation of synthetic substance P on the isolated guinea-pig ileum. *Clin. Exp. Pharmacol. Physiol.*, **4**, 453–61.

Byers, M.R. (1984) Dental sensory receptors. *Int. Rev. Neurobiol.*, **25**, 39–94.

Cantalamessa, F., De Caro, G. and Perfumi, M. (1975) Effects of chronic administration of eledoisin or physalaemin on the rat salivary glands. *Pharmacol. Rev. Commun.*, **7**, 259–71.

Cascieri, M.A. and Liang, T. (1983) Characterization of the substance P receptor and the inhibition of radioligand binding by guanine nucleotides. *J. Biol. Chem.*, **258**, 5158–64.

Cascieri, M.A., Chicchi, G.G., Friedinger, R.M. *et al.* (1986) Conformationally constrained tachykinin analogs which are selective ligands for the eledoisin binding site. *Mol. Pharmacol.*, **29**, 34–8.

Charlton, C.G. and Helke, C.J. (1985) Autoradiographic localization and chracterization of spinal cord substance P binding sites: high densities in sensory, autonomic, phrenic and Onuf's motor nuclei. *J. Neurosci.*, **5**, 1653–61.

Charlton, C.G. and Helke, C.J. (1986) Ontogeny of substance P receptors in rat spinal cord: quantitative changes in receptor number and differential expression in specific loci. *Dev. Brain Res.*, **29**, 81–91.

Charnay, Y., Paulin, C., Chayvialle, J.A. and Dubois, P.M. (1983) Distribution of substance P-like immunoreactivity in the spinal cord and dorsal root ganglia of the human foetus and infant. *Neuroscience*, **10**, 41–55.

Costa, M., Cuello, A.C., Furness, J.B. and Frano, R. (1980) Distribution of enteric neurons showing immunoreactivity for substance P in the guinea pig ileum. *Neuroscience*, **5**, 323–31.

Cuello, A.C. and Kanazawa, I. (1978) The distribution of substance P immunoreactive fibers in the rat central nervous system. *J. Comp. Neurol.*, **178**, 129–56.

Cuello, A.C., Priestley, J.V. and Matthews, M.R. (1982) Localization of substance P in neuronal pathways. *Ciba Found. Symp.*, **9**, 55–83.

Dalsgaard, C.-J., Risling, M. and Cuello, C. (1982) Immunohistochemical localization of substance P in the lumbrosacral spinal pia matter and ventral roots of the cat. *Brain Res.*, **246**, 168–71.

Dalsgaard, C.-J., Haegerstrand, A., Theodorsson-Norheim, E. *et al.* (1985) Neurokinin A-like immunoreactivity in rat primary sensory neurons: coexistence with substance P. *Histochemistry*, **83**, 37–39.

Dalsgaard, C.-J., Hultgardh-Nilsson, A., Haegerstrand, A. and Nilsson, J. (1989) Neuropeptides as growth factors. Possible roles in human diseases. *Regul. Pept.*, **25**, 1–9.

Dam, T.V. and Quirion, R. (1986) Pharmacological characterization and autoradiographic localization of substance P receptors in guinea pig brain. *Peptides*, **7**, 855–64.

Dam, T.V., Escher, E. and Quirion, R. (1986) Tachykinins and phosphatidyl inositol turnover in rat brain, in *Substance P and Neurokinins* (eds J.

Henry, R. Couture, A.C. Cuello *et al.*) Springer-Verlag, New York, pp. 75–7.

Dam, T.V., Escher, E. and Quirion, R. (1988) Evidence for the existence of three classes of neurokinin receptors in brain. Differential ontogeny of neurokinin-1, neurokinin-2 and neurokinin-3 binding sites in rat cerebral cortex. *Brain Res.*, **453**, 372–6.

Dam, T.V., Escher, E. and Quirion, R. (1990a) Visualization of neurokinin-3 receptor sites in rat brain using the highly selective ligand [³H] senktide. *Brain Res.*, **506**, 175–9.

Dam, T.V., Martinelli, B. and Quirion, R. (1990b) Autoradiographic distribution of brain neurokinin-1/substance P receptors using a highly selective ligand [³H]-Sar⁹,Met(O₂)-substance P. *Brain Res.*, **531**, 333–7.

Dam, T.V., Takeda, Y., Krause, J.E. *et al.* (1990c) Gamma-preprotachykinin (72–92)-peptide amide: an endogenous preprotachykin-I gene-derived peptide that preferentially binds to neurokinin-2 receptors. *Proc. Natl. Acad. Sci. USA*, **87**, 246–50.

Davies, J. and Dray, A. (1976) Substance P in the substantia nigra. *Brain Res.*, **107**, 623–7.

De Felipe, M.C., Molinero, M.T. and Del Rio, J. (1989) Long-lasting neurochemical and functional changes in rats induced by neonatal administration of substance P antiserum. *Brain Res.*, **485**, 301–8.

Del Fiacco, M., Dessi, M.L. and Leranti, M.C. (1984) Topographical localization of substance P in the human postmortem brainstem. An immunohistochemical study in the newborn and adult tissue. *Neuroscience*, **12**, 591–611.

Deutch, A.Y., Maggio, J.E., Bannon, M.J. *et al.* (1985) Substance K and substance P differentially modulate mesolimbic and mesocortical systems. *Peptides*, **6** (Suppl.2), 113–22.

Dick, E., Miller, R.F. and Behbehani, M.M. (1980) Opioids and substance P influence ganglion cells in amphibian retina. *Invest. Opthalmol. Visual Sci.*, **19** (ARVO Suppl.) 132.

Diez-Guerra, F.J., Veira, J.A., Augood, S. and Emson, P.C. (1989) Ontogeny of the novel tachykinins neurokinin A, neurokinin B and neuropeptide K in the rat central nervous system. *Regul. Pept.*, **25**, 87–97.

D'Orléans-Juste, P., Dion, S., Mizrahi, J. and Regoli, D. (1985) Effects of peptides and non-peptides on isolated arterial smooth muscles: role of endothelium. *Eur. J. Pharmacol.*, **114**, 9–21.

D'Orléans-Juste, P., Dion, S., Drapeau, G. and Regoli, D.(1986) Different receptors are involved in the endothelium-mediated relaxation and the smooth muscle contraction of the rabbit pulmonary artery in response to substance P and related neurokinins. *Eur. J. Pharmacol.*, **125**, 37–44.

Duckles, S.P. and Buck, S.H. (1982) Substance P in the cerebral vasculature: depletion by capsaicin suggests a sensory role. *Brain Res.*, **245**, 171–4.

Edvinsson, L. and Uddman, R. (1982) Immunohistochemical localization and dilatory effect of substance P on human cerebral vessels. *Brain Res.*, **232**, 466–71.

Edvinsson, L., McCulloch, J. and Uddman, R. (1981) Substance P: immunohistochemical localization and effect upon cat pial arteries *in vitro* and *in situ*. *J. Physiol. (Lond.)*, **318**, 251–8.

Erjavec, F., Lembeck, F. and Florjanc-Irman, T. (1981) Release of histamine by substance P. *N.-S. Arch. Pharmacol.*, **317**, 67–70.

Erspamer, G.F., Erspamer, V. and Piccinnelli, D. (1980) Parallel bioassay of physalaemin and kassinin, a tachykinin dodecapeptide from the skin of the African frog *Kassina senegalensis*. *N.-S. Arch. Pharmacol.*, **311**, 61–5.

Evans, P.D., Reale, V., Merzan, R.M. and Villegas, J. (1990) Substance P modulation of the membrane potential of the Schwann cell of the squid giant nerve fibre. *Glia*, **3**, 393–404.

Fitzgerald, M. (1987) Prenatal growth of fine diameter primary afferents into the rat spinal cord: a transganglionic tracer sensory study. *J. Comp. Neurol*, **261**, 98–104.

Fitzgerald, M. and Gibson, S. (1984) The postnatal physiological and neurochemical development of peripheral sensory C fibres. *Neuroscience*, **13**, 933–44.

Flood, J.F., Baker, M.L., Hernandez, E.N. and Morley, J.E. (1990) Modulation of memory retention by neuropeptide K. *Brain Res.*, **520**, 284–90.

Freidin, M. and Kessler, J.A. (1990) Cytokine regulation of substance P expression in sympathetic neurons. *Proc. Natl. Acad. Sci. USA*, **88**, 3200–3.

Furness, J.B., Papka, R.F., Della, N.G. *et al.* (1982) Substance P-like immunoreactivity in nerves associated with the vascular system in guinea pigs. *Neuroscience*, **7**, 447–59.

Gamse, R., Holzer, P. and Lembeck, F. (1980) Decrease of substance P in primary afferent neurons and impairment of neurogenic plasma extravasation by capsaicin. *Br. J. Pharmacol.*, **68**, 207–13.

Gilbert, R.F.T. and Emson, P.C. (1979) Substance P in rat CNS and duodenum during development. *Brain Res.*, **171**, 166–70.

Gilbey, M.P., McKenna, K.E. and Schramm, L.P. (1983) Effects of substance P on sympathetic preganglionic neurons. *Neurosci. Lett.*, **41**, 157–9.

Glickman, R.D., Adolph, A.R. and Dowling, J.E. (1980) Does substance P have a physiological role in the carp retina? *Invest. Opthalmol. Visual Sci.*, **19** (ARVO Suppl.), 281.

Guard, S., Watling, K.J. and Watson, S.P. (1988) Neurokinin-3 receptors are linked to inositol phospholipid hydrolysis in the guinea pig ileum longitudinal muscle-myenteric plexus preparation. *Br. J. Pharmcol.*, **94**, 148–54.

Handelmann, G.E., Selsky, J.H. and Helke, C.J. (1984) Substance P administration to neonatal rats increases adult sensitivity to substance P. *Physiol. Behav.*, **33**, 297–300.

Handelmann, G.E., Shults, C. and O'Donohue, T.L. (1987) A developmental influence of substance P on its receptors. *Int. J. Dev. Neurosci.*, **5**, 11–15.

Hanley M.R. and Jackson, T. (1987) Substance K receptor: return of the magnificent seven. *Nature*, **329**, 766–7.

Hartung, H.-P., Wolters, K. and Toyka, K.V. (1986) Substance P: binding properties and studies on cellular responses in guinea pig macrophages. *J. Immunol.*, **136**, 3856–63.

Hassessian, H., Drapeau, G. and Couture, R. (1988) Spinal action of neurokinins producing cardiovascular responses in the conscious freely moving rat: evidence for a NK-1 receptor mechanism. *N.-S. Arch. Pharmacol.*, **338**, 649–54.

Haxhiu, M.A., Haxhiu-Poskurica, B., Moracic, V. *et al.* (1990) Reflex and chemical responses of tracheal submucosal glands in piglets. *Respir. Physiol.*, **82**, 267–77.

Hecht, K., Oehme, P., Poppei, M. and Hect, T. (1979) Conditioned-reflex learning of normal juvenile and adults rats exposed to action of substance P and of an SP analogue. *Pharmazie*, **34**, 419–23.

Helke, C.J., Neil, J.J., Massari, V.J. and Loewy, A.D. (1982) Substance P neurons project from the ventral medulla to the intermediolateral cell column and ventral horn in the rat. *Brain Res.*, **243**, 147–52.

Helke, C.J. , Charlton, C.G. and Wiley, R.G. (1985) Suicide transport of ricin demonstrates the presence of substance P receptors on medullary somatic and autonomic motor neurons. *Brain Res.*, **328**, 190–5.

Helke, C.J., Charlton, C.G. and Wiley, R.G. (1986) Studies on the cellular localization of spinal cord substance P receptors. *Neuroscience*, **19**, 523–33.

Helke, C.J., Krause, J.E., Mantyh, P.W. *et al.* (1990) Diversity in mammalian tachykinin peptidergic neurons: multiple peptides, receptors and regulatory mechanisms. *FASEB J.*, **4**, 1606–15.

Hershey, A.D. and Krause, J.E. (1990) Molecular characterization of a functional cDNA encoding the rat substance P receptor. *Science*, **247**, 958–62.

Hinsey, J.C. and Gasser, H.S. (1930) The component of the dorsal root mediating vasodilation and the Sherrington contraction. *Am. J. Physiol.*, **92**, 679–789.

Holm, I., Thulin, L. and Hellgren, M (1978) Anticholeretic effect of substance P in anesthetized dogs. *Acta Physiol. Scand.*, **102**, 274–80.

Holzer, P. Lembeck, F. (1980) Neurally mediated contraction of ileal longitudinal muscle by substance P. *Neurosci. Lett.*, **17**, 101–5.

Holzer, P., Gamse, R. and Lembeck, F. (1980) Distribution of SP in the gastrointestinal tract: lack of effect of capsaicin pretreatment. *Eur. J. Pharmacol.*, **61**, 303–7.

Holzer, P., Bucsics, A. and Lembeck, F. (1982) Distribution of capsaicin-sensitive nerve fibers containing immunoreactive substance P in cutaneous and visceral tissues of the rat. *Neurosci. Lett.*, **31**, 253–7.

Huston, J.P. and Staubli, U. (1979) Post-trial injection of substance P into lateral hypothalamus and amygdala, respectively, facilitates and impairs learning. *Behav. Neural Biol.*, **27**, 244–8.

Ihara, H. and Nakanishi, S. (1990) Selective inhibition of expression of the substance P receptor mRNA in pancreatic acinar AR42J cells by glucocorticoids. *J. Biol. Chem.*, **265**, 22441–5.

Improta, G. and Broccarde, M. (1990) Tachykinin: effects on gastric secretion and emptying in rats. *Pharmacol. Res.*, **22**, 605–10.

Inagaki, S., Sakanaka, M., Shiosaka, S. *et al.* (1982a) Ontogeny of substance P-containing neuron system of the rat: immunohistochemical analysis. I. Forebrain and upper brainstem. *Neuroscience* , **7**, 251–77

Inagaki, S., Sakanaka, M., Shiosaka, S. *et al.* (1982b) Experimental and immunohistochemical studies of the cerebellar substance P of the rat: localization, postnatal ontogeny and ways of entry to the cerebellum. *Neuroscience*, **7**, 639–45.

Inman, R.D., Chiu, B. and Marshall, K.W. (1986) Substance P and arthritis analysis of plasma and synovial fluid levels. *Arthritis Rheum.*, **29**, Suppl. 1, S9.

Iversen, S.D. (1982) Behavioral effects of substance

P through dopaminergic pathways in the brain. *Ciba Found. Symp.*, **91**, 307–24.

Iversen, L.L., Foster, A.C., Watling, K.J. *et al.* (1987) Multiple receptors and binding sites for tachykinins, in *Substance P and Neurokinins* (eds J.L. Henry, R. Couture, A.C. Cuello *et al.*), Springer-Verlag, New York, pp. 40–3.

Jacques, L. and Couture, R. (1990) Studies on the vascular permeability induced by intrathecal substance P and bradykinin in the rat. *Eur. J. Pharmacol.*, **184**, 9–20.

Johansson, O., Hökfelt, T., Pernow, B. *et al.* (1981) Immunohistochemical support for three putative transmitters in one neuron: coexistence of 5-hydroxytryptamine-, substance P- and thyrotropin releasing hormone-like immunoreactivity in medullary neurons projecting to the spinal cord. *Neuroscience*, **6**, 1857–81.

Jonsson, G. and Hallman, H. (1982a) Substance P counteracts neurotoxin damage on norepinephrine neurons in rat brain during ontogeny. *Science*, **215**, 75–7.

Jonsson, G. and Hallman, H. (1982b) Substance P modifies the 6-hydroxydopamine induced alteration of postnatal development of central noradrenaline neurons. *Neuroscience*, **7**, 2909–18.

Jonsson, G. and Hallman, H. (1983) Effect of substance P on the 5,7-dihydroxytryptamine induced alteration of postnatal development of central serotonin neurons. *Med. Biol.*, **61**, 105–12.

Kafetzopoulos, E., Holzhauer, M.S. and Houston, J.P. (1986) Substance P injected into the region of the nucleus basalis magnocellularis facilitates performance of an inhibitory avoidance task. *Psychopharmacology*, **90**, 281–3.

Kawaguchi, Y., Hoshimaru, M., Nawa, H. and Nakanishi, S. (1986) Sequence analysis of cloned cDNA for rat substance P precursor: existence of a third substance P precursor. *Biochem. Biophys. Res. Commun.*, **139**, 1040–6.

Keeler, J.R., Charlton, C.G. and Helke, C.J. (1985) Cardiovascular effects of spinal cord substance P: studies with a stable receptor agonist. *J. Pharmacol. Exp. Ther.*, **233**, 755–60.

Kelley, A.E., Stinus, L. and Iversen, S.D. (1979) Behavioral activation induced in the rat by substance P infusion into ventral tegmental area: implication of dopaminergic A10 neurones. *Neurosci. Lett.*, **11**, 335–9.

Kessler, J.A. and Black, I.B. (1980) Nerve growth factor stimulates the development of substance P in sensory ganglia. *Proc. Natl. Acad. Sci. USA*, **77**, 649–52.

Kessler, J.A., Adler, J.E., Jonakait, G.M. and Black, I.B. (1984) Target organ regulation of substance P in sympathetic neurons in culture. *Dev. Biol.*, **103**, 71–9.

Konturek, S.J., Jaworek, J., Tasler, J. *et al.* (1981) Effect of substance P and its C-terminal hexapeptide on gastric and pancreatic secretion in the dog. *Am. J. Physiol.*, **21**, G74–G81.

Kotani, H., Hoshimaru, M., Nawa, H. and Nakanishi, S. (1986) Structure and gene organization of bovine neuromedin K precursor. *Proc. Natl. Acad. Sci. USA*, **83**, 7074–8.

Krause, J.E., Chirgwin, J.M., Carter, M.S. *et al.* (1987) Three rat preprotachykinin mRNAs encode the neuropeptides substance P and neurokinin A. *Proc. Natl. Acad. Sci. USA*, **84**, 881–5.

Krause, J.E., Cremins, J.D., Carter, M.S. *et al.* (1988) Solution hybridization-nuclease protection assays for sensitive detection of differentially spliced substance P- and neurokinin A-encoding messenger ribonucleic acids. *Methods Enzymol.*, **168**, 634–52.

Krause, J.E., MacDonald, M.R. and Takeda, Y. (1989) The polyprotein nature of substance P precursors. *BioEssays*, **10**, 62–9.

Kroegel, C., Giembycz, M.A. and Barnes, P.J. (1990) Characterization of eosinophil cell activation by peptides. Differential effects of substance P, melittin and FMET-Leu-Phe. *J. Immunol.*, **145**, 2581–7.

Lamour, Y., Dutar, P. and Jobert, A. (1983) Effects of neuropeptides on rat cortical neurons: laminar distribution and interaction with the effect of acetylcholine. *Neuroscience*, **10**, 107–17.

Laufer, R., Wormser, U., Friedman, Z.Y. *et al.* (1985) Neurokinin B is a preferred agonist for a neuronal substance P receptor and its action is antagonized by enkephalin. *Proc. Natl. Acad. Sci. USA*, **82**, 7444–8.

Lee, C.-M. and Cheng, W.T. (1988) Effects of neonatal monosodium glutamate treatment on substance P binding sites in the rat retina. *Neurosci. Lett.*, **92**, 310–14.

Lee, C.-M., Campbell, N.J., Williams, B.J. and Iversen, L.L. (1986) Multiple tachykinin binding sites in peripheral tissues and in brain. *Eur. J. Pharmacol.*, **130**, 209–17.

Lee, C.-M., Kum, W., Cockram, C.S. *et al.* (1989) Functional substance P receptors on a human astrocytoma cell line (U-373 MG). *Brain Res.*, **488**, 328–31.

Lembeck, F. and Donnerer, J. (1981) Time course of capsaicin-induced functional impairments in comparison with changes in neuronal substance P content. *N.-S. Arch. Pharmacol.*, **316**, 240–3.

Lembeck, F. and Gamse, R. (1982) Substance P in peripheral sensory processes. *Ciba Found. Symp.* **91**, 35–54.

Lembeck, F. and Holzer, P. (1979) Substance P as a neurogenic mediator of antidromic vasodilation and neurogenic plasma extravasation. *N.-S. Arch. Pharmacol.*, **310**, 175–83.

Levine, J.D., Clark, R., Devor, M. *et al.* (1984) Intraneuronal substance P contributes to the severity of experimental arthritis. *Science*, **226**, 547–9.

Lindefors, N.E., Brodin, E., Theodorsson-Norheim, E. and Ungerstedt, U. (1985) Calcium-dependent potassium-stimulated release of neurokinin A and neurokinin B from rat brain regions *in vitro*. *Neuropeptides*, **6**, 453–61.

Ljungdahl, A., Hökfelt, T. and Nilsson, G. (1978) Distribution of substance P like immunoreactivity in the central nervous system of the rat. I. Cell bodies and nerve terminals. *Neuroscience*, **3**, 861–943.

Lockerbie, R.O., Beaujouan, J.-C., Saffroy, M. and Glowinski, J. (1988) An isolated growth cone-enriched fraction from developing rat brain has substance P binding sites. *Dev. Brain Res.*, **40**, 1–9.

Lotz, M., Carson, D.A. and Vaughan, J.H. (1987) Substance P activation of rheumatoid synoviocytes: neural pathway in pathogenesis of arthritis. *Science*, **235**, 893–5.

Lundberg, J.M., Hökfelt, T., Arggård, A. *et al.* (1980) Peripheral peptide neurons: distribution, axonal transport and some aspects on possible function, in *Neural Peptides and Neuronal Communication* (eds E. Costa and M. Trabucchi), Raven Press, New York, pp. 25–36.

MacDonald, M.R., McCourt, D.W. and Krause, J.E. (1988) Posttranslational processing α, β and γ-preprotachykinins: cell-free translation and early posttranslational processing events. *J. Biol. Chem.*, **263**, 15176–83.

Maggio, J.E. (1988) Tachykinins. *Annu. Rev. Neurosci.*, **11**, 13–28.

Maggio, J.E. and Hunter, J.C. (1984) Regional distribution of kassinin-like immunoreactivity in rat central and peripheral tissues and the effect of capsaicin. *Brain Res.*, **307**, 370–3.

Mansson, B., Nilsson, B.-O. and Ekstrom, J. (1990) Effects of repeated infusions of substance P and vasoactive intestinal peptide on the weights of salivary glands subjected to atrophying influences in rats. *Br. J. Pharmacol.*, **101**, 853–8.

Mantyh, P.W., Hunt, S.P. and Maggio, J.E. (1984a) Substance P receptors; localization by light microscopic autoradiography in rat brain using [^{3}H] SP as the radioligand. *Brain Res.*, **307**, 147–65.

Mantyh, P.W., Pinnock, R.D., Downes, C.P. *et al.* (1984b) Correlation between inositol phospholipid hydrolysis and substance P receptors in rat CNS. *Nature*, **309**, 795–7.

Mantyh, P.W., Gates, T., Mantyh, C.R. and Maggio, M.D. (1989a) Autoradiographic localization and characterization of tachykinin receptor binding sites in the rat brain and peripheral tissues. *J. Neurosci.*, **9**, 258–79.

Mantyh, P.W., Johnson, D.J., Boehmer, C.G. *et al.* (1989b) Substance P receptor binding sites are expressed by glia *in vivo* after neuronal injury. *Proc. Natl. Acad. Sci. USA*, **86**, 5193–7.

Mastrangelo, D., Mathison, R., Huggel, H.J. *et al.* (1987) The rat isolated portal vein: a preparation sensitive to neurokinins, particularly neurokinin B. *Eur. J. Pharmacol.*, **134**, 321–6.

Masu, Y., Nakayama, K., Tamaki, H. *et al.* (1987) cDNA cloning of bovine substance K through oocyte expression system. *Nature*, **329**, 836–8.

Mayberg, M., Langer, R.S., Zervas, N.T. and Moskowitz, M.A. (1981) Perivascular meningeal projections from cat trigeminal ganglia: possible pathway for vascular headaches in man. *Science*, **213**, 228–30.

Miller, A., Costa, M., Furness, J.B. and Chubb, I.W. (1981) Substance P immunoreactive sensory nerves supply the rat iris and cornea. *Neurosci. Lett.*, **23**, 243–9.

Minamino, N., Kangawa, K., Fukuda, A. and Matsuo, H. (1984) Neuromedin L: a novel mammalian tachykinin identified in porcine spinal cord. *Neuropeptides*, **4**, 157–66.

Morin-Surun, M.P., Jordan, D., Champagnat, J. *et al.* (1984) Excitatory effects of iontophoretically applied substance P on neurons in the nucleus tractus solitarius of the cat: lack of interaction with opiates and opioids. *Brain Res.*, **307**, 388–92.

Mousli, M., Bronner, C., Landry, Y. *et al.* (1990a) Direct activation of GTP-binding regulatory protein (G proteins) by substance P and compound 48/80. *FEBS Lett.*, **259**, 260–2.

Mousli, M., Bueb, J.L., Bronner, C. *et al.* (1990b) G-protein activation: a receptor-independent mode of action for cationic amphiphilic neuropeptides and venom peptides. *Trends Pharmacol. Sci.*, **11**, 358–62.

Murray, M., Saffroy, M., Torrens, Y. *et al.* (1988) Tachykinin binding sites in the interpeduncular nucleus of the rat: normal distribution, postnatal development and the effects of lesions. *Brain Res.*, **459**, 76–92.

Nagashima, A., Takano, Y., Tateisha, K. *et al.* (1989) Central pressor actions of neurokinin B: increases in neurokinin B contents in discrete nuclei in spontaneously hypertensive rats. *Brain Res.*, **499**, 198–203.

Nakai, K. and Kasamatsu, T. (1984) Accelerated regeneration of central catecholamine fibers in cat occipital cortex: effects of substance P. *Brain Res.*, **323**, 374–9.

Nakanishi, S. (1987) Substance P precursor and kininogen: their structures, gene organizations and regulation. *Physiol. Rev.*, **67**, 1117–42.

Namura, H., Shiosaka, S., Inagaki, S. *et al.* (1982) Distribution of substance P-like immunoreactivity in the lower brainstem of the human fetus: an immunohistochemical study. *Brain Res.*, **252**, 315–25.

Narumi, S. and Fujita, T. (1978) Stimulatory effects of substance P and nerve growth factor (NGF) on neurite outgrowth in embryonic chick dorsal root ganglia. *Neuropharmacology*, **17**, 73–6.

Narumi, S. and Maki, Y. (1978) Stimulatory effects of substance P on neurite extension and cyclic AMP levels in cultured neuroblastoma cells. *J. Neurochem.*, **30**, 1321–6.

Nawa, H., Hirose, T., Takashima, H. *et al.* (1983) Nucleotide sequences of cloned cDNA for two types of bovine brain substance P precursor. *Nature*, **306**, 32–6.

Nawa, H., Kotani, H. and Nakanishi, S. (1984) Tissue specific generation of two prepro-tachykinin mRNAs from one gene by alternative RNA splicing. *Nature*, **312**, 729–34.

Ni, L. and Jonakait, G.M. (1988) Development of substance P-containing neurons in the central nervous system in mice: an immunocytochemical study. *J. Comp. Neurol.*, **275**, 493–510.

Nilsson, J., von Euler, A.M. and Dalsgaard, C.J. (1985) Stimulation of connective tissue cell growth by substance P and substance K. *Nature*, **315**, 61–3.

Nishimoto, T., Akai, M., Inagaki, S. *et al.* (1982) On the distribution and origins of substance P in the papillae of the rat tongue: an experimental and immunohistochemical study. *J. Comp. Neurol.*, **207**, 85–92.

Nomura, H., Shiosaka, S., Inagaki, S. *et al.* (1982) Distribution of substance P-like immunoreactivity in the lower brainstem of the human fetus: an immunohistochemical study. *Brain Res.*, **252**, 315–25.

O'Donohue, T.L., Helke, C.J., Shults, C.W. *et al.* (1990) Tachykinin receptors, in *Handbook of Chemical Neuroanatomy* Vol. **9**. *Neuropeptides in the CNS*, Part II (eds T. Hökfelt and M.J. Kuhar), Elsevier, Amsterdam, pp. 395–442.

O'Dowd, B.F., Lefkowitz, R.J. and Caron, M.G. (1989) Structure of the adrenergic and related receptors. *Annu. Rev. Neurosci.*, **12**, 67–83.

Ogawa, T., Kanazawa, I. and Kimura, S. (1985) Regional distribution of substance P, neurokinin A and neurokinin B in rat spinal cord, nerve roots and dorsal root section or spinal transection. *Brain Res.*, **359**, 152–7.

Olgart, L., Gazelius, B., Brodin, E. and Nilsson, G. (1977) Release of substance P-like immunoreactivity from the dental pulp. *Acta Physiol. Scand.*, **101**, 510–12.

Otsuka, M. and Konishi, S. (1976) Release of substance P-like immunoreactivity from isolated spinal cord of newborn rat. *Nature*, **264**, 83–4.

Otsuka, M. and Yanagisawa, M. (1980) The effects of substance P and baclofen on motorneurons of isolated spinal cord of the newborn rat. *J. Exp. Biol.*, **89**, 201–14.

Otten, U., Rüegg, U.T., Hill, R.C. *et al.* (1982) Correlation between substance P content of primary sensory neurones and pain sensitivity in rats exposed to antibodies to nerve growth factor. *Eur. J. Pharmacol.*, **85**, 351–3.

Paulin, C., Charnay, Y., Dubois, P.M. and Chayvialle, J.A. (1986) Localisation de substance P dans le système nerveux du foetus humain: résultats préliminaires. *C.R. Acad. Sci. (Paris)*, **29**, 253–60.

Payan, D.G., Brewster, D.R. and Goetzl, E.J. (1983) Specific stimulation of human T lymphocytes by substance P. *J. Immunol.*, **131**, 1613–15.

Payan, D.G., Brewster, D.R. and Goetzl, E.J. (1984) Stereospecific receptors for substance P on cultured human IM-9 lymphoblasts. *J. Immunol.*, **133**, 3260–5.

Payan, D.G., McGillis, J.P. and Organist, M.L. (1986) Binding characteristics and affinity labelling of protein constituents of the human IM-9 lymphoblast receptor for substance P. *J. Biol. Chem.*, **261**, 14321–9.

Pearson, J., Brandeis, L. and Cuello, A.C (1982) Depletion of substance P-containing axons in the substantia gelatinosa of patients with diminished pain sensitivity. *Nature*, **295**, 61–3.

Pernow, B. (1983) Substance P. *Pharmacol. Rev.*, **35**, 85–141.

Phillis, J.W. and Limacher, J.J. (1974) Substance P

excitation of cerebral cortical Betz cells. *Brain Res.*, **69**, 158–63.

Quirion, R. (1985) Multiple tachykinin receptors. *Trends Neurosci.*, **8.**, 183–5.

Quirion, R. and Dam, T.V. (1985) Multiple tachykinin receptors in guinea pig brain. High densities of substance K (neurokinin A) binding sites in the substantia nigra. *Neuropeptides*, **6**, 191–204.

Quirion, R. and Dam, T.V. (1986) Ontogeny of substance P receptor binding sites in rat brain. *J. Neurosci.*, **6**, 2187–99.

Quirion, R. and Dam, T.V. (1988) Multiple neurokinin receptors: recent developments. *Regul. Pept.*, **22**, 18–25.

Quirion, R., Shults, C., Moody, T. *et al.* (1983) Autoradiographic distribution of substance P receptors in rat central nervous system. *Nature*, **303**, 714–16.

Randic, M. and Miletic, V. (1977) Effects of substance P in cat dorsal horn neurons activated by noxious stimuli. *Brain Res.*, **128**, 164–9

Regoli, D., Drapeau, G., Dion, S. and Couture, R. (1988) New selective agonists for neurokinin receptors: pharmacological tools for receptors characterization. *Trends Pharmacol. Sci.*, **9**, 290–5.

Regoli, D., Dion, S., Rhaleb, N.E. *et al.* (1989) Selective agonists for receptors of substance P and related neurokinins. *Biopolymers*, **28**, 81–90.

Rovero, P., Pestellini, V., Rhaleb, N.E. *et al.* (1989) Structure–activity studies of neurokinin A. *Neuropeptides*, **13**, 263–70.

Ruff, M.R., Wahl, S.M. and Pert, C.B. (1985) Substance P receptor-mediated chemotaxis of human monocytes. *Peptides*, **6**, 107–11.

Saffroy, M., Beaujouan, J.C., Torrens, Y. *et al.* (1988) Localization of tachykinin binding sites (NK1, NK2, NK3 ligands) in the rat brain. *Peptides*, **9(2)**, 227–41.

Sakanaka, M., Inagaki, S., Shiosaka, S. *et al.* (1982) Ontogeny of substance P-containing neuron system of the rat: immunohistochemical analysis. II Lower brainstem. *Neuroscience*, **7**, 1097–126.

Sastry, B.R. (1979) Substance P effects on spinal nociceptive neurons. *Life Sci.*, **24**, 2169–77.

Savard, P., Mërand, Y., Bëdard, P. *et al.* (1983) Comparative effects of neonatal hypothyroidism and euthyroidism on TRH and substance P content of lumbar spinal cord in saline and PCPA-treated rats. *Brain Res.*, **277**, 263–8.

Schini, V.B., Katusic, Z.S. and Vanhoutte, P.M. (1990) Neurohypophyseal peptides and tachykinins stimulate the production of cyclic GMP in cultured porcine aortic endothelial cells. *J. Pharmacol. Exp. Ther.*, **255**, 994–1000.

Senba, E., Shiosaka, S., Hara, Y. *et al.* (1982) Ontogeny of the peptidergic system in the rat spinal cord: immunohistochemical analysis. *J. Comp. Neurol.*, **208**, 54–66.

Serra, M.C., Bazzoni, F., Della Bianca, V. *et al.* (1988) Activation of human neutrophils by substance P. Effect on oxidative metabolism, exocytosis, cytosolic Ca^{2+} concentration and inositol phosphate formation. *J. Immunol.*, **141**, 2118–24.

Sharkey, K.A. and Templeton, D. (1984) Substance P in the rat parotid gland: evidence for a dual origin from the otic and trigeminal ganglia. *Brain Res.*, **304**, 392–6.

Shigematsu, K., Niwa, M., Kurihara, M. *et al.* (1987) Alterations in substance P binding in brain nuclei of spontaneously hypertensive rats. *Am. J. Physiol.*, **252**, H301–6.

Shigemoto, R., Yokota, Y., Tsuchida, K. and Nakanishi, S. (1990) Cloning and expression of a rat neuromedin K receptor cDNA. *J. Biol. Chem.*, **265**, 623–8.

Shortland, P., Molander, C., Woolf, C.J. and Fitzgerald, M. (1990) Neonatal capsaicin treatment induces invasion of the substantia gelatinosa by the terminal arborizations of hair follicle afferents in the rat dorsal horn. *J. Comp. Neurol.*, **296**, 23–31.

Shults, C.W., Quirion, R., Chronwall, B. *et al.* (1984) A comparison of the anatomical distribution of substance P and substance P receptors in the rat central nervous system. *Peptides*, **5**, 1097–128.

Sivam, S.P., Krause, J.E., Breese, G.R. and Hong, J.S. (1991) Dopamine-dependent postnatal development of enkephalin and tachykinin neurons of rat basal ganglia. *J. Neurochem.*, **56**, 1499–508.

Snyder, S.H. (1979) Peptide and neurotransmitter receptors in the brain: regulation by ions and guanyl nucleotides, in *Central Regulations of the Endocrine System* (eds K. Fuxe, T. Hökfelt and R. Luft), Plenum, New York, pp. 109–17.

Snyder-Keller, A.M. (1991) Development of striatal compartmentalization following pre- or postnatal dopamine depletion. *J. Neurosci.*, **11**, 810–21.

Soder, O. and Hellstrom, P.M. (1989) The tachykinins neurokinin A and physalaemin stimulate murine thymocyte proliferation. *Int. Arch. Allergy Appl. Immunol.*, **90**, 91–6.

Somogyi, P., Priestley, J.V., Cuello, A.C. et al. (1982) Synaptic connections of substance P immunoreactive nerve terminals in the substantia nigra of the rat: a correlated light and electron microscopic study. *Cell Tissue Res.*, **223**, 469–86.

Stanisz, A.M., Befus, D. and Bienenstock, J. (1986) Differential effects of vasoactive intestinal peptide, substance P and somatostatin on immunoglobulin synthesis and proliferations by lymphocytes from Peyer's patches, mesenteric lymph nodes and spleen. *J. Immunol.*, **136**, 152–6.

Staubli, U. and Huston, J.P. (1980) Facilitation of learning by post-trial injection of substance P into the medial septal nucleus. *Behav. Brain Res.*, **1**, 245–55.

Stjarne, P., Lundblad, L., Anggård, A. *et al.* (1989) Tachykinins and calcitonin gene-related peptide: co-existence in sensory nerves of the nasal mucosa and effects on blood flow. *Cell Tissue Res.*, **256**, 439–46.

Stoessl, A.J. and Hill, D.R. (1990) Autoradiographic visualization of NK-3 tachykinin binding sites in the rat brain, utilizing [^{3}H] senktide. *Brain Res.*, **534**, 1–7.

Stoessl, A.J., Dourish, C.T. and Iversen, S.D. (1988) The NK-3 tachykinin agonist senktide elicits 5-HT mediated behaviour following central or peripheral administration in mice and rats. *Br. J. Pharmacol.*, **94**, 285–7.

Takano, Y., Nagashima, A., Masui, H. *et al.* (1986) Distribution of substance K (neurokinin A) in the brain and peripheral tissues of rat. *Brain Res.*, **369**, 400–4.

Takeda, Y. and Krause, J.E. (1989a) Neuropeptide K potently stimulates salivary gland secretion and potentiates substance P-induced salivation. *Proc. Natl. Acad. Sci. USA*, **86**, 392–6.

Takeda, Y. and Krause, J.E. (1989b) γ-preprotachykin-(72-92)-peptide amide potentiates substance P-induced salivation. *Eur. J. Pharmacol.*, **161**, 267–71.

Takeda, Y., Takeda, J., Smart, B. and Krause, J.E. (1990) Regional distribution of neuropeptide γ, and other tachykinin peptides derived from the substance P gene in the rat. *Regul. Pept.*, **28**, 323–33.

Tanaka, T., Ikai, K. and Imamura, S. (1988) Effects of substance P and substance K on the growth of cultured keratinocytes. *J. Invest. Dermatol.*, **90**, 399–401.

Tatemoto, K., Lundberg, J.M., Jörnvall, H. and Mutt, V. (1985) Neuropeptide K: isolation, structure and biological activities of a novel brain tachykinin. *Biochem. Biophys. Res. Commun.*, **128**, 947–53.

Tervo, K., Tervo, T., Erenko, L. *et al.* (1982) Effect of sensory and sympathetic denervation on substance P immunoreactivity in nerve fibers of the rabbit eye. *Exp. Eye Res.*, **34**, 577–85.

Too, H.-P., Cordova, J.L. and Maggio, J.E. (1989) A novel radioimmunoassay for neuromedin K. I. Absence of neuromedin K-like immunoreactivity in guinea pig ileum and urinary bladder. II. Heterogeneity of tachykinins in guinea pig tissues. *Regul. Pept.*, **26**, 93–105.

Torrens, Y., Beaujouan, J. L., Saffroy, M. *et al.* (1986) Substance P receptors in primary cultures of cortical astrocytes from the mouse. *Proc. Natl. Acad. Sci. USA*, **83**, 9216–20.

Turski, L., Klockgether, T., Schwarz, M. *et al.* (1990) Substantia nigra: a site of action of muscle relaxant drugs. *Ann. Neurol.*, **28**, 341–8.

Unger, T., Rascher, W., Schuster, C. *et al.* (1981) Central blood pressure effects of substance P and angiotensin II: role of sympathetic nervous system and vasopressin. *Eur. J. Pharmacol.*, **71**, 33–42.

Viger, A., Beaujouan, J.C., Torrens, Y. and Glowinski, J. (1983) Specific binding of a ^{125}I-substance P derivative to rat brain synaptosomes. *J. Neurochem.*, **40**, 1030–9.

Walker, P.D., Green, T.L., Jonakait, G.M. and Hart, R.P. (1991) A comparison of substance P peptide and preprotachykinin mRNA levels during development of rat medullary raphe and neostriatum. *Int. J. Dev. Neurosci.*, **9**, 47–55.

Wall, P.D., Fitzgerald, M., Nussbaumer, J.C. *et al.* (1982) Somatotrophic maps are disorganized in adult rodents treated neonatally with capsaicin. *Nature*, **295**, 691–3.

Watson, S.P., Sandberg, B.E.B., Hanley, M.R. and Iversen, L.L. (1983) Tissue selectivity of substance P alkyl esters: suggesting multiple receptors. *Eur. J. Pharmacol*, **87**, 77–84.

Wyke, B. (1981) The neurology of joints: a review of general principles. *Clin. Rheum. Dis.*, **7**, 233–39.

Yamasaki, H., Kubota, Y. and Tohyama, M. (1985) Ontogeny of substance P-containing fibers in the taste buds and the surrounding epithelium. I Light microscopic analysis. *Dev. Brain Res.*, **18**, 301–5.

Yashpal, K., Dam, T.V. and Quirion, R. (1990) Quantitative autoradiographic distribution of multiple neurokinin binding sites in rat spinal cord. *Brain Res.*, **506**, 259–66.

Yashpal, K., Dam, T.V. and Quirion, R. (1991a) Effects of dorsal rhizotomy on neurokinin receptor sub-types in the rat spinal cord: a quantitative autoradiographic study. *Brain Res.*, **552**, 240–7.

Yashpal, K., Kar, S., Quirion, R. and Henry, J.L. (1991b) Noxious cutaneous stimulation produces analgesia in the tail-flick test and decreases binding of substance P in the dorsal horn. *Soc. Neurosci. (Abst.)*, **17**, 1006.

Yew, D.T., Luo, C.B., Zheng, D.R. *et al.* (1990) Development and localization of enkephalin and substance P in the nucleus of tractus solitarius in the medulla oblongata of human fetuses. *Neuroscience*, **34**, 491–8.

Yokota, Y., Sasai, Y., Tanaka, K. *et al.* (1989) Molecular characterization of a functional cDNA for rat substance P receptor. *J. Biol. Chem.*, **264**, 17649–52.

Young, D., Waitches, G., Birchmeier, C. *et al.* (1986) Isolation and characterization of a new cellular oncogene encoding a protein with multiple potential transmembrane domains. *Cell*, **15**, 711–19.

Zderic, S.A., Duckett, J.W., Wein A.J. *et al.* (1990) Development factors in the contractile response of the rabbit bladder to both autonomic and non-autonomic agents. *Pharmacology*, **41**, 119–23.

Ziche, M., Morbidelli, C., Pacini, M. *et al.* (1990) Substance P stimulates neovascularization *in vivo* and proliferation of cultured endothelial cells. *Microvasc. Res.*, **40**, 264–78.

Ziche, M., Morbidelli, L., Geppetti, P. *et al.* (1991) Substance P induces migration of capillary endothelial cells: a novel NK-1 selective receptor mediated activity. *Life Sci.*, **48**, PL7–11.

Paul H. Robinson, John D. Stephenson and Timothy H. Moran

5.1 INTRODUCTION

The term cholecystokinin (CCK) was first used to describe a hormonal factor identified in extracts from the intestine which caused gall bladder contraction (Ivy and Oldberg, 1928). Later, Harper and Raper (1943) demonstrated that intestinal extracts could also stimulate pancreatic secretion and termed the responsible factor, pancreozymin. Further experiments (Jorpes and Mutt, 1973) showed that pancreozymin and CCK were identical. CCK was first characterized as a 33-amino acid peptide in which the C-terminal pentapeptide was identical to that found in gastrin. Subsequently, several molecular forms of CCK have been identified including CCK-39, CCK-33, CCK-8 and CCK-4 (Dockray, 1981). These have been shown to have a number of pharmacological actions in the gut, the physiological significance of which remains to be determined.

CCK was originally identified in the central nervous systems of rats (Vanderhaegen *et al.*, 1975) and non-human primates (Strauss and Yalow, 1978) as gastrin-like immunoreactivity and it was subsequently shown that the majority of brain CCK was in the form of the octapeptide (Dockray *et al.*, 1985). Little is known about the specific functions of CCK in the brain although it has been proposed to function both as a neurotransmitter or a neuromodulator in peripheral and central neurons (Larsson and Rehfeld, 1979).

Neuronal CCK can be detected in both cell bodies and terminals but occurs mainly in synaptic vesicles (Emson *et al.*, 1980). Its biosynthetic pathway, in both cortical and subcortical rat brain regions, was described by Goltermann (1982) and Goltermann *et al.* (1980, 1981). CCK has been shown to be released from brain slices in response to a depolarizing stimulus (Emson *et al.*, 1980) and to be inactivated by deamination (Williams *et al.*, 1981). Receptors for CCK have been identified in brain (Innis and Snyder, 1980; Saito *et al.*, 1980) and their differential distribution mapped in autoradiographic studies (Zarbin *et al.*, 1983; Moran and McHugh, 1990). Finally, iontophoretically applied CCK increases neuronal firing in a variety of brain regions (Dodd and Kelly, 1981; Bunney *et al.*, 1985). Together, these findings support the proposed function of CCK as a transmitter or modulator within the CNS.

5.2 CCK RECEPTOR SUBTYPES

There are two types of CCK receptors (Moran *et al.*, 1986). Type A receptors are present in the pancreas (Innis and Snyder, 1980), pylorus (Smith *et al.*, 1984) and vagal afferents

Receptors in the Developing Nervous System Vol. 2: Neurotransmitters. Edited by Ian S. Zagon and Patricia J. McLaughlin. Published in 1993 by Chapman & Hall. ISBN 0 412 49400 0. Vols. 1 and 2 (set) ISBN 0 412 54520 9.

(Moran *et al.*, 1990). In the rat CNS, type A receptors are restricted to the nucleus tractus solitarius, area postrema, interpeduncular nucleus and other discrete brain structures (Moran *et al.*, 1986; Hill *et al.*, 1987). In contrast, the type B CCK receptor is widely distributed in the brain (Moran *et al.*, 1986; Moran and McHugh, 1990). The basis for the distinction between the subtypes is pharmacological and was originally based on differences in the relative affinities of various CCK agonists for the two receptor subtypes. Thus, sulfated forms of CCK bind with higher affinity to the CCK-A receptors than unsulfated CCK, pentagastrin, CCK-4 and gastrin. In contrast, sulfated and non-sulfated forms of CCK and gastrin show smaller differences in their binding affinities to the CCK-B receptor (Innis and Snyder, 1980; Moran *et al.*, 1986). The development of selective CCK receptor antagonists has confirmed this receptor differentiation (Hill *et al.*, 1987; Dourish and Hill, 1989; Chang and Lotti, 1986; Lotti and Chang, 1989). Antagonists at central CCK-B receptors also have a high affinity for peripheral gastrin receptors suggesting that CCK-B and gastrin receptors may be structurally related (Yu *et al.*, 1990).

5.3 AUTORADIOGRAPHIC RECEPTOR DISTRIBUTION

The distribution of CCK receptors in the adult rat brain has been mapped using autoradiographic techniques by a variety of investigators (Zarbin *et al.*, 1983; Moran and McHugh, 1990; Van Dijk *et al.*, 1984). We have already discussed the central distribution of type A CCK receptors. Although they have been identified autoradiographically in only a restricted number of sites, electrophysiological and functional data suggest that they may be more widely distributed although at such a low density that their presence is difficult to detect by standard receptor techniques (Crawley, 1988; Wang *et al.*, 1988). Type B CCK receptors are widely distributed

in the rat brain. Within the olfactory bulb, high densities are evident in the internal granular layer, the inferior olfactory nucleus and caudate putamen. CCK receptors are also found in a variety of limbic structures including the hippocampal formation, the hippocampus, subiculum, dentate gyrus, inferior amygdala and nucleus accumbens (especially in its medial aspects). Binding is found throughout the depth of the cortex, but is concentrated in cortical laminae, II to IV. The highest densities of cortical CCK receptors are found throughout the full rostral-caudal extent of the cingulate cortex.

Only a few areas within the diencephalon contain significant densities of CCK binding. These include the reticular thalamic nuclei, paraventricular thalamic nucleus, ventral medial hypothalamus and the supraoptic nucleus.

Within the midbrain, low levels of binding are found in the dorsal raphe and the several components of the substantia nigra. Type B CCK receptors were only found at a few sites in the brainstem. Low levels of binding are evident in the superior spinal and medial vestibular nucleus, suprageniculate nuclei and cuneate nuclei. Lateral aspects of the nucleus tractus solitarius also contain type B CCK receptors. Low densities of CCK binding have also been demonstrated in the spinal tract of the trigeminal nerve. CCK receptors are present in the cerebellum of the guinea pig and primate, but not in the rat cerebellum.

5.4 DEVELOPMENT OF CENTRAL CCK AND ITS RECEPTORS

Studies of the development of levels of CCK suggest that increase of the peptide is primarily postnatal in altricial subprimate species such as the rat, whereas in precocial species such as the guinea pig, there is significant prenatal development of CCK levels (Goldman *et al.*, 1985).

Although the distribution of CCK receptors and the classification of receptor subtypes

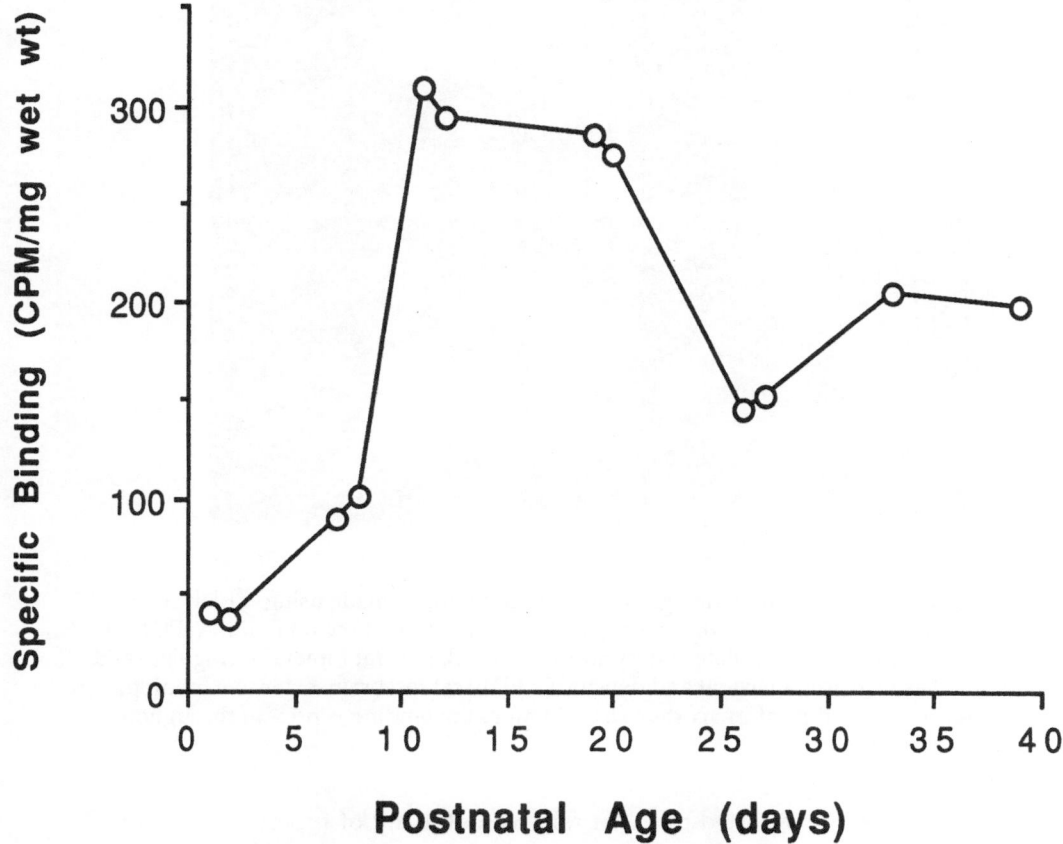

Fig. 5.1.   Age-associated changes in CCK receptor numbers in developing rat brain homogenates. Reproduced from Hays *et al.* (1981).

have been made in the adult, the ontogeny of CCK receptors in the CNS has been examined only rarely. Hays *et al.* (1981), examining binding to rat brain homogenates, demonstrated that the ontogeny of CCK receptors was primarily postnatal. Figure 5.1 (adapted from Hays *et al.* (1981)) shows that specific binding of CCK to membrane preparations of the rat forebrain increased from birth until it reached a plateau between postnatal day (PD) 12 and PD20. After this time, levels fell and then gradually approached adult levels. Autoradiographic studies have confirmed that the primary development of CCK receptors in the rat brain occurs between birth and PD20.

Pelaprat *et al.* (1988) demonstrated that, at birth, CCK receptors were absent from most regions of the rat brain and were confined to selected areas, namely ventromedial thalamic nucleus, endopyriform cortex and medial nucleus of amygdala. Labeling gradually increased during the first three weeks of life to above levels found in the adult and thereafter declined to adult levels over the fourth week.

The distribution of CCK receptors in the forebrain of a 1-day-old rat is shown in Fig. 5.2A. This shows that binding of ^{125}I-labeled Bolton Hunter CCK-8 (Garcha *et al.*, 1987) is limited to the central cingulate and pyriform

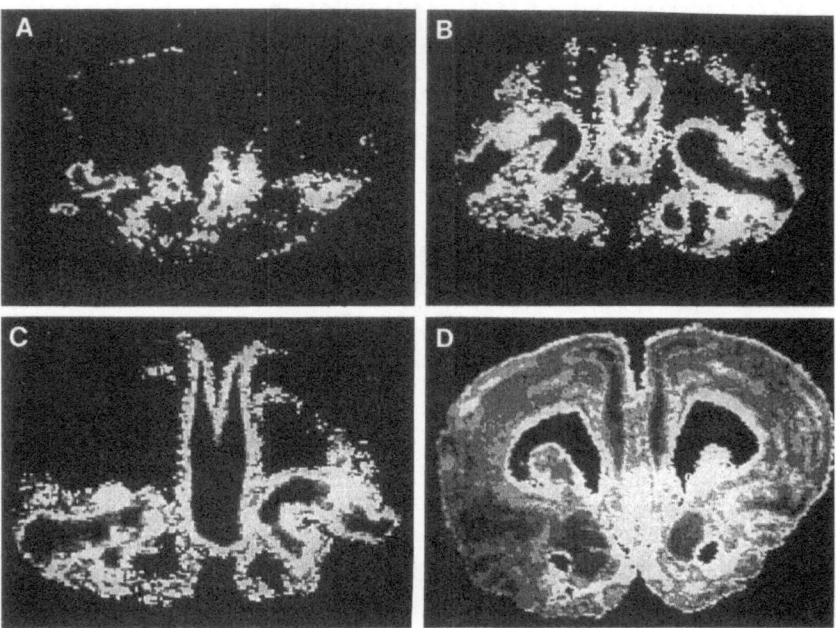

Fig. 5.2.   Computerized microdensitometry images from autoradiographs made using ^{125}I-labeled Bolton Hunter-CCK-8, to demonstrate the ontogeny of specific CCK binding in the rat brain. A: PD1 rat forebrain, showing specific binding in cingulate and pyriform cortex. B: PD6 rat forebrain: cingulate and pyriform cortex binding has increased in extent and density. C: PD10 rat forebrain: extensive binding, still confined to cingulate cortex and pyriform cortex. D: adult rat brain: binding is present throughout the cortex and in subcortical structures.

cortices. In contrast to the results reported by Pelaprat *et al.* (1988), binding was not observed in all pups. This difference may be attributed to small variations in the maturity of one-day-old pups during a period of rapid receptor development. By PD6 binding was consistently observed and had increased in both of these areas (Fig. 5.2B).

As rat pups mature, cortical binding becomes more extensive but is still concentrated in limbic areas. Up to PD10, CCK receptors remained confined to paleocortical areas, in which they increased in density (Fig. 5.2C). Between PD10 and PD15 there was rapid development of receptors in the neocortex which, by PD15, adopted approximately the distribution observed in the adult (Fig. 5.2D).

To date, no studies specifically examining

the relative ontogeny of types A and B CCK receptors have been conducted.

5.5   DEVELOPMENT OF CCK-CONTAINING NEURONS IN RAT CNS

CCK neuron systems develop contemporaneously with CCK receptors (Noyer *et al.*, 1980; Cho *et al.*, 1983; Kiyama *et al.*, 1983; Duchemin *et al.*, 1987). CCK neurons first appear in the developing ventral tegmental area and in the primordium of the medial forebrain bundle around gestational age 15. Thereafter, CCK cell numbers increase in the forebrain and the upper brainstem until birth and continue to rise until around PD10. In the ventral tegmental area, however, maximum levels of CCK neurons are found in the late gestational period and then fall in the adult, suggesting that CCK processing occurs largely in terminal regions. Moreover, in the hypothalamic dorsal medial nucleus high levels are found up to PD5 and disappear by PD10. This temporal pattern of development of CCK containing neurons is similar to that of CCK receptors in the rodent brain.

5.6 OTHER CCK RECEPTOR SYSTEMS

The ontogeny of CCK receptors in the brain can be contrasted with their development in the upper gastrointestinal tract. Previous studies have shown that the locus of binding sites of CCK in the circular muscle of the rat pyloric sphincter is evident at 17 days of gestation and persists as a dense area of binding through development into adulthood (Robinson *et al.*, 1988). CCK binding sites are also transiently found in other areas of the developing rat stomach. The gastric mucosa has extensive binding in the fetus that persists until PD 10–PD 15. The antral circular muscle is also densely packed with CCK binding sites in fetal and early postnatal life. In both these cases binding declines in both density and area with increasing age. Thus, in contrast to the brain, CCK receptors in the GI tract may play a role in the functional ontogeny of the GI tract.

5.7 CONCLUSIONS

CCK is one of a number of gastrointestinal peptides which is found in the rat CNS. We have shown that at birth, receptors for CCK are confined to the pyriform and cingulate cortical regions. The major increase in receptor density and distribution occurred during the first two weeks of postnatal development, with the 15-day-old rat showing a pattern of CCK distribution similar to that seen in the adult. This developmental pattern accords with those described by Hays *et al.* (1981) and Pelaprat *et al.* (1988) and parallel changes in CCK concentration (Cho *et al.*, 1983). The last

authors showed that CCK immunoreactivity appeared first in the ventral tegmental area and pyriform cortex one week after birth and rapidly spread to other forebrain areas. In contrast to CCK, other neuropeptides, e.g. somatostatin (Shiosaka *et al.*, 1982), are present in high concentration in the fetal rat nervous system. However, the late appearance of CCK in the rat nervous system contrasts with its early prenatal development in non-primate precocial species, such as the guinea pig and chicken (Goldman *et al.*, 1985) and in non-human primate, the macaque monkey (Hayashi *et al.*, 1989). The lack of CCK binding in the developing rat brain also contrasts with its early appearance in the stomach, which showed extensive binding four days before birth (Robinson *et al.*, 1988). This suggests that in the rat, CCK may be involved in gut, but not brain, development.

Little is known about the CCK fiber system, but the observations that CCK first appears in the ventral tegmental area and pyriform cortex (Cho *et al.*, 1983; Kiyama *et al.*, 1983) and coexists with dopamine in several brain regions, suggest that the CCK tract from the mesencephalon to the limbic cortex (Hokfelt *et al.*, 1980) may be identical to the dopaminergic mesolimbic pathway. This co-localization provides the rationale for investigating CCK function in psychiatric states such as schizophrenia in which abnormal dopamine function has been postulated.

CCK has been proposed to be an endogenous regulator of satiety, although its site of action remains the subject of debate. The demonstration that CCK induces satiety in rats at PD1, when CCK receptors are abundant in the stomach but not in the CNS (Robinson *et al.*, 1987, 1988), suggests that either gastric mechanisms are most important for regulating CCK-induced satiety in the neonate or that the effect is mediated through highly localized brain regions in which CCK receptors are present at birth but are not easily visualized.

REFERENCES

Bunney, B.S., Chiodo, L.A. and Freeman, A.S. (1985) Further studies on the specificity of proglumide as a selective CCK antagonist in the central nervous system. *Ann. NY Acad. Sci.*, **448**, 345–51.

Chang, R.S.L. and Lotti, V.J. (1986) Biochemical and pharmacological characterization of an extremely potent and selective non-peptide cholecystokinin antagonist. *Proc. Natl. Acad. Sci. USA*, **83**, 4923–6.

Cho, H.J., Shiotani, Y., Shiosaka, S. *et al.* (1983) Ontogeny of cholecystokinin-8 containing neuron system of the rat: an immunohistochemical analysis. I. Forebrain and upper brainstem. *J. Comp. Neurol.*, **218**, 25–41.

Crawley, J.N. (1988) Behavioral analysis of antagonists of the peripheral and central effects of CCK, in *Cholecystokinin Antagonists* (eds R. Y. Wang and R. Schoenfeld), Alan R. Liss, New York, pp. 243–62.

Dockray, G.J. (1981) Cholecystokinin, in *Gut Hormones*, 2nd edn (eds S. R. Bloom and J. M. Polak), Churchill Livingstone, Edinburgh, pp. 228–39.

Dockray, G.J., Desmond, H., Gayston, R.J. *et al.* (1985) Cholecystokinin and gastrin forms in the central nervous system. *Ann. NY Acad. Sci.*, **448**, 32–43.

Dodd, J. and Kelly, J.S. (1981) The actions of cholecystokinin and related peptides in pyramidal neurons of the mammalian hippocampus. *Brain Res.*, **205**, 337–50.

Dourish, C.T. and Hill, D.R. (1987) Classification and function of CCK receptors. *Trends Pharmacol. Sci.*, **8**, 207–8.

Duchemin, A.M., Quach, T.T., Iadarola, M.J. *et al.* (1987) Expression of the cholecystokinin gene in rat brain during development. *Dev. Neurosci.*, **9**, 61.

Emson, P.C., Lee, C.M. and Rehfeld, J.F. (1980) Cholecystokinin octapeptide: vesticular localization and calcium dependent release from rat brain *in vitro*. *Life Sci.*, **26**, 2157–63.

Garcha, G.S., McHugh, P.R., Moran, T.H. *et al.* (1987) Development of cholecystokinin receptors and suppression of ingestion and gastric emptying in the rat. *J. Physiol. (Lond.)*, **387**, 93.

Goldman, S.A., Monahan, J.W. and Schneider, B. S. (1985). The regional and subcellular development of cholecystokinin immunoreactivity in vertebrate brain. *Brain Res.*, **354**, 237–46.

Goltermann, N.R. (1982) *In vivo* synthesis of cholecystokinin in rat cerebral cortex: identification of COOH-terminal peptides with labelled amino acids. *Peptides*, **3**, 733–7.

Goltermann, N.R., Rehfeld, J.F. and Roigaard-Petersen, H. (1980) In vivo biosynthesis of cholecystokinin in rat cerebral cortex. *J. Biol. Chem.*, **255**, 6181–5.

Goltermann, N.R., Stengaart-Petersen, V., Rehfeld, J.F. *et al.* (1981) Newly synthesized cholecystokinin in subcellular fractions of the rat brain. *J. Neurochem.*, **36**, 959–65.

Harper, A.A. and Raper, H.S. (1943) Pancreozymin a stimulant of the secretion of pancreatic enzymes in extracts of the small intestine. *J. Physiol. (Lond.)*, **102**, 115–25.

Hayashi, M., Yamashita, A., Shimizu, K. and Oshima, K. (1989) Ontogeny of cholecystokinin-8 and glutamic acid decarboxylase in cerebral neocortex of the macaque monkey. *Exp. Brain Res.*, **74**, 249–55.

Hays, S.E., Goodwin, F.K. and Paul, S.M. (1981) Cholecystokinin receptors in brain: effect of obesity, drug treatment and lesions. *Peptides*, **2**, 21–6.

Hill, D.R., Campbell, N.J., Shaw, T.M. *et al.* (1987) Autoradiographic localization and biochemical characterization of peripheral type CCK receptors in rat CNS using highly selective nonpeptide CCK antagonists. *J. Neurosci.*, **7**, 2967–76.

Hökfelt, T.L., Skirboll, L., Rehfeld, J.F. *et al.* (1980) A subpopulation of mesencephalic dopamine neurons projecting to limbic areas contains a cholecystokinin-like peptide: evidence from immunohistochemistry combined with retrograde tracing. *Neuroscience*, **5**, 2093–124.

Innis, R.B. and Snyder, S.-H. (1980) Distinct cholecystokinin receptors in brain and pancreas. *Proc. Natl. Acad. Sci. USA*, **77**, 6919–21.

Ivy, A.C. and Olberg, E. (1982) A hormone mechanism for gall-bladder contraction and evacuation. *Am. J. Physiol.*, **86**, 599–613.

Jorpes, J.E. and Mutt, V. (1973) Secretin and cholecystokinin (CCK), in *Secretin, Cholecystokinin, Pancreozymin and Gastrin. Handbook of Experimented Pharmacology*, Vol. **34** (eds J. E. Jorpes and V. Mutt), Springer-Verlag, Berlin, pp. 1–179.

Kiyama, H., Shiosaka, S., Kubota, Y. *et al.* (1983) Ontogeny of cholecystokinin-8 containing neuron system of the rat: an immunohistochemical analysis. II. Lower brainstem. *Neuroscience*, **10**, 1341–59.

Larsson, L.I. and Rehfeld, J.F. (1979) Localization

and molecular heterogeneity of cholecystokinin in the central and peripheral nervous system. *Brain Res.*, **165**, 201–18.

Lotti, V.J. and Chang, R.S.L. (1989) A new potent and selective non-peptide gastrin antagonist and brain cholecystokinin receptor (CCK-B) ligand: L-365,260. *Eur. J. Pharmacol.*, **162**, 273–80.

Moran, T.H. and McHugh, P.R. (1990) Cholecystokinin receptors, in *Handbook of Chemical Neuroanatomy*: Vol. **9**, *Neuropeptides in the CNS*, Part II, (eds A. Bjorklund, T. Hokfeld and M.J. Kuhar), Elsevier, Amsterdam, pp. 455–76.

Moran, T.H., Robinson, P.H., Goldrich, M.S. *et al.* (1986) Two brain CCK receptors: implications for behavioral actions. *Brain Res.*, **362**, 175–9.

Moran, T.H., Norgren, R., Crosby, R.J. and McHugh, P.R. (1990) Central and peripheral vagal transport of cholecystokinin binding sites occurs in afferent fibers. *Brain Res.*, **526**, 95–102.

Noyer, M., Diem Bui, N., Deschodt-Lanckman, M. *et al.* (1980) Postnatal development of the cholecystokinin–gastrin family of peptides in the brain and gut of rat. *Life Sci.*, **27**, 2197–203.

Pelaprat, D., Dusart, I. and Peschanski, M. (1988) Postnatal development of cholecystokinin binding sites in the rat forebrain and midbrain: an autoradiographic study. *Dev. Brain Res.*, **44**, 119–32.

Robinson, P.H., Moran, T.H., Goldrich, M. and McHugh, P.R. (1987) Development of cholecystokinin receptors in rat upper gastrointestinal tract. *Am. J. Physiol.*, **252**, G529–G534.

Robinson, P.H., Moran, T.H. and McHugh, P.R. (1988) Cholecystokinin inhibits independent ingestion in neonatal rats. *Am. J. Physiol.*, **255**, R14–R20.

Saito, A., Sankaran, H., Goldfine, I.D. *et al.* (1980) Cholecystokinin receptors in brain: characterization and distribution. *Science*, **208**, 1155–6.

Shiosaka, S., Takatsuki, K., Sakanaka, M. *et al.* (1982) Ontogeny of somatostatin containing neuron system in the rat. Immunohistochemical observations. II. Forebrain and diencephalon. *J. Comp. Neurol.*, **204**, 211–24.

Smith, G.T., Moran, T.H., Coyle, J.T. *et al.* (1984) Anatomical localization of cholecystokinin receptors to the pyloric sphincter. *Am. J. Physiol.*, **246**, R127–130.

Strauss, E. and Yalow, R.S. (1978) Species specificity of cholecystokinin in gut and brain of several mammalian species. *Proc. Natl. Acad. Sci. USA*, **75**, 486–9.

Vanderhaeghen, J.J., Signeau, J.C. and Gepts, W. (1975) New peptide in vertebrate CNS reacting with antigastrin antibodies. *Nature*, **251**, 604–5.

Van Dijk, A., Richards, J.G., Trzeciak, A. *et al.* (1984) Cholecystokinin receptors: biochemical demonstration and autoradiographical localization in rat brain and pancreas using [^{3}H]cholecystokinin-8 as a radioligand. *J. Neurosci.*, **4**, 1021–33.

Wang, R.Y., Kasser, R.J. and Hu, X.-T. (1988) Cholecystokinin receptor subtypes in the rat nucleus accumbens, in *Cholecystokinin Antagonists* (eds R. Y. Wang and R. Schoenfeld), Alan R. Liss, New York, pp. 199–216.

Williams, R.G., Goyku, R.J., Zoo, N.Y. *et al.* (1981) Changes in brain cholecystokinin octapeptide following lesions of the media forebrain bundle. *Brain Res.*, **212**, 221–30.

Yu, D.H., Huang, S.C., Wank, S.A. *et al.* (1990) Pancreatic receptors for cholecystokinin: evidence for three receptor classes. *Am. J. Physiol.*, **258**, G86–G95.

Zarbin, M.A., Innis, R.B., Wamsley, J.K. *et al.* (1983) Autoradiographic localization of cholecystokinin receptors in rodent brain. *J. Neurosci.*, **3**, 877–906.

GABA$_A$/BENZODIAZEPINE RECEPTORS IN THE DEVELOPING MAMMALIAN BRAIN

Angel Luis de Blas

6.1 THE GABA$_A$ BENZODIAZEPINE RECEPTORS

GABA (γ-aminobutyric acid) is the main inhibitory transmitter in the brain. Up to 30–40% of the brain synapses use GABA as a neurotransmitter (Bloom and Iversen, 1971). When GABA binds to the GABA$_A$ receptor (GABAR) it opens the associated Cl$^-$ channel allowing the flow of this anion inside the neuron, therefore hyperpolarizing the neuronal membrane and making the cell less reactive to excitatory neurotransmitters. Benzodiazepines (BZDs) are the most widely prescribed drugs. They are potent anxiolytic, antiepileptic and muscle relaxing agents. The psychotropic effects produced by the BZDs (i.e. librium and valium) result from their binding to brain benzodiazepine receptors (BZDRs) that are present in the membranes of many brain neurons (Bräestrup and Squires, 1977; Möhler and Okada, 1977). The BZDR is part of a protein complex that also includes the GABAR and a chloride channel. Barbiturates and some steroids also bind to these receptor-channel protein complexes affecting the efficacy of GABA in opening the Cl$^-$ channel. The BZDs potentiate the inhibitory effect of GABA by increasing the frequency of channel opening events induced by GABA. In this way, the total Cl$^-$ current flowing through the channel is increased. For reviews of the pharmacological, biochemical and molecular biology of the GABAR/BZDR see Olsen and Tobin (1990) and Schwartz (1988).

There are other drugs that specifically bind to the various elements of the GABAR complex. Thus, muscimol (MUS) is an agonist and bicuculline is an antagonist of GABA. Picrotoxinin and TBPS (butylbicyclophosphorothionate) bind to the Cl$^-$ channel and block it. There are also specific drugs for the BZDR which are agonists (e.g. diazepam), antagonists (e.g. Ro15-1788) and inverse-agonist (e.g. β-carboline carboxylate ethyl ester, β-CCE). The inverse agonists decrease the effectiveness of GABA in opening the Cl$^-$ channel. Therefore they are proconvulsant and anxiogenic substances with effects opposite to diazepam and other agonist benzodiazepines. The antagonists block the effects of both agonists and inverse agonists but they do not affect in one way or the other the GABA-induced opening of the Cl$^-$ channel.

In addition to the GABA$_A$ receptors, the brain also has GABA$_B$ receptors with different pharmacological and molecular properties. Thus (–)baclofen is a specific agonist for the GABA$_B$ receptors whereas bicuculline is not a ligand for this receptor. In addition, the GABA$_B$ receptor is not a Cl$^-$ channel. It belongs to the family of receptors coupled to G proteins. The

Receptors in the Developing Nervous System Vol. 2: Neurotransmitters. Edited by Ian S. Zagon and Patricia J. McLaughlin. Published in 1993 by Chapman & Hall. ISBN 0 412 49400 0. Vols. 1 and 2 (set) ISBN 0 412 54520 9.

activation of this receptor regulates the activity of K$^+$ and Ca^{2+} channels. There is little information available on the developmental aspects of the GABA$_B$ receptors. This chapter will discuss exclusively the development of the GABA$_A$/benzodiazepine receptor complex.

The GABAR/BZDR complex has been solubilized with various detergents and purified in several laboratories by using ligand affinity chromatography (AC) on immobilized BZDs (Sigel and Barnard, 1984; Taguchi and Kuriyama, 1984; Vitorica *et al.*, 1988). In addition, there has been recent development of a novel immunoaffinity chromatography (IAC) purification method of the receptor (Park *et al.*, 1991; Park and De Blas, 1991a,b) by using one of the monoclonal antibodies (62-3G1) to the GABAR/BZDR complex that was developed in our laboratory (Vitorica *et al.*, 1988). The receptor preparations purified by either method (AC or IAC) showed several peptides in the range 50 000–60 000 daltons (Fig. 6.1; Sigel and Barnard, 1984; Vitorica *et al.*, 1988; Park *et al.*, 1991; Park and De Blas, 1991a,b). In addition to the M_r, the peptides can be identified by: (1) photoaffinity labeling (PAL) with [^{3}H]flunitrazepam (FNZ) or [^{3}H]muscimol (Fig. 6.2); (2) reactivity with subunit specific monoclonal and polyclonal antibodies; (3) peptide mapping and (4) extent of glycosylation (Park *et al.*, 1991; Park and De Blas, 1991a,b). Recent cloning studies and receptor subunit expression after mRNA injection in *Xenopus laevis* oocytes or transfection into mammalian cells indicate that the neuronal receptor can be functionally reconstituted with a combination of α, β and γ subunits (Pritchett *et al.*, 1989a; Ymer *et al.*, 1989; Verdoorn *et al.*, 1990). To date, six variants of the α subunit (α$_1$–α$_6$), three of the β (β$_1$–β$_3$), two of the γ (γ$_1$–γ$_2$), one δ and one ρ have been cloned from mammalian brain (Schofield *et al.*, 1987; Olsen and Tobin, 1990; Cutting *et al.*, 1991). Combinations of four or five of these subunits form a receptor–channel complex. These membrane-spanning subunits are arranged such that they form a transmembrane Cl$^-$ channel in a fashion similar to the nicotinic acetylcholine and glycine receptors. All the subunits are similarly organized having a long extracellular amino-terminal peptide and four hydrophobic transmembrane α-helix domains (Schofield *et al.*, 1987). All the subunits of the complex participate in the formation of the transmembrane Cl$^-$ channel. The reconstitution of the receptor func-

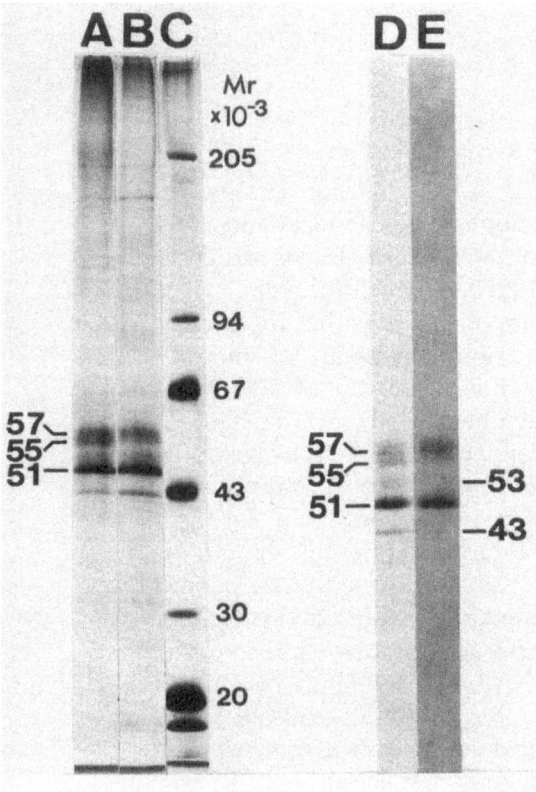

Fig. 6.1. SDS-PAGE of the GABAR/BZDR complex. The bovine brain receptor was purified by immunoaffinity chromatography (IAC) on mAb 62-3G1 (lanes A and D) or by Ro 7-1986/1 affinity chromatography (AC) (lanes B and E). Each lane contained 1 µg of the purified receptor. SDS-PAGE was in a 5–20% polyacrylamide gradient (lanes A–C) or 10% polyacrylamide (lanes D and E). The gels were stained by a silver method. Each lane shows a different receptor preparation. Lane C has M_r markers. Reprinted from Park *et al.* (1991), with permission.

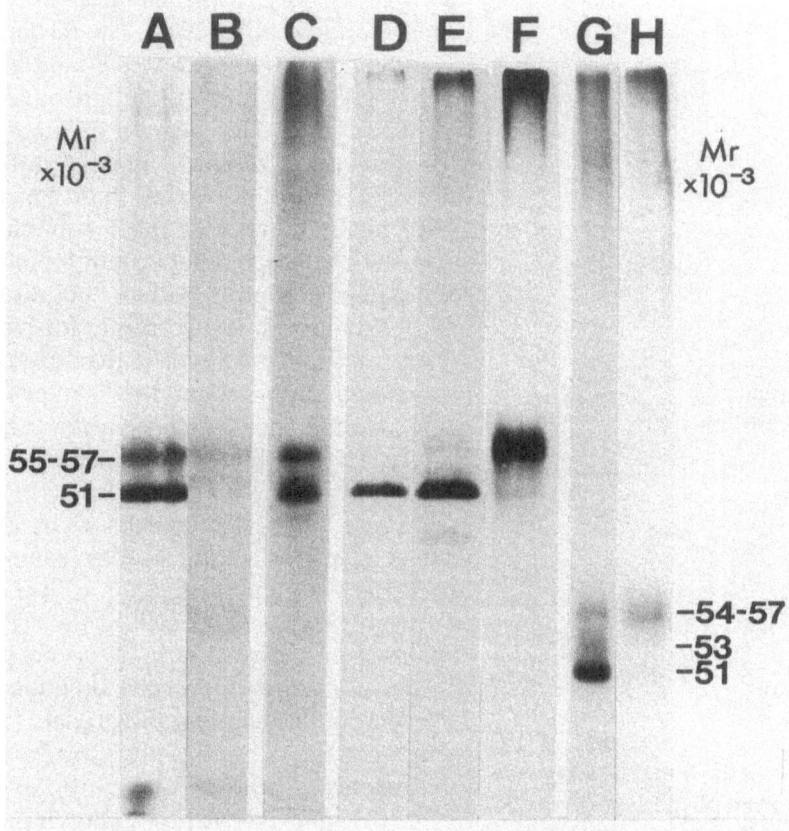

Fig. 6.2. PAL of the purified GABAR/BZDR complex from bovine cerebral cortex (fluorographs). Lanes A, C, D, E, and G show [³H]FNZ labeling, whereas lanes B, F, and H show [³H]MUS labeling. Lanes A and B contain membranes and lanes C–H purified receptors. In lane C, the membrane receptor was PAL, solubilized, and purified by IAC on mAb 62-3G1. In lanes D–F, the receptor complex was purified by Ro 7-1986/1 AC, followed by PAL. In lanes G and H, the receptor was immunoadsorbed on the mAb 62-3G1 gel beads, washed, and then PAL. The x-ray films were exposed for 26 days (lanes A and B), 20 days (lane C), 10 days (lanes D, F, and G), or 55 days (lane H). Lane E is a 60-day exposed fluorograph identical to lane D. SDS-PAGE was in 5–20% polyacrylamide gradient (lanes A and B) or 10% polyacrylamide (lanes C–H). Lanes G and H were from SDS-PAGE in which samples were run for a longer time in order to resolve the PAL peptides better. Reprinted from Park *et al.* (1991), with permission.

tion by expression studies together with the biochemical studies on the purified receptor suggest that the α_1 subunits bind BZDs whereas the β subunits bind muscimol and GABA. The γ_1 and γ_2 subunits seem to be necessary for the functional coupling of the GABAR and BZDR as well as for the binding of BZDs to the α subunits (Pritchett *et al.*, 1989a; Verdoorn *et al.*, 1990; Ymer *et al.*, 1990).

The cloning studies have also shown a high degree of heterogeneity in the subunit composition of the GABAR/BZDR complex. Earlier studies had also suggested receptor heterogeneity. Thus photoaffinity labeling (PAL) with [³H]FNZ had revealed differences in the peptide composition of the BZDRs in brain, cerebellum and hippocampus (Fig. 6.3; Sieghart and Karobath, 1980; Sieghart and

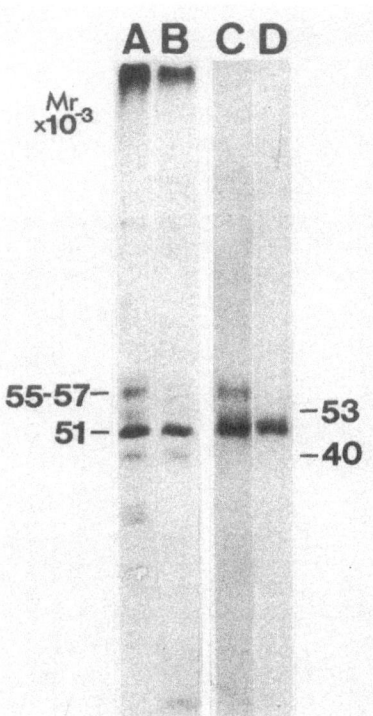

Fig. 6.3. IAC-purification of the [³H]FNZ PAL membrane receptors (fluorographs). Lanes A and B show the PAL membranes and lanes C and D show the mAb 62-3G1 IAC purified receptors after being PAL in membranes. Lanes A and C represent receptors from bovine cerebral cortex, and lanes B and D from bovine cerebellum. Reprinted from Park and de Blas (1991b), with permission.

Drexler, 1983; Park and De Blas, 1991b). Ligand binding studies had previously shown that CL218-872, β-CCE and zolpidem could distinguish two BZDR types (named type I and type II). The cerebellar BZDR is mostly type I whereas in the cerebral cortex both types I and II BZDRs co-exist (Klepner *et al.*, 1979; Sieghart *et al.*, 1983). The cloning and expression of the receptors in mammalian cells have shown that the alpha subunits dictate whether the BZDR is type I or II. The type I BZDR is associated with the α_1 subunit whereas α_2, α_3 and α_5 are associated with type II. In addition, α_5 has lower affinities for zolpidem than α_2 and α_3 indicating heterogeneity of

the type II BZDR (Pritchett *et al.* 1989b; Pritchett and Seeburg, 1990). The recombinant receptors composed of α_6 (α_6, β_2 and γ_2) bind both [³H]MUS and the benzodiazepine [³H]R015– 4513. However, it neither binds other benzodiazepines nor β-carbolines (Lüddens *et al.*, 1990). In addition, Ymer *et al.* (1990) have shown that the γ subunits also affect the pharmacology of the receptors. The recombinant receptors with γ_1 subunit (α_1, β_1 and γ_1) have much lower affinity for the antagonists and inverse agonists than the recombinants with the γ_2 subunit (α_1, β_1 and γ_2).

The significance of this subunit heterogeneity is not yet known but functional implications have been suggested from the aforementioned pharmacological specificities as well as from localization studies using *in situ* hybridization techniques (Shivers *et al.*, 1989) and from functional studies of the receptors expressed either in *Xenopus laevis* oocytes or transfected mammalian cells (Levitan *et al.*, 1988; Verdoorn *et al.*, 1990; Sigel *et al.*, 1990). It seems that the subunit heterogeneity allows the existence of a large combinatorial variety of GABAR and BZDR types showing differential sensitivities to GABA, benzodiazepines and other ligands.

The localization of the GABAR/BZDR in the brain was first revealed by radioligand autoradiography such as [³H]FNZ, [³H]MUS and [³⁵S]TBPS which are specific for the BZDR, GABAR and Cl⁻ channel respectively. In this way complete brain maps for each ligand have been generated (Palacios *et al.*, 1981; McCabe and Wamsley, 1986). These studies have shown that there is no total co-localization of these three binding sites (McCabe and Wamsley, 1986). These results have been interpreted by others as an indication that not all the GABAR and BZDR are physically coupled to one another (Unnerstall *et al.*, 1981). This interpretation is supported by the recent cloning and functional reconstitution studies which indicate the necessity of the γ_2 subunits (in addition to α and β subunits) for: (1) the binding of BZDs to the

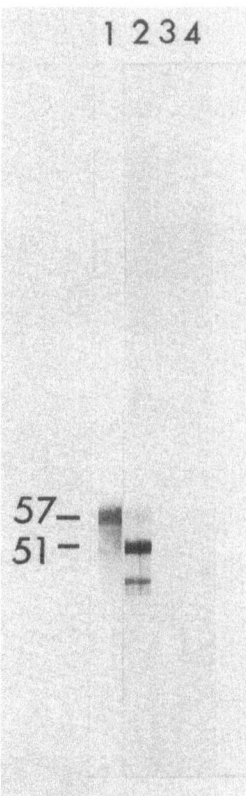

Fig. 6.4.   Immunoblots with mAbs of AC-purified GABAR/BZDR complex. Lanes 1–4 are the mAb 62-3G1, a mouse antiserum, the mAb 62-5F6, and the mAb 62-2G4 respectively. Reprinted from Vitorica *et al.* (1988), with permission.

study the receptor localization at the cellular and subcellular level by both light and electron microscopy immunocytochemistry (De Blas *et al.*, 1988; Juiz *et al.*, 1989; Yazulla *et al.*, 1989). Similar localization has been found by using mAbs 62-3G1 (De Blas *et al.*, 1988) and bd-17 (Richards *et al.*, 1987) which specifically recognize both the β_2 and β_3 subunits of the GABAR (Ewert *et al.*, 1990, 1992) or the mAb bd-24 (Richards *et al.*, 1987) which is specific for α_1 subunit (Ewert *et al.*, 1990). These studies complement and expand the earlier radioligand autoradiography mapping and have allowed the subcellular localization of the GABAR/BZDR complex in the brain to be revealed (Figs. 6.5–6.7). Given the subunit heterogeneity of the receptors and the selectivity of the available antibodies for few (although the most abundant) subunits, such as α_1, β_2, and β_3, the immunocytochemical maps of the various receptor types are still incomplete. New antibodies specific for all the other subunits need to be developed for completing the mapping.

The mapping of the various receptor subunit mRNAs has also been done with *in situ* hybridization techniques by using specific cDNA or cRNA probes (Siegel, 1988; Wisden *et al.*, 1988, 1989; Khrestchatisky *et al.*, 1989; Pritchett *et al.*, 1989a; Shivers *et al.*, 1989; Lüddens *et al.*, 1990; Malherbe *et al.*, 1990; Olsen and Tobin, 1990). These studies complement the radioligand autoradiography and immunocytochemistry mapping studies. They have revealed which subunits co-exist in the same cells, which presumably indicates which subunits are made and assembled into mature receptors by a particular cell. The *in situ* hybridization studies have some limitations, such as the almost exclusive labeling of the cell bodies, not showing where in the neuron the receptors are localized. It is conceivable that the same cell might assemble different combinations of receptor subunits in different synapses or cell areas (i.e. dendrites vs axons). It is also possible that an expressed

α_1 subunit of the receptor complex and (2) for the BZD-dependent stimulation of the GABA-induced opening of the Cl$^-$ channel. In the absence of the γ subunits, the α and β subunits form GABA-gated Cl$^-$ channels but are insensitive to BZDs (Pritchett *et al.*, 1989a; Verdoorn *et al.*, 1990; Ymer *et al.*, 1990). Therefore, the ligand-binding mismatch as well as the heterogeneity of the receptors regarding the ligand-binding properties indicate that the radioligand binding maps cannot discriminate all the receptor types defined at the mRNA level.

Monoclonal (mAb) and polyclonal antibodies raised in our laboratory (Fig. 6.4; Vitorica *et al.*, 1987, 1988) have been used to

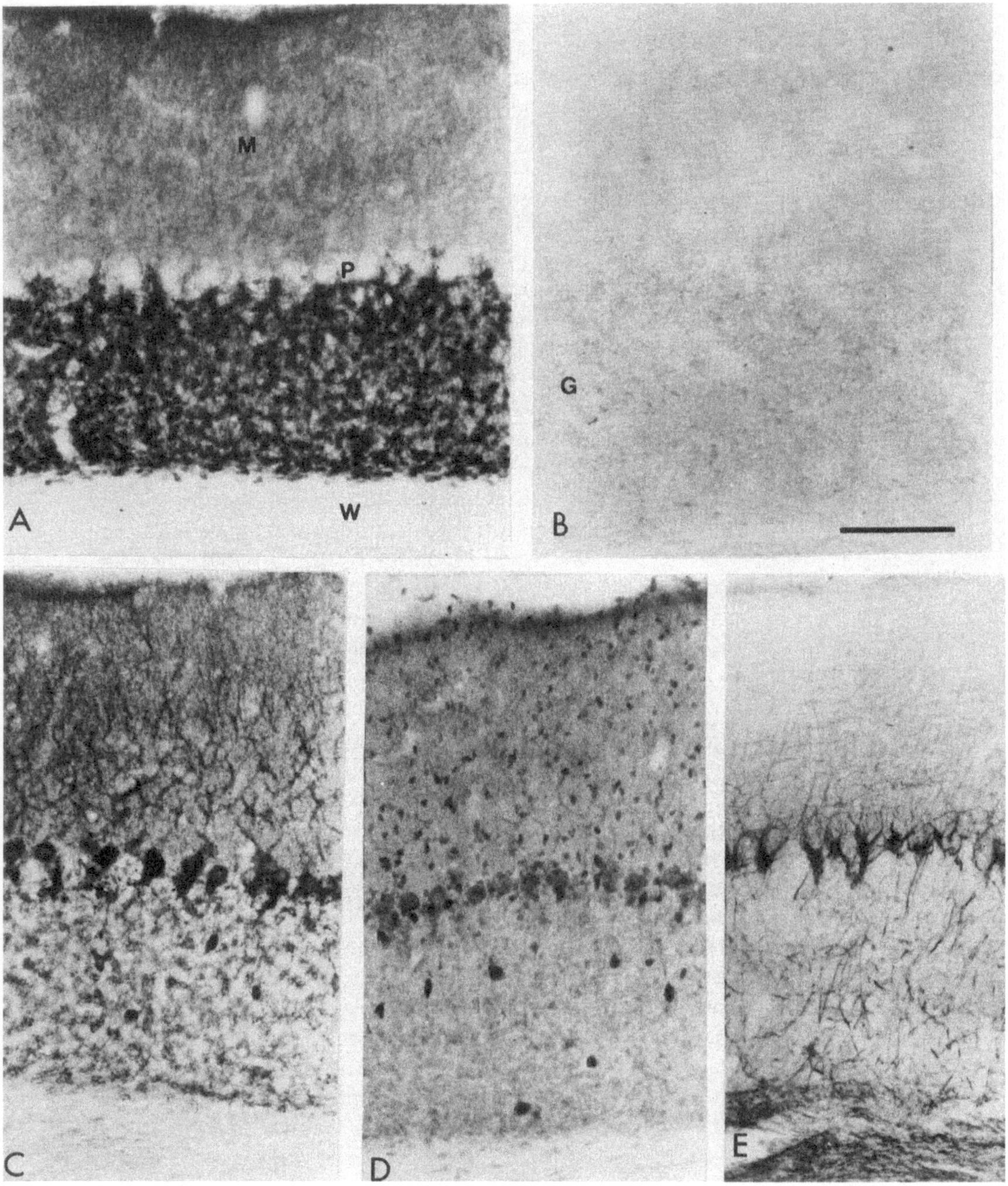

Fig. 6.5.   Rat cerebellum immunocytochemistry with mAb 62-3G1 (A, B), anti-GAD (C), anti-GABA (D), and anti-NFP (E). B, The mAb 62-3G1 (1 ml) was incubated with 9 μg of the AC-purified GABAR/BZDR complex previous to the incubation with the brain tissue. M, molecular layer; P, Purkinje cell layer; G, granule layer; W, white matter. Notice the large GABA-containing Golgi II cells in the granule layer (C, D). Bar, 100 μm. Reprinted from de Blas *et al.* (1988), with permission.

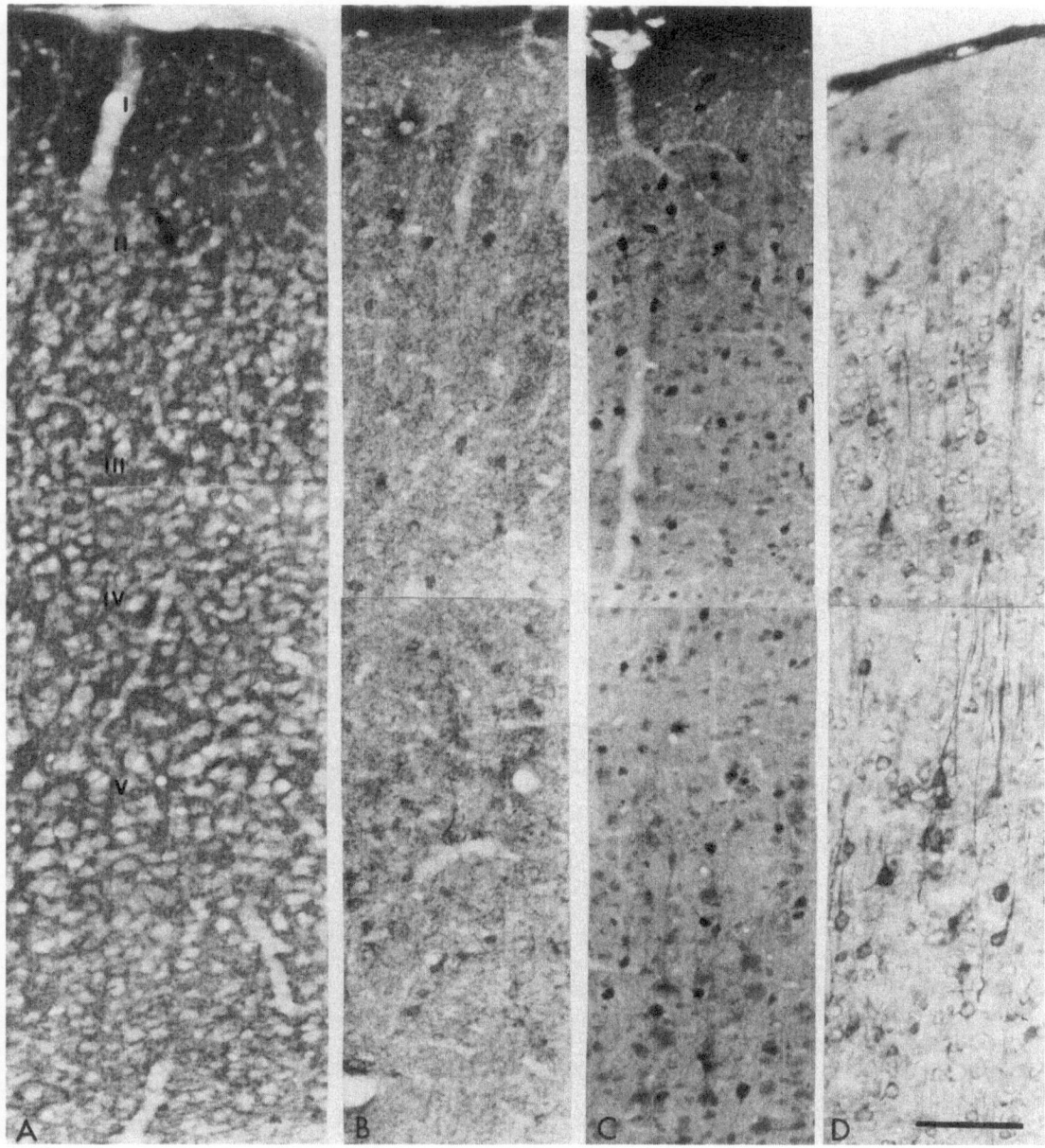

Fig. 6.6. Rat cerebral cortex immunocytochemistry with 62-3G1 (A), anti-GAD (B), anti-GABA (C), and the mAb 8-6A2 (D). Notice that the GABAR, GAD, and GABA immunoreactivities are distributed through all layers. The mAb 8-6A2 was used as a marker for neurons and for the identification of the pyramidal cells (De Blas *et al.*, 1984). Bar, 100 μm. Reprinted from De Blas *et al.* (1988), with permission.

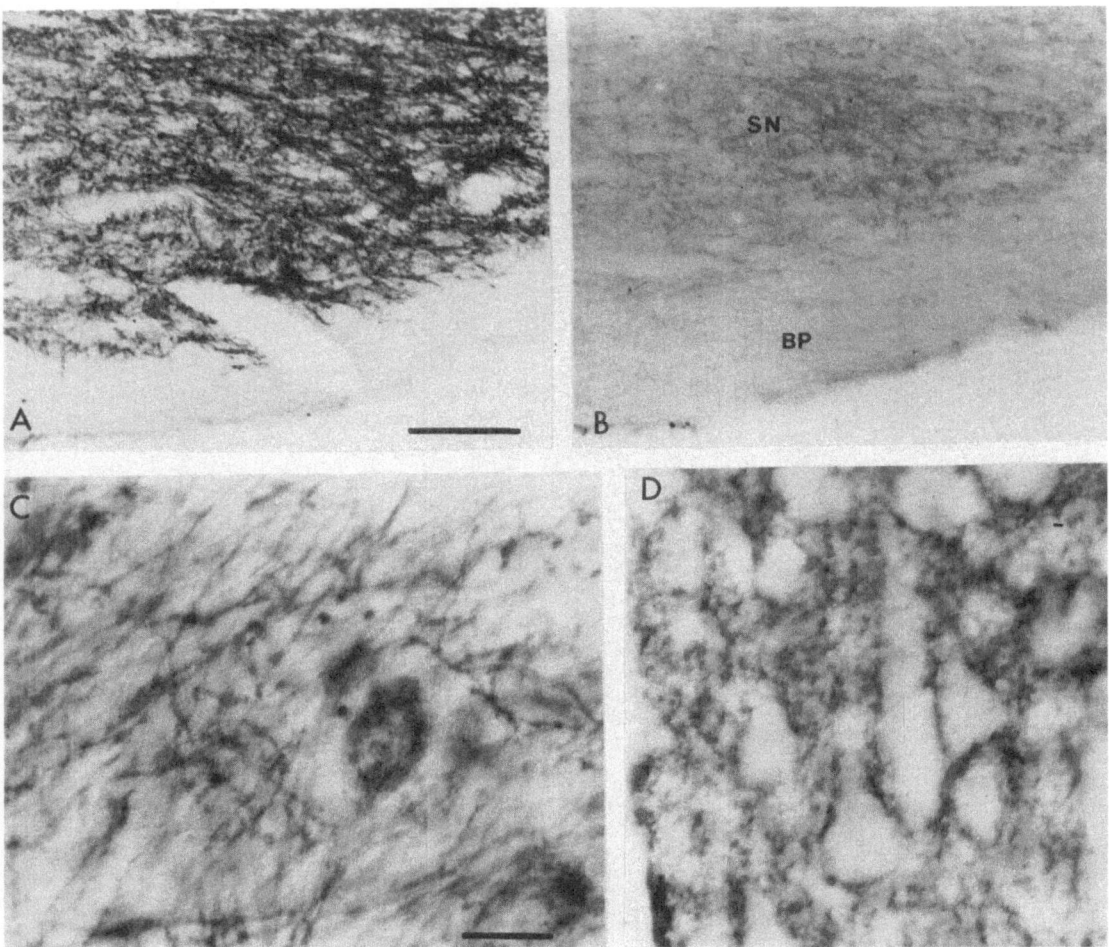

Fig. 6.7. Rat substantia nigra (A–C) and cerebral cortex (D) immunocytochemistry with 62-3G1. B, Control in which mAb was incubated with 9 μg of purified GABAR/BZDR complex previous to the incubation with the tissue. SN, substantia nigra; BP, basis pedunculus. Notice in (C) the presence of receptor clusters on dendrites and cell bodies, and in (D) the association of the reaction product with the neuronal surface of the cortical neurons. Bar, 100 μm (A, B); 20 μm (C, D). Reprinted from De Blas *et al.* (1988), with permission.

mRNA for a specific subunit might not be assembled into a mature receptor in the absence of other subunits. In addition, the different relative abundance of the subunit mRNAs and the limited sensitivity of the assays also make it difficult to interpret correctly the qualitative *in situ* hybridization results.

6.2 THE GABAR/BZDR DURING MAMMALIAN BRAIN DEVELOPMENT

6.2.1 RADIOLIGAND BINDING STUDIES

The first studies on the ontogenesis of the GABAR in the rat brain were done by Coyle and Enna (1976), who reported that GABARs were already present in the embryo at embryonic day (ED)15 although they remained at low levels during birth (25% of the adult values) and up to postnatal day (PD)8, when GABAR levels increased dramatically. The time courses for the GABA synthesizing enzyme, glutamate decarboxylase (GAD), and for the GABAR were similar. However, the neurotransmitter GABA developed earlier than both GABAR and GAD. At birth, the levels of GABA were 50% of the adult values reaching maximum levels around ten days earlier than GABAR or GAD. These results, which have been observed by other authors (see below) suggest that either the GABA turnover in embryos is slower than in adults or that there are alternative synthetic pathways for GABA. This early development of the GABA system suggests that tonic inhibition predominates over excitation during the early development of the brain. The results also suggest a role for the GABA system in the trophic and synaptic interactions that occur during development.

The development of the BZDR in the rat brain has been studied by several groups (Bräestrup and Nielsen, 1978; Palacios *et al.*, 1979; Candy and Martin, 1979; Lippa *et al.*, 1981; Chisholm *et al.*, 1983; Vitorica *et al.*, 1990a). The development of the BZDR in the spinal cord (Bruning *et al.*, 1990) and in the mouse brain (Garrett and Tabakoff, 1985) has also been studied. In the rat, maximum [³H] FNZ binding was observed by the third postnatal week whereas maximum [³H]MUS binding was reached by the fourth postnatal week. The [³H]MUS binding developed more slowly than the [³H]FNZ binding. This developmental mismatch (as well as the distribution mismatch discussed above) has been interpreted by Palacios *et al.* (1979) as evidence that not all the BZDR and GABAR co-exist. The developmental studies have also shown that at birth most of the BZDRs are type II which rapidly increase during the first postnatal week reaching the adult levels by the third or fourth week. The type I receptors however, appear later, increasing after the first week and

reaching adult levels by the second postnatal week. These observations probably reflect the different time courses of expression of the various receptor subunits associated with types I and II BZDRs as discussed above. The time course of GABAR expression during development determined by the radioligand binding assays coincides with the results obtained with a different assay based on the expression of GABAR in *Xenopus laevis* oocytes after the injection of polyA⁺ mRNA obtained from rats of various ages (Carpenter *et al.*, 1988).

6.2.2 PHOTOAFFINITY LABELING, IMMUNOPRECIPITATION AND NORTHERN BLOT STUDIES

In new born animals, the photoaffinity labeling of the membrane bound GABAR/BZDR with [³H]FNZ reveals two peptides of 59 and 55 kDa, whereas in adults, the labeled peptides are of 55 and 51 kDa (Eichinger and Sieghart, 1986). We have investigated the [³H]FNZ photolabeled peptides during the development of the rat brain (Vitorica *et al.*, 1990a) by immunoprecipitation of the solubilized and photolabeled receptor using subunit specific antibodies: the monoclonal antibody (62-3G1) which is specific for both β_2 and β_3 subunits (Ewert *et al.*, 1992) and an antiserum to the α_1 subunit. Both antibodies to the purified GABAR/BZDR were raised and characterized in our laboratory (Vitorica *et al.*, 1987, 1988). Age-dependent heterogeneity of the [³H]FNZ photolabeled subunits was revealed. Peptides of 59, 57, 53 and 51 kDa were immunoprecipitated from the brain of newborn rats, the higher M_r peptides being the most abundant. During early development there is a progressive decrease of the higher M_r peptides and an increase of the 51 kDa peptide. By PD20 the pattern of immunoprecipitated photolabeled peptides had reached the characteristics of the adult animal in which the main photolabeled peptide is the 51 kDa peptide (Fig. 6.8). During the different stages of development, the photolabeling of

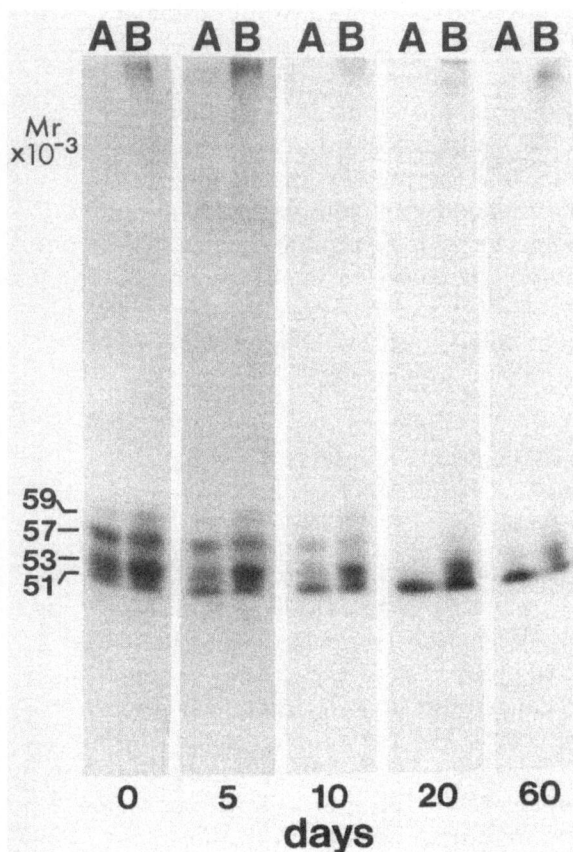

AB AB AB AB AB

Mr
×10⁻³

59
57
53
51

0   5   10   20   60
days

Fig. 6.8. Fluorographs of the [³H]FNZ photoaffinity-labeled GABAR/BZDR during the postnatal development of the rat cerebral cortex. The immunoprecipitates of the triton X-100 (TX)-SDS-solubilized GABAR/BZDR by mAb 62-3G1 and by rabbit antiserum A are shown in lanes A and B, respectively. Reprinted from Vitorica *et al.* (1990a), with permission.

the 51 kDa peptide, but not the others, could be inhibited by Cl218-872 indicating that this peptide was associated with type I BZDR (Fig. 6.9). In support of the results obtained with binding assays already discussed above, the results with photolabeling and the immunoprecipitation also indicated that at birth only type II BZDR were present in the brain and that type I receptors appeared later. These results also showed that the antibodies recognize both types I and II BZDRs.

The differential expression of receptor subunits during the development of the rat brain has also been observed by Garrett *et al.* (1990) using Northern blot hybridization with ³²P-random primed α_1 and β_1 cDNA probes. The β_1 mRNA was highest at birth, then decreased to adult levels by PD5–PD7. In contrast, the α_1 mRNA was low at birth but increased thereafter reaching the adult levels by PD14– PD25. Levitan *et al.* (1988) have shown that the amount of α_3 mRNA is prominent in the cerebellum of the 12-day-old calf, then declines with age.

The various studies using subunit specific probes (photoaffinity labeling, antibodies or DNA) have shown that during development there is differential regulation of the expression of the various GABAR/BZDR subunits. The coordinated expression of some subunits and their assembly into functional receptors are areas that will be explored in the near future as the appropriate tools become available.

6.2.3   RADIOLIGAND AUTORADIOGRAPHY AND *IN SITU* HYBRIDIZATION

The prenatal development of the BZDR in the rat CNS was described by Schlumpf *et al.* (1983) using [³H]FNZ and [³H]Ro15-1788 autoradiography. The BZDR was first detected at ED14 in spinal cord and lower brainstem progressing in a caudorostral gradient. At ED14 and ED15 the expression was extended to the brainstem, mesencephalon and parts of the diencephalon. At ED16, the BZDR was also expressed in the corpus striatum, olfactory bulb and the frontoventral part of the neocortex. By ED21 the receptor was expressed in the remaining neocortex. In the neocortex, the BZDR appeared first in the superficial layer. Then between ED16 and ED18 the receptor was also expressed in the subcortical plate and at ED21 the BZDR appeared in the cortical plate with an inside–out density gradient. Based on these results the authors postulate that the expression of the BZDR is related to

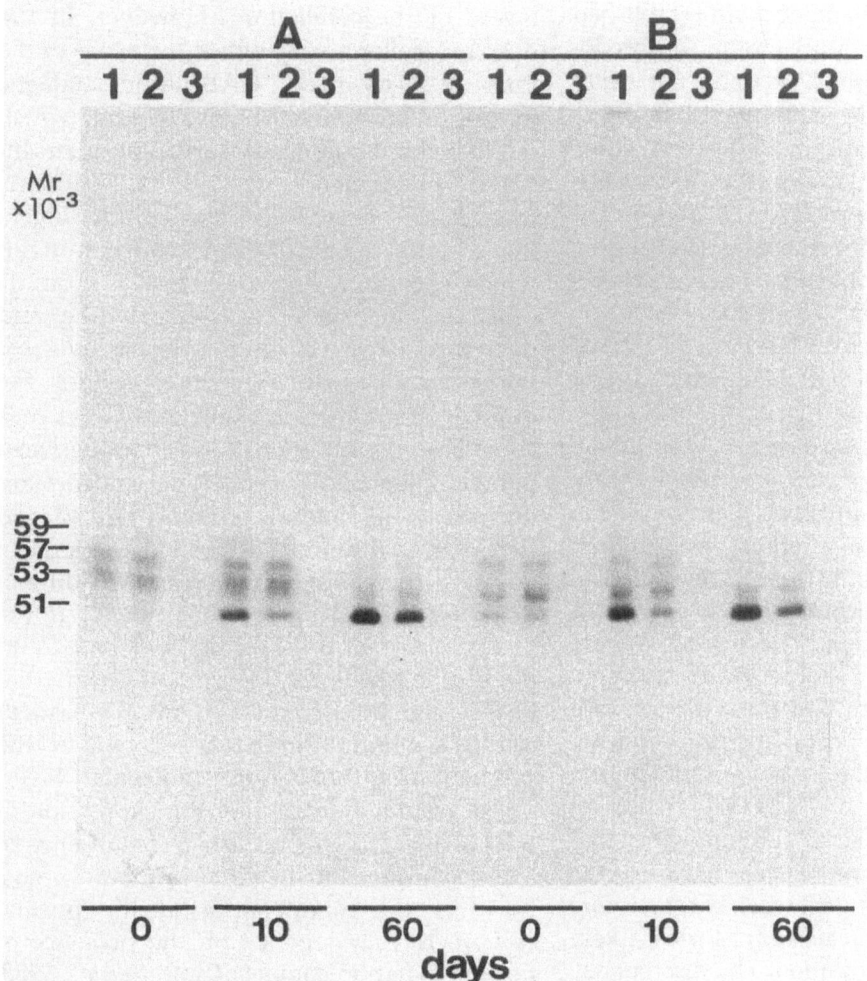

Fig. 6.9.   Fluorographs of [³H]FNZ photoaffinity-labeled and immunoprecipitated GABAR/BZDR from 0-, 10- and 60-day-old rats. The membranes were photoaffinity-labeled with [³H]FNZ, solubilized, and immunoprecipitated with either the mAb 62-3G1 (A) or rabbit antiserum A (B). Lanes 1, 2, and 3 show photoaffinity labeling in controls and in the presence of 300 nM Cl218-872 and 300 nM clonazepam, respectively. Reprinted from Vitorica *et al.* (1990a), with permission.

cell differentiation rather than to synaptogenesis. With the exception of the subplate of the developing cerebral cortex (see below), most authors have noticed that the developmental expression of the GABAR/BZDR is not directly related to synaptogenesis. Some examples are derived from developmental studies on the GABAR and BZDR in the cerebellum (Palacios and Kuhar, 1982; Rotter and Frostholm, 1986; Frostholm and Rotter, 1987; Zdilar *et al.*, 1992). In rats and mice the cerebellum develops postnatally. The dividing external granule cell layer, from which the granule cells are derived, remains unlabeled with either [³H]FNZ or [³H]MUS. The binding of the latter to the granule cell GABAR occurs for the first time during their migration from the external granule layer to their final position in the granule cell layer, that is before afferent and efferent synaptic contacts are established by these cells. In the adult brain, the predominant binding of [³H]FNZ is to the molecular layer. In contrast, the predominant binding of [³H]MUS is to the granule cell layer. These results show one more example of the localization mismatch between GABAR and BZDR binding sites as mentioned above. The binding of [³H]FNZ to the molecular layer increases dramatically during PD11–PD15. Most likely this binding represents the binding of [³H]FNZ to the α₁ subunit of the GABAR/BZDR. *In situ* hybridization studies (Zdilar *et al.*, 1992) indicate that the expression of the α₁ subunit in Purkinje cells appears very early in development between PD1 and PD5, that is, prior to the establishment of afferent and efferent synaptic connections in these cells. Later, at PD11–PD13 the α₁ mRNA is also expressed in stellate and basket cells. These studies were done with a riboprobe for the α₁ subunit. These results indicate that the expression of the GABAR/BZDR occurs before synaptogenesis. However, the *in situ* hybridization studies by Gambarana *et al.* (1990), using an oligonucleotide for the α₁ subunit, show that the α₁ subunit is synthesized between the second and third postnatal week and therefore the expression of this subunit is most likely related to synapse formation. The discrepancy between the two studies might be explained by the increased sensitivity of the *in situ* hybridization assays done in the former study in which a full-length radiolabeled riboprobe was used instead of the end-labeled 40-mer oligonucleotides used in the latter study.

The relationship between the formation and maintenance of synaptic contacts and GABAR/BZDR expression has also been investigated by using mouse neurological mutants: Purkinje cell degeneration (PKCD), weaver, staggerer and reeler (Rotter *et al.*, 1988; Rotter and Frostholm, 1988) as well as in the rat dystonic mutant (Beales *et al.*, 1990). In the PKCD mutant in which the Purkinje cells have disappeared by PD45, there is a reduction in [³H]FNZ binding in all the cell layers of the cerebellum. However, in the deep cerebellar nuclei, where the axons of the Purkinje cells make GABAergic contacts, there is an increase of [³H]FNZ binding which might be the result of a denervation supersensitivity phenomenon. In the PKCD mutants there is also decreased [³H]MUS binding to the granule cell layer. The weaver mutant, which shows an important loss of granule cells, has increased [³H]FNZ binding and decreased [³H]MUS binding in all layers of the vermis. In the staggerer where the number of granule, Purkinje and Golgi cells as well as the number of synapses between parallel fibers and Purkinje cell dendrites are decreased, the binding of both [³H]FNZ and [³H]MUS is diminished in all the layers. In addition, there is also an increased binding of [³H]FNZ to the deep cerebellar nuclei. In the reeler where all cells are malpositioned and the Purkinje cells are deprived of the afferent basket cell inputs, there is an increase of [³H]FNZ binding to all the layers of the cerebellum including the molecular layer. These results indicate that the expression of the GABAR/BZDRs occurs in the absence of a full complement of GABAergic afferents, however the maintenance of the normal receptor levels depends on the presence of normal synaptic contacts (Rotter *et al.*, 1988; Rotter and Frostholm, 1988). In the dystonic rat mutant there is a concomitant increase of GAD activity in the deep cerebellar nuclei with a decrease in [³H]MUS binding. These results suggest the existence of receptor down regulation in these nuclei (Beales *et al.*, 1990). It seems that the synaptic contacts are involved in the regulation of the normal expression of the receptors. Nevertheless, the cells also express receptors in the absence of synapses.

6.2.4 IMMUNOCYTOCHEMISTRY

Our monoclonal antibody 62-3G1 to β₂ and β₃ subunits of the GABAR/BZDR has been used in several developmental studies of the expression of the GABAR/BZDR. Thus, Huntley *et al.* (1990) have studied the development of the sensory–motor cortex of the macaque monkey. The GABAR/BZDR immunoreactivity with mAb 62-3G1 appeared at ED121 in layers III and IV and subplate. During development there are laminar changes in the distribution of the GABAR, reaching the adult distribution by PD1.5. In the adult the highest level of immunoreactivity occurs in layers II and III (Fig. 6.10). The changes in the laminar distribution of the immunoreactivity parallelled the distribution of the GABA neurons which indicated the existence of early GABA and GABAR interactions. The immunoreactivity in the subplate supports the notion that this layer is a temporary target for ingrowing cortical afferents during early development. Later in development this layer disappears and some cells become interdispersed in the white matter.

In the rat neocortex (Cobas *et al.*, 1991) the GABAR/BZDR immunoreactivity with mAb 62-3G1 appeared first at ED14 in the external primordial plexiform layer. By ED16 the receptor was present in lamina I and subplate. At ED18 the receptor appeared in the lower part of the cortical plate. In the latter, the GABAR/BZDR immunoreactivity and the distribution of GABA neurons show a similar spatiotemporal and inside–out gradient. Nevertheless, the expression of the GABAR/BZDR precedes both the axogenesis of cortical interneurons and the formation of symmetrical synapses. Therefore, the GABA system during early embryogenesis might have a morphogenetic rather than synaptic role. However, in the rat brain subplate as indicated above in the case of the monkey, the GABA system might be involved in transient synaptic connectivity.

In the rat thalamus (Bentivoglio *et al.*, 1991) the GABAR/BZDR immunoreactivity with mAb 62-3G1 undergoes rearrangements during the first postnatal weeks. During this time the levels of GABAR/BZDR decreased in the reticular nucleus while they increased in the dorsal thalamus. The receptor was expressed at ED14, that is as early as the GABAergic innervation of these nuclei. The study, however, does not show whether GABAergic synaptic contacts have been established by that time.

Lauder *et al.* (1986) have studied the development of the GABA neurons in the rat brain by immunocytochemistry with an antibody to GABA. These authors have shown that GABA appeared first at ED13 in the brainstem, mesencephalon and diencephalon. At ED16 the cells were also found in the basal forebrain and cortex. The temporal and spatial distribution of the GABA immunoreactivity was similar to that of the BZDR revealed by radioligand autoradiography (Schlumpf *et al.*, 1983) as discussed earlier. Nevertheless, the GABA immunoreactivity appeared one day earlier than the BZDR. These results also support a morphogenetic and/or trophic function of GABA during embryogenesis as proposed by Chronwall and Wolff (1980). Although some authors favor the notion that GABARs are formed following synaptogenesis (Gambarana *et al.*, 1990; Meinecke and Rakic, 1990), most of the studies discussed earlier indicate that GABAR/BZDR expression does not follow synaptogenesis. In most cases, GABA, GAD and GABAR are expressed earlier than synaptogenesis (Aoki *et al.*, 1989; Fairén *et al.*, 1986). In fact, in certain cases the GABAR function might modulate synaptogenesis as will be discussed below.

6.3 THE GABAR/BZDR AND BRAIN PLASTICITY

We have already mentioned the plasticity of the GABAR/BZDR during the abnormal development of the brain in studies with neurological mutants. Experimental denervation has also revealed plasticity of the GABAR/BZDR. The denervation of the striatonigral fibers results in an increase in the density of type I and a decrease of type II BZDRs in the substantia nigra (Lo *et al.*, 1983). These results suggest a postsynaptic

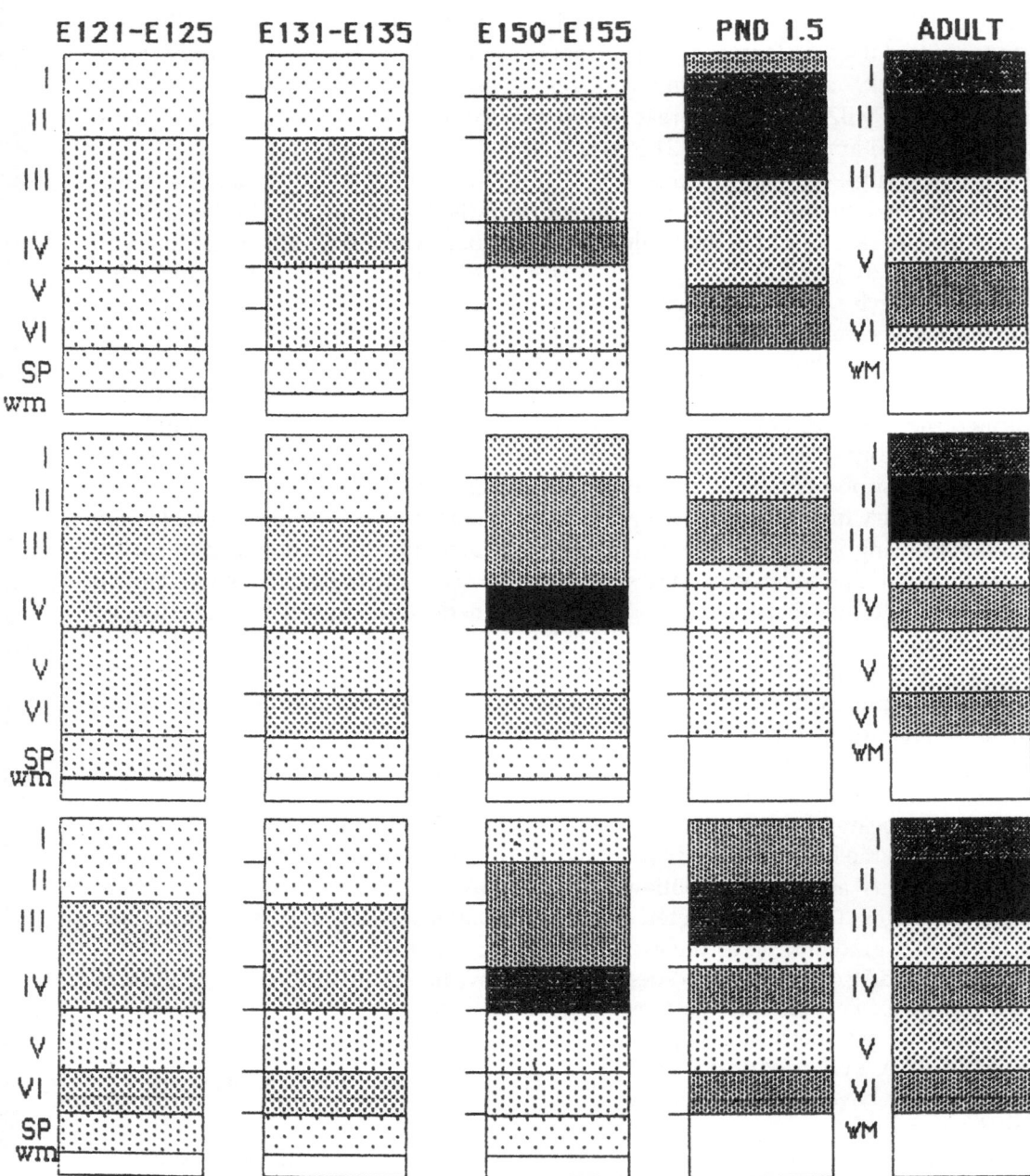

Fig. 6.10. Schematic diagram illustrating the major changes in laminar distribution and staining intensity of receptor immunoreactivity in areas 4, 3a, 3b, 1 and 2 from ED125 to adulthood in the sensory–motor cortex of the macaque monkey. Increasing density of stippling represents increasing intensity of the immunostaining. Note that the laminar boundaries depicted are schematic, and are not meant to reflect changes in laminar thickness with increasing age. Reprinted from Huntley *et al.* (1990), with permission.

localization of BZDR type I and a presynaptic localization of BZDR type II in this structure. They also reveal the possible existence of a phenomenon of denervation supersensitivity as in the case of the PKCD and staggerer mutants already discussed above. In post-mortem human brains from patients with unilateral lesions of the optic nerve, Palacios *et al.* (1987) have found a selective decrease of GABAR/ BZDR in the lateral geniculate body in the layers innervated by the damaged nerve. The previous studies were done by using radioligand autoradiography techniques.

Immunocytochemical techniques with the mAb 62-3G1 to the GABA/BZDR have revealed that in macaque monkey, either intravitreal injection of the sodium channel blocker tetrodotoxin (TTX) or monocular enucleation are followed by a decrease of GABAR/BZDR immunoreactivity in the columns of the visual cortex (areas 17 and 18) dominated by the deprived eye (Fig. 6.11, Hendry *et al.*, 1990). In the same study, similar results were obtained by using [³H]FNZ and [³H]MUS radioligand autora-diography techniques. These results indicate that the density of the GABAR/BZDR in the monkey visual cortex is regulated not only by denervation but also by the electrical activity. Similar activity-dependent regula-tion of GAD and GABA also occurs in the visual cortex (Hendry and Jones, 1988). These results suggest that decreased GABA-ergic activity in the columns driven by the

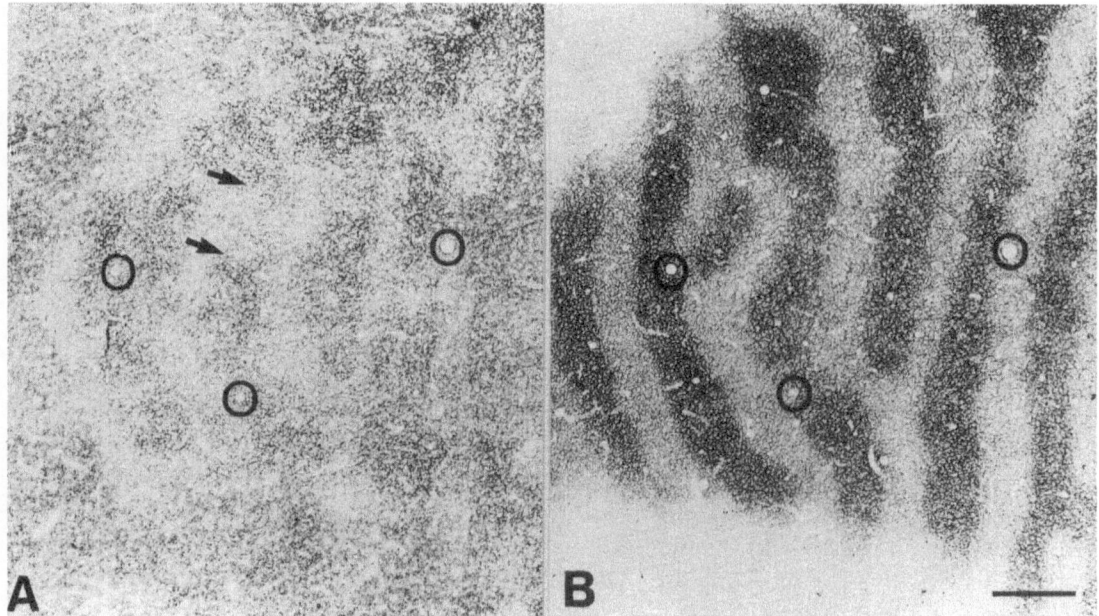

Fig. 6.11. Photomicrographs of tangential sections through layer IVC of a monkey visual cortex injected intravitreally with TTX 5 days before death. A, GABA$_A$ receptor immunostaining in layer IVC consists of stripes of intense immunostaining that alternate with stripes of lighter immunostaining. Individual stripes pass through both layer IVCβ and the more lightly stained layer IVCα (between arrows). B, Section adjacent to A stained for CO. Intense staining of normal-eye dominance stripes and light staining for injected-eye stripes are evident. Comparison of the profiles of radially oriented blood vessels (circles) shows that the receptor immunostaining in the injected-eye columns is reduced to that in the normal-eye columns. Scale bar, 500 μm. Reprinted from Hendry *et al.* (1990), with permission.

deprived eye might contribute to the functional expansion of the ocular dominance columns driven by the other eye.

In monocularly deprived kittens, changes in the GABAR/BZDR expression was opposite to those observed in monkeys since the density of GABAR increased in area 17 among other regions of the brain (Skangiel-Kramska and Kossut, 1984; Shaw and Cynader, 1988). Therefore, the plastic changes of the GABAR that follow denervation seem to be heterogeneous and adapted to the functional characteristics of the system. This functional heterogeneity might result from a combination of factors: the molecular heterogeneity of the GABAR/BZDR and the differential regulation of the subunit expression as discussed above. No effect on either [³H]FNZ (Shaw *et al.*, 1987) or [³H]MUS (Mower *et al.*, 1988) binding in the cat visual cortex was found after rearing these animals in the dark. Therefore, it seems that the GABAergic system develops normally in the cat visual cortex in the absence of visual input from both eyes. Nevertheless, this phenomenon is not common to all brain structures. Schliebs *et al.* (1986) have reported a decrease in the binding of [³H]FNZ to the lateral geniculate nucleus and superior colliculus of rats reared in the dark.

Evidence of the importance of the role of the GABAR in the synaptic organization of the ocular dominance columns has been provided by Reiter and Stryker (1988). When muscimol is injected in the visual cortex of the cat during the critical period and the eye is monocularly closed, the inputs from the closed eye become dominant. Thus, the less active inputs became dominant when the cortical cell discharges were inhibited by the GABAR/BZDR agonist muscimol. This result was the opposite of what happened when muscimol was not injected. In this case, the open eye was dominant. These results show the pivotal role that the GABAergic transmission, mediated by GABAR/BZDR, has on brain plasticity and synaptogenesis.

6.4 DEVELOPMENT OF THE GABAR/BZDR IN NEURONAL CULTURES

Rat neuronal cultures from cerebral cortex (McCarthy and Harden, 1981; White *et al.*, 1981) and cerebellar granule cells (Meier *et al.*, 1984) as well as neuronal cultures from other species (Huang *et al.*, 1980; Borden *et al.*, 1984; Kuriyama *et al.*, 1987; Mehta and Ticku, 1988) express GABAR/BZDR complexes. Syapin *et al.* (1985) have shown that in embryonic neuronal cultures, most of the BZDR are type II. We have discussed above that during normal development, the early postnatal BZDR is type II. The type I BZDR appears later and increases dramatically during the first and second postnatal weeks coinciding with the expression of the 51 kDa peptide that is PAL by [³H]FNZ as discussed earlier. The development of the GABAR/BZDR complex in primary neuronal cultures from rat brain embryos has been investigated by Vitorica *et al.* (1990b). The receptor complexes of the cultured neurons were photoaffinity labeled with [³H]FNZ, solubilized with detergent and immunoprecipitated with the subunit specific mAb 62-3G1 (to β_2 and β_3) and a polyclonal antibody specific for the α_1 subunit. The results have shown that the cultured neurons express five different photoaffinity labeled peptides of 51, 53, 54, 57 and 59 kDa which are similar in size to the ones found in the brain of newborn rats (Fig. 6.12). However, during *in vitro* development of the neuronal cultures, the transition of the receptors from the embryonic to the mature form does not occur (Fig. 6.8 vs. Fig. 6.13). These findings contrast with the important changes in subunit composition observed during the normal development of the rat brain, mainly regarding the progressive enrichment in the 51 kDa subunit during the early postnatal brain maturation. The neuronal cultures might be missing a cell type and/or soluble factor that regulates the normal expression of the GABAR/BZDR subunits (i.e., the 51 kDa peptide).

Other studies have shown that granule cell cultures from rat cerebellum have high affinity GABAR regardless of the age of the cultures. However, when the cells were cultured during eight days in the presence of 50 μM GABA or muscimol or 150 μM 4,5,6,7-tetrahydroisooxazolo[5-4-c]pyridin-3-ol (THIP) the cultures developed low affinity GABA receptors as well as an increased number of neurites. These

effects could be blocked by bicuculline whereas (–)baclofen had no effect (Meier *et al.*, 1984; Hansen *et al.*, 1987). A similar effect on the expression of the low affinity GABAR was produced by taurine (Abraham and Schousboe, 1989). These studies suggest that GABA and taurine exert trophic actions on the neurons, including the induction of the expression of the low affinity GABAR. The trophic

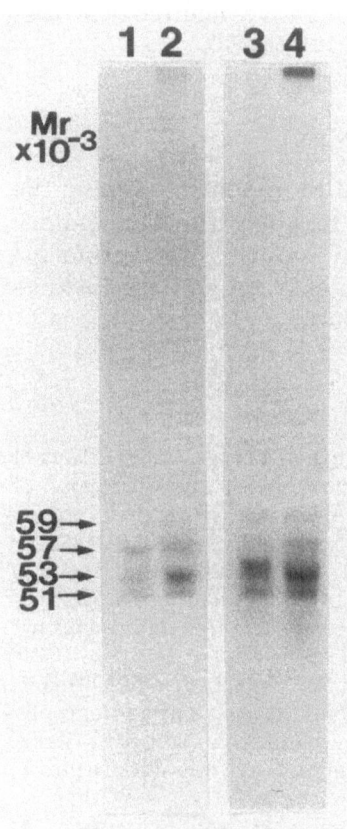

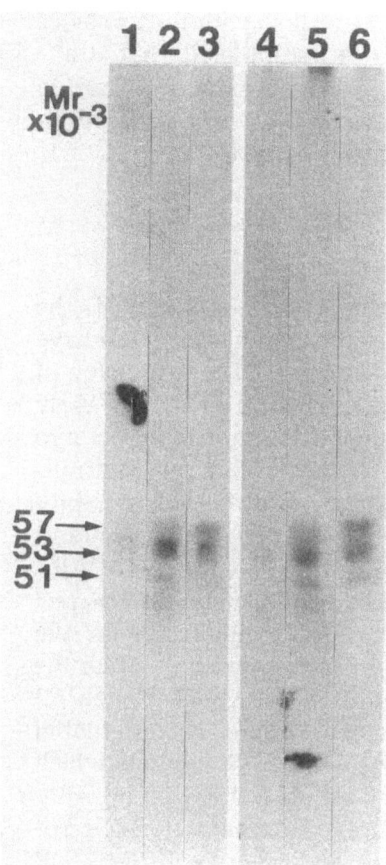

Fig. 6.12. Fluorographs of the [³H]FNZ photolabeled and immunoprecipitated GABAR/BZDR receptor. Lanes 1 and 2 are immunoprecipitates of newborn rat brain membranes and lanes 3 and 4 show the immunoprecipitated receptor from cells maintained in culture for 14 days. mAb 62-3G1 was used in lanes 1 and 3 and rabbit antiserum A in lanes 2 and 4. Reprinted from Vitorica *et al.* (1990b), with permission.

Fig. 6.13. Fluorographs of the immunoprecipitated [³H]FNZ photoaffinity labeled GABAR/BZDR from rat neuronal cultures. The photoaffinity labeled receptor was solubilized and immunoprecipitated with either mAb 62-3G1 (lanes 1–3) or rabbit antiserum A (lanes 4–6). Cells maintained in culture for 5 days (lanes 1 and 4), 14 days (lanes 2 and 5) and 21 days (lanes 3 and 6) are shown. Reprinted from Vitorica *et al.* (1990b), with permission.

effect results from the binding of GABA and other agonists to the high affinity GABAR. It has been proposed that the hyperpolarization of the membrane which follows the opening of the GABA-gated chloride channel is the signal that mediates the trophic and differentiating actions of GABA both *in vivo* and in culture (Belhage *et al.*, 1990). This hypothesis is consistent with the very early expression of GABA and GABAR during development as indicated above, as well as with the existence of non-vesicular K^+-stimulated but Ca^{2+}-independent release of GABA and taurine from the growth cones (Taylor and Gordon-Weeks, 1989; Taylor *et al.*, 1990).

6.5 SUMMARY

The studies on the development of the GABAR/BZDR in the mammalian brain have indicated the following. (1) The expression of GABA, GAD and GABAR/BZDR is an early developmental event that occurs in the embryo before the development of other transmitter–receptor systems and before synaptogenesis. (2) GABA seems to have a trophic and morphogenetic role mediated by GABAR/BZDR during early development. (3) Synaptic connectivity seems to be important for the consolidation and maintenance but not for the initial expression of the GABAR/BZDR. (4) During development, there is differential expression of GABAR/BZDR subunits which seems to be regulated at least in part by cellular interactions and/or trophic factors, electrical activity, GABA and other GABAR/BZDR modulators. The differential expression of various GABAR/BZDR types and subunits during development might reflect the adaptation to the various functional roles that the GABAR/BZDR might play during brain development (i.e., morphogenesis, plasticity, synaptogenesis etc.). (5) There is plasticity in the expression of these receptors that is dependent on neuronal activity and innervation. The GABAergic transmission mediated by GABAR

plays an important role in regulating neuronal plasticity and synaptic reorganization in experimental denervation and sensory deprivation.

These results suggest that the GABAR/BZDR might be involved in organizing and maintaining the connectivity and synaptic function of certain brain areas during the normal development, maturation and aging as well as in controlling adaptive synaptic changes in response to injury, sensory deprivation and degenerative neurological disorders.

ACKNOWLEDGEMENT

I thank Ms Laura Morissette for typing the manuscript. The research in my laboratory was supported by grant NS17708 from the National Institute of Neurological Disorders and Stroke.

REFERENCES

Abraham, J.H. and Schousboe, A. (1989) Effects of taurine on cell morphology and expression of low affinity GABA receptors in cultured cerebellar granule cells. *Neurochem. Res.*, **14**, 1031–8.

Aoki, E., Semba, R. and Kashiwamata, S. (1989) When does GABA-like immunoreactivity appear in the rat cerebellar GABAergic neurons? *Brain Res.*, **502**, 245–51.

Beales, M., Lorden, J.F., Walz, E. and Oltmans, G.A. (1990) Quantitative autoradiography reveals selective changes in cerebellar GABA receptors of the rat mutant dystonic. *J. Neurosci.*, **10**, 1874–5.

Belhage, B., Hansen, G.H. and Schousboe, A. (1990) GABA agonist induced changes in ultrastructure and GABA receptor expression in cerebellar granule cells is linked to hyperpolarization of the neurons. *Int. J. Dev. Neurosci.*, **8**, 473–9.

Bentivoglio, M., Spreafico, R., Alvarez-Bolado, G. *et al.* (1991) Differential expression of the $GABA_A$ receptor complex in the dorsal thalamus and reticular nucleus: an immunohistochemical study in the adult and developing rat. *Eur. J. Neurosci.*, **3**, 118–25.

Bloom, F. E. and Iversen, L. L. (1971) Localizing ^{3}H-GABA in nerve terminals of rat cerebral cortex by electron microscopic autoradiography. *Nature*, **229**, 628–30.

Borden, L.A., Czajkowski, C., Chan, C.Y., and Farb, D.H. (1984) Benzodiazepine receptor synthesis and degradation by neurons in culture. *Science*, **226**, 857–60.

Bräestrup, C. and Nielsen, M. (1978) Ontogenic development of benzodiazepine receptors in the rat brain. *Brain Res.*, **147**, 170–3.

Bräestrup, C. and Squires, F. (1977) Specific benzodiazepine receptors in rat brain characterized by high-affinity [^{3}H]diazepam binding. *Proc. Natl. Acad. Sci. USA*, **74**, 3805–9.

Brüning, G., Bäuer, R. and Baumgarten, H.G. (1990) Postnatal development of [^{3}H]flunitrazepam and [^{3}H]strychnine binding sites in rat spinal cord localized by quantitative autoradiography. *Neurosci. Lett.*, **110**, 6–10.

Candy, J.M. and Martin, I.L. (1979) The postnatal development of the benzodiazepine receptor in the cerebral cortex and cerebellum of the rat. *J. Neurochem.*, **32**, 655–8.

Carpenter, M.V., Parker, I. and Miledi, F.R.S. (1988) Expression of GABA and glycine receptors by messenger RNAs from the developing rat cerebral cortex. *Proc. R. Soc. Lond.*, **234**, 159–70.

Chisholm, J., Kellogg, C. and Lippa, A. (1983) Development of benzodiazepine binding subtypes in three regions of rat brain. *Brain Res.*, **267**, 388–91.

Chronwall, B. and Wolff, J.R. (1980) Prenatal and postnatal development of GABA accumulating cells in the occipital neocortex of rat. *J. Comp. Neurol.*, **190**, 187–208.

Cobas, A., Fairén, A., Alvarez-Bolado, G. and Sánchez, M.P. (1991) Prenatal development of the intrinsic neurons of the rat neocortex: a comparative study of the distribution of GABA immunoreactive cells and the GABA$_A$ receptor. *Neuroscience*, **40**, 375–97.

Coyle, J.T. and Enna, S.J. (1976) Neurochemical aspects of the ontogenesis of GABAenergic neurons in the rat brain. *Brain Res.*, **111**, 119–33.

Cutting, G.R., Lu, L., O'Hara, B.F. *et al.*(1991) Cloning of the γ-aminobutyric acid (GABA)$_{ρ1}$ cDNA: a GABA receptor subunit highly expressed in the retina. *Proc. Natl. Acad. Sci. USA*, **88**, 2673–7.

de Blas, A.L., Kuljis, R.O. and Cherwinski, H.M. (1984) Mammalian brain antigens defined by monoclonal antibodies. *Brain Res.*, **322**, 277–87.

de Blas, A.L., Vitorica, J. and Friedrich, P. (1988) Localization of GABA$_A$ receptor in the rat brain with a monoclonal antibody to the 57 000 M$_r$ peptide of GABA$_A$ receptor/benzodiazepine receptor/Cl$^-$ channel complex. *J. Neurosci.*, **8**, 602–14.

Eichinger, A. and Sieghart, W. (1986) Postnatal development of proteins associated with different benzodiazepine receptors. *J. Neurochem.*, **46**, 173–80.

Ewert, M., Shivers, B., Lüddens, H. *et al.* (1990) Subunit selectivity and epitope characterization of monoclonal antibodies directed against the GABA$_A$/benzodiazepine receptor. *J. Cell Biol.*, **110**, 2043–8.

Ewert, M., de Blas, A.L., Möhler, H. and Seeburg, P.H. (1992) A prominent epitope on GABA$_A$ receptors is recognized by two different monoclonal antibodies. *Brain Res.*, **569**, 57–62.

Fairén, A., Cobas, A. and Fonseca, M. (1986) Times of generation of glutamic acid decarboxylase immunoreactive neurons in mouse somatosensory cortex. *J. Comp. Neurol.*, **251**, 67–83.

Frostholm, A. and Rotter, A. (1987) The ontogeny of [^{3}H]muscimol binding sites in the C57BL/6J mouse cerebellum. *Dev. Brain Res.*, **37**, 157–66.

Gambarana, C., Pittman, R. and Siegel, R.E. (1990) Developmental expression of the GABA$_A$ receptor alpha-1 subunit mRNA in the rat brain. *J. Neurobiol.*, **21**, 1169–79.

Garrett, K. M. and Tabakoff, B. (1985) The development of type 1 and type 2 benzodiazepine receptors in the mouse cortex and cerebellum. *Pharmacol. Biochem. Behav.*, **22**, 985–92.

Garrett, K. M., Saito, N., Duman, R.S. *et al.* (1990) Differential expression of GABA$_A$ receptor subunits. *Mol. Pharmacol.*, **37**, 652–7.

Hansen, G.H., Belhage, B., Schousboe, A. and Meier, E. (1987) Temporal development of GABA agonist induced alterations in ultrastructure and GABA receptor expression in cultured cerebellar granule cells. *Int. J. Neurosci.*, **5**, 263–9.

Hendry, S.H.C. and Jones, E.G. (1988) Activity-dependent regulation of GABA expression in the visual cortex of adult monkeys. *Neuron*, **1**, 701–12.

Hendry, S.H.C., Fuchs, J., de Blas, A.L. and Jones, E.G. (1990) Distribution and plasticity of immunocytochemically localized GABA$_A$ receptors in adult monkey visual cortex. *J. Neurosci.*, **10**, 2438–50.

Huang, A., Barker, J.L., Paul, S.M. *et al.* (1980) Characterization of benzodiazepine receptors in primary cultures of fetal mouse brain and spinal cord neurons. *Brain Res.*, **190**, 485–91.

Huntley, G.W., de Blas, A.L. and Jones, E.G. (1990) GABA_A receptor immunoreactivity in adult and developing monkey sensory-motor cortex. *Exp. Brain Res.*, **82**, 519–35.

Juiz, J.M., Helfert, R.H., Wenthold, R.J. *et al.* (1989) Immunocytochemical localization of the GABA/benzodiazepine receptor in the guinea pig cochlear nucleus: evidence for receptor localization heterogeneity. *Brain Res.*, **504**, 173–9.

Khrestchatisky, M., MacLennan, A.J., Chiang, M.-Y. *et al.* (1989) A novel α subunit in rat brain GABA_A receptors. *Neuron*, **3**, 745–53.

Klepner, C.A., Lippa, A.S., Benson, P.I. *et al.* (1979) Resolution of two biochemically and pharmacologically distinct benzodiazepine receptors. *Pharmacol. Biochem. Behav.*, **11**, 457–62.

Kuriyama, K., Tomono, S., Kishi, M. *et al.* (1987) Development of gamma-aminobutyric acid (GABA)ergic neurons in cerebral cortical neurons in primary culture. *Brain Res.*, **416**, 7–21.

Lauder, J.M., Han, V.K.M., Henderson, P. *et al.* (1986) Prenatal ontogeny of the GABAergic system in the rat brain: an immunocytochemical study. *Neuroscience*, **19**, 465–93.

Levitan, E.S., Schofield, P.R., Burt, D.R. *et al.* (1988) Structural and functional basis for GABA_A receptor heterogeneity. *Nature*, **335**, 76–9.

Lippa, A.S., Beer, B., Sano, M. C. *et al.* (1981) Differential ontogeny of type 1 and type 2 benzodiazepine receptors. *Life Sci.*, **28**, 2343–7.

Lo, M.M.S., Niehoff, D.L., Kuhar, M.J. and Snyder, S.H. (1983) Differential localization of type I and type II benzodiazepine binding sites in substantia nigra. *Nature*, **306**, 57–60.

Lüddens, H., Pritchett, D.B., Köhler, M. *et al.* (1990) Cerebellar GABA_A receptor selective for a behavioral alcohol antagonist. *Nature*, **346**, 648–51.

Malherbe, P., Sigel, E., Baur, R. *et al.* (1990) Functional characteristics and sites of gene expression of the α_1, β, γ_2-isoform of the rat GABA_A receptor. *J. Neurosci.*, **10**, 2330–7.

McCabe, R.T. and Wamsley, J.K. (1986) Autoradiographic localization of subcomponents of the macromolecular GABA receptor complex. *Life Sci.*, **39**, 1937–46.

McCarthy, K.D. and Harden, T.K. (1981) Identification of two benzodiazepine binding sites on cells cultured from rat cerebral cortex. *J. Pharmacol. Exp. Ther.*, **216**, 183–91.

Mehta, A.K. and Ticku, M.K. (1988) Developmental aspects of benzodiazepine receptors and GABA-gated chloride channels in primary cultures of spinal cord neurons. *Brain Res.*, **454**, 156–63.

Meier, E., Drejer, J. and Schousboe, A. (1984) GABA induces functionally active low-affinity GABA receptors on cultured cerebellar granule cells. *J. Neurochem.*, **43**, 1737–44.

Meinecke, D.L. and Rakic, P. (1990) Developmental expression of GABA and subunits of the GABA_A receptor complex in an inhibitory synaptic circuit in the rat cerebellum. *Dev. Brain Res.*, **55**, 73–86.

Möhler, H. and Okada, T. (1977) Benzodiazepine receptor: demonstration in the central nervous system. *Science*, **198**, 849–51.

Mower, G.D., Rustad, R. and Frost White, W. (1988) Quantitative comparisons of GABA neurons and receptors in the visual cortex of normal and dark-reared cats. *J. Comp. Neurol.*, **272**, 293–302.

Olsen R.W. and Tobin, A.J. (1990) Molecular biology of GABA_A receptors. *FASEB J.*, **4**, 1469–80.

Palacios, J.M. and Kuhar, M. (1982) Ontogeny of high affinity GABA and benzodiazepine receptors in the rat cerebellum: an autoradiographic study. *Dev. Brain Res.*, **2**, 531–9.

Palacios, J.M., Niehoff, D. L. and Kuhar, M. J. (1979) Ontogeny of GABA and benzodiazepine receptors: effects of Triton X-100, bromide and muscimol. *Brain Res.*, **179**, 390–5.

Palacios, J.M., Wamsley, J.K. and Kuhar, M.J. (1981) High affinity GABA receptors – autoradiographic localization. *Brain Res.*, **222**, 285–307.

Palacios, J.M., Cortés, R. and Probst, A. (1987) Receptor plasticity in the human brain: some autoradiographic studies. *J. Recept. Res.*, **7**, 581–97.

Park, D. and de Blas, A.L. (1991a) Peptide subunits of γ-aminobutyric acid_A/benzodiazepine receptors from bovine cerebral cortex. *J. Neurochem.*, **56**, 1972–9.

Park, D. and de Blas, A.L. (1991b) Peptide heterogeneity of GABA_A/benzodiazepine receptors in bovine cerebral cortex and cerebellum. *Brain Res.*, **550**, 279–86.

Park, D., Vitorica, J., Tous, G. and de Blas, A.L. (1991) Purification of the γ-aminobutyric acid_A/benzodiazepine receptor complex by immunoaffinity chromatography. *J. Neurochem.*, **56**, 1962–71.

Pritchett, D.B., Sontheimer, H., Shivers, B.D. *et al* (1989a) Importance of a novel GABA_A receptor subunit for benzodiazepine pharmacology. *Nature*, **338**, 582–5.

Pritchett, D.B., Lüddens, H. and Seeburg, P. (1989b) Type I and type II GABA_A-benzodiazepine receptors produced in transfected cells. *Science*, **245**, 1389–92.

Pritchett, D.B. and Seeburg, P.H. (1990) γ-Aminobutyric acid_A receptor α_5-subunit creates novel type II benzodiazepine receptor pharmacology. *J. Neurochem.*, **54**, 1802-4.

Reiter, H.O. and Stryker, M.P. (1988) Neural plasticity without postsynaptic action potentials: less-active inputs become dominant when kitten visual cortical cells are pharmacologically inhibited. *Proc. Natl. Acad. Sci. USA*, **85**, 3623–7.

Richards, J.G., Schoch, P., Haring, P. *et al.* (1987) Resolving GABA$_A$/benzodiazepine receptors: cellular and subcellular localization in the CNS with monoclonal antibodies. *J. Neurosci.*, **7**, 1866–86.

Rotter, A. and Forstholm, A. (1986) Cerebellar benzodiazepine receptor distribution: an autoradiographic study of the normal C57BL/6J and Purkinje cell degeneration mutant mouse. *Neurosci. Lett.*, **71**, 66–71.

Rotter, A. and Frostholm, A. (1988) Cerebellar benzodiazepine receptors: cellular localization and consequences of neurological mutations in mice. *Brain Res.*, **444**, 133–46.

Rotter, A., Gorenstein, C. and Frostholm, A. (1988) The localization of GABA$_A$ receptors in mice with mutations affecting the structure and connectivity of the cerebellum. *Brain Res.*, **439**, 236–48.

Schliebs, R. and Rothe, T. (1988) Development of GABA$_A$ receptor in the cortical visual structures of rat brain. Effect of visual pattern deprivation. *Gen. Physiol. Biophys.*, **7**, 281–92.

Schliebs, R., Rothe, T. and Bigl, V. (1986) Dark rearing affects the development of benzodiazepine receptors in the central visual structures of rat brain. *Dev. Brain Res.*, **24**, 179–85.

Schlumpf, M., Richards, J.G., Lichtensteiger, W. and Möhler, H. (1983) An autoradiographic study of the prenatal development of benzodiazepine-binding sites in rat brain. *Neuroscience*, **3**, 1478–87.

Schofield, P.R., Darlison, M.G., Fujita, M. *et al.* (1987) Sequence and functional expression of the GABA$_A$ receptor shows a ligand-gated receptor super-family. *Nature*, **328**, 221–7.

Schwartz, R.D. (1988) The GABA$_A$ receptor-gated ion channel: biochemical and pharmacological studies of structure and function. *Biochem. Pharmacol.*, **37**, 3369–75.

Shaw, C. and Cynader, M. (1988) Unilateral eyelid suture increases GABA$_A$ receptors in cat visual cortex. *Dev. Brain Res.*, **40**, 148–53.

Shaw, C., Aoki, C., Wilkinson, M. *et al.* (1987) Benzodiazepine ([³H]flunitrazepam) binding in cat visual cortex: ontogenesis of normal characteristics and the effects of dark rearing. *Dev. Brain Res.*, **37**, 67–76.

Shaw, C., Needler, M.C., Wilkinson, M. *et al.* (1984) Alterations in the receptor number, affinity and laminar distribution in cat visual cortex during the critical period. *Prog. Neuro-Psychopharmacol. Biol. Psychiat.*, **8**, 627–34.

Shaw, C., Needler, M.C., Wilkinson, M. *et al.* (1986) Modification of neurotransmitter receptor sensitivity in cat visual cortex during the critical period. *Dev. Brain Res.*, **22**, 67–73.

Shivers, B.D., Killisch, I., Sprengel, R. *et al.* (1989) Two novel GABA$_A$ receptor subunits exist in distinct neuronal subpopulations. *Neuron*, **3**, 327–37.

Siegel, R.E. (1988) The mRNA encoding GABA$_A$/benzodiazepine receptor subunits are localized in different cell populations of the bovine cerebellum. *Neuron*, **1**, 579–84.

Sieghart, W. and Drexler, G. (1983) Irreversible binding of [³H]flunitrazepam to different proteins in various brain regions. *J. Neurochem.*, **41**, 47–55.

Sieghart, W. and Karobath, M. (1980) Molecular heterogeneity of benzodiazepine receptor. *Nature*, **286**, 285–7.

Sieghart, W., Mayer, A. and Drexler, G. (1983) Properties of [³H]flunitrazepam binding to different benzodiazepine binding proteins. *Eur. J. Pharmacol.*, **88**, 291–9.

Sigel, E. and Barnard, E.A. (1984) A gamma-aminobutyric acid/benzodiazepine receptor complex from bovine cerebral cortex. Improved purification with preservation of regulatory sites and their interaction. *J. Biol. Chem.*, **259**, 7219–23.

Sigel, E., Baur, R., Trube, G. *et al.* (1990) The effect of subunit composition of rat brain GABA$_A$ receptors on channel function. *Neuron*, **5**, 703–11.

Skangiel-Kramska, J. and Kossut, M. (1984) Increase of GABA receptor binding activity after short lasting monocular deprivation in kittens. *Acta Neurobiol. Exp.*, **44**, 33–9.

Squires, R.F., Saederup, E., Damgaard, I. and Schousbae, A. (1990) Development of benzodiazepine and pictrotopin (*t*-butylbicyclophorothionate) binding sites in rat cerebellar granule cells in culture. *Neurochem.*, **54**, 473–8.

Syapin, P.J., Cole, R., de Vellis, J. *et al.* (1985) Benzodiazepine binding characteristics of embryonic rat brain neurons grown in culture. *J. Neurochem.*, **45**, 1797–801.

Taguchi, J.-I. and Kuriyama, K. (1984) Purification of γ-aminobutyric acid (GABA) receptor from rat brain by affinity column chromatography using a new benzodiazepine 1012-S, as an immobilized ligand. *Brain Res.*, **323**, 219–26.

Taylor, J. and Gordon-Weeks, P.R. (1989) Developmental changes in the calcium dependency of γ-aminobutyric acid release from isolated growth cones: correlation with growth cone morphology. *J. Neurochem.*, **53**, 834–43.

Taylor, J., Docherty, M. and Gordon-Weeks, P.R. (1990) GABAergic growth cones: release of endogenous GABA precedes the expression of synaptic vesicle antigens. *J. Neurochem.*, **54**, 1689–99.

Unnerstall, J.R., Kuhar, M.J., Niehoff, D.L. and Palacios, J.M. (1981) Benzodiazepine receptors are coupled to a subpopulation of GABA receptors: evidence from a quantitative autoradiographic study. *J. Pharmacol. Exp. Ther.*, **218**, 797–804.

Verdoorn, T.A., Draguhn, A., Ymer, S. *et al.* (1990) Functional properties of recombinant rat GABA$_A$ receptors depend upon subunit composition. *Neuron*, **4**, 919–28.

Vitorica, J., Park, D. and de Blas, A.L. (1987) Immunochemical localization of the GABA$_A$ receptor in the rat brain. *Eur.J. Pharmacol.*, **136**, 451–3.

Vitorica, J., Park, D., Chin, G. and de Blas, A.L. (1988) Monoclonal antibodies and conventional antisera to the γ-aminobutyric acid$_A$/benzodiazepine receptor/Cl⁻ channel complex. *J. Neurosci.*, **8**, 615–22.

Vitorica, J., Park, D., Chin G. and de Blas, A.L. (1990a) Characterization with antibodies of the GABA/benzodiazepine receptor complex during development of the rat brain. *J. Neurochem.*, **54**, 187–94.

Vitorica, J., Park, D. and de Blas, A.L. (1990b) The GABA$_A$/benzodiazepine receptor complex in rat brain neuronal cultures. Characterization by immunoprecipitation. *Brain Res.*, **537**, 209–15.

White, W. F., Dichter, M. A. and Snodgrass, S. N. (1981) Benzodiazepine binding and interactions with the GABA receptor complex in living cultures of rat cerebral cortex. *Brain Res.*, **215**, 162–76.

Wisden, W., Moris, B.J., Darlison, M. G. *et al.* (1988) Distinct GABA$_A$ receptor alpha-subunit mRNAs show differential patterns of expression in bovine brain. *Neuron*, **1**, 937–47.

Wisden, W., Moris, B. J., Darlison, M. G. *et al* (1989) Localization of GABA$_A$ receptor alpha-subunit mRNAs in relation to receptor subtypes. *Mol. Brain Res.*, **5**, 305–10.

Yazulla, S., Studholme, K.M., Vitorica, J. and De Blas, A.L. (1989) Immunochemical localization of GABA$_A$ receptors in goldfish and chicken retinas. *J. Comp. Neurol.*, **280**, 15–26.

Ymer, S., Schofield, P.R., Draguhn, A. *et al.* (1989) GABA$_A$ receptor beta subunit heterogeneity: functional expression of cloned cDNAs. *EMBO J.*, **8**, 1665–70.

Ymer, S., Draguhn, A., Wisden, W. *et al.* (1990) Structural and functional characterization of the γ$_1$ subunit of GABA/benzodiazepine receptors. *EMBO J.*, **9**, 3261–7.

Zdilar, D., Rotter, A. and Frostholm, A. (1992) Expression of GABA$_A$/benzodiazepine receptor α$_1$-subunit mRNA and [³H]flunitrazepam binding sites during postnatal development of the mouse cerebellum. *Dev. Brain Res.*, **61**, 63–71.

GABA AND BENZODIAZEPINE RECEPTORS IN THE DEVELOPING VISUAL SYSTEM

Reinhard Schliebs and Thomas Rothe

7.1 INTRODUCTION

For transmitting visual information from the retina via the thalamus (lateral geniculate nucleus, superior colliculus) to the cortex, visual pathways use excitatory synaptic signals (Schiller, 1986). γ-Aminobutyric acid (GABA), acting on both type A and type B receptors, is the major inhibitory transmitter in the visual system, and mediates fast synaptic inhibition by activating a chloride channel. The type A GABA receptor belongs to a super-family of ligand-gated ion channels including glycine, glutamate and nicotinic acetylcholine receptors and is a hetero-oligomeric complex which comprises binding sites for GABA and for its allosteric modulators (benzodiazepines, barbiturates) together with the integral chloride ion channel. The native $GABA_A$ receptor is postulated to have a pentameric structure composed of various combinations of at least four subunits: α, β, γ and δ. Further diversity within each class (α_1, ..., α_6; β_1, ..., β_3; γ_1, γ_2) has been reported, and thus regional variations in the expression of distinct $GABA_A$ subunits have been observed (for review, see Silvilotti and Nistri, 1991). An additional GABA receptor subunit, termed GABA p_1-receptor, has recently been identified that demonstrates unique pharmacological properties and is primarily expressed in the retina (Cutting *et al.*, 1991).

Baclofen, a GABA receptor agonist, binds selectively to a subpopulation of GABA recognition sites which are termed $GABA_B$ receptors. An unusual characteristic of $GABA_B$ binding is the absolute requirement for Ca^{2+}. The main effect of baclofen via $GABA_B$ receptors is to inhibit presynaptically the release of excitatory transmitters by depressing Ca^{2+} currents probably mediated via a GTP-binding protein, whereas the postsynaptic action of baclofen is an activation of K^+ currents. However, the postsynaptic action of baclofen is nearly negligible when compared to its presynaptic action (Silvilotti and Nistri, 1991).

In most areas of the brain there is a predominance of $GABA_A$ sites over $GABA_B$ ones (exceptionally the molecular layer of cerebellum and interpeduncular nucleus of mesencephalon), although the ratio of $GABA_A$ and $GABA_B$ receptors varies between different brain regions. The functional significance of various ratios of $GABA_A$ and $GABA_B$ sites in the brain is still unclear.

Electrophysiological measurements disclosed an unusual GABA receptor in the visual system of frog and guinea pig. Recordings of

Receptors in the Developing Nervous System Vol. 2: Neurotransmitters. Edited by Ian S. Zagon and Patricia J. McLaughlin. Published in 1993 by Chapman & Hall. ISBN 0 412 49400 0. Vols. 1 and 2 (set) ISBN 0 412 54520 9.

excitatory postsynaptic potentials (EPSPs) in the frog optic tectum by optic nerve stimulation showed that GABA enhances excitatory mechanisms in this brain area (Nistri and Silvilotti, 1985). Application of a very high concentration of GABA in guinea-pig superior colliculus resulted in inhibition, whereas an excitatory effect was observed after application of much lower doses. The sustained excitatory response to GABA displays unusual pharmacological characteristics which suggest the existence of a novel receptor type distinct from the types A and B (for review, see Silvilotti and Nistri, 1991).

The presence of GABAergic local circuit neurons within the visual thalamic nuclei and the visual cortex (VC) of mammals has been disclosed in a number of studies (Ribak, 1978; Sterling and Davis, 1980; McDonald *et al.*, 1981; Hendrickson *et al.*, 1983; Ohara *et al.*, 1983; Somogyi *et al.*, 1983; Fitzpatrick *et al.*, 1984; Montero and Singer, 1984; Penny *et al.*, 1984; Giolli *et al.*, 1985; Gabbott *et al.*, 1986; Somogyi, 1989). GABA-releasing structures are assumed to gate and modify the responses evoked at the postsynaptic cell by the excitatory signals. This is emphasized by the findings that GABA in the visual system is involved in such fundamental phenomena as center-surround inhibition, directional and orientation selectivity, and binocular interaction (Kanno and Okada, 1988; Somogyi, 1989).

Neurotransmitter receptors are one of the decisive links in the chain of synaptic information processing. Due to their adaptive properties in response to altered levels of neurotransmitter as a consequence of changes in neuronal activity (Schwartz *et al.*, 1983), they are of particular importance for functional-adaptive processes occurring during the maturation of the visual system with respect to pattern recognition. GABA$_A$ receptors represent the target structures which receive GABAergic input and are partly associated with benzodiazepine receptors (Stephenson, 1988). The benzodiazepine receptors are thought to play a neuromodulatory role by modifying the activities of a certain population of GABA receptors (Costa, 1988). On the basis of the regulatory properties of benzodiazepines and barbiturates on GABA binding, GABA$_A$ receptors have been further subdivided: GABA$_{A1}$ receptors can be modulated by benzodiazepines and barbiturates, whereas GABA$_{A2}$ receptors are insensitive to the action of these agents (Friedman and Redburn, 1990).

The activation of GABA$_A$ receptors, which are localized mainly on the postsynaptic cell, leads to an inhibition of the excitatory response on the postsynaptic cell. This precise interaction of excitatory and inhibitory signals seems to be a major step in efficient processing of visual information. Thus the question arises as to whether the functional maturation of the GABA receptors in the visual system during an early period of postnatal life is driven by the presence of adequate visual experience. Therefore, in the following section the maturation of GABA$_A$ and benzodiazepine receptors in the visual regions is considered followed by a short description of the effects of visual deprivation on the GABA and benzodiazepine receptor ontogeny.

7.2 GABA AND THE VISUAL SYSTEM

7.2.1 RETINA

GABA is thought to function as a neurotransmitter in the retina. Significant levels of GABA have been detected in the inner plexiform layer where terminals of bipolar, amacrine and ganglion cells are present. There is now evidence that a subset of amacrine cells utilizes GABA as their transmitter (Voaden *et al.*, 1980), and possesses a high affinity uptake and release system for GABA (Brandon *et al.*, 1979). GABAergic amacrine cells constitute a major portion of the inhibitory interneurons, and are reported to make pre- and postsynaptic contacts with bipolar cells and axon terminals and to form synapses onto amacrine cells (including the GABAergic subset) and ganglion cells. They participate in the establishment of receptive field properties like size, velocity and directional selectivity of ganglion cells and seem to be involved in forming both ON and OFF circuitries of light stimulation (e.g. Murashima *et al.*, 1990).

Binding and physiological studies suggest that both $GABA_A$ and $GABA_B$ receptors as well as benzodiazepine receptors are present in the retina, mainly localized within the inner plexiform layer. It is assumed that the p_1 GABA receptor subunit recently detected and highly expressed in the retina plays an essential role in retinal neurotransmission (Cutting *et al.*, 1991), but further functional and physiological studies must be awaited. $GABA_A$ receptors may be involved in regulation of both basal and light-dependent acetylcholine release. But $GABA_A$ receptors that are involved in the regulation of acetylcholine release do not seem to be coupled to benzodiazepine and barbiturate modulatory sites (Friedman and Redburn, 1990). $GABA_B$ receptors may participate in the regulation of GABA release in the circuitry for acetylcholine release.

In contrast to the large number of neurochemical and physiological studies of the adult retina, there are fewer data on age-related changes of the GABA system in the mammalian retina (Drago *et al.*, 1989). GABA concentration and glutamic acid decarboxylase (GAD) activity (used as a marker enzyme for GABAergic neurons) in the mouse retina were found to increase from birth until two weeks of age followed by a slight decrease at 20 weeks. GABA and GAD levels reached peaks when the maturation of the retina was complete (Murashima *et al.*, 1990). Similarly, developmental changes in GABA and benzodiazepine binding sites in rat retina have been described (Guarnieri *et al.*, 1982). However, the property of benzodiazepine receptors to be modulated by GABA was steadily lost with age indicating age-dependent functional changes of GABA and benzodiazepine receptors in the retina. It has been shown in the rat retina that $GABA_A$ binding sites increase markedly until postnatal day (PD) 14 followed by a considerable loss of receptor sites until PD 35 by which time the adult value is reached (Fig. 7.1). However, benzodiazepine binding sites rise markedly from PD10 to PD37, persist until PD50 and then slightly decrease until adulthood (Fig. 7.1). The comparison of the developmental appearance of [³H]muscimol and [³H]flunitrazepam binding sites in the rat retina shows that the age-related increase in benzodiazepine binding sites is delayed as compared to the developmental course of $GABA_A$ receptor binding. The distinct temporal appearance of $GABA_A$ and benzodiazepine binding sites might indicate a separate, uncoupled, development of both receptor types or an age-related change in the functional properties of the $GABA_A$–benzodiazepine–chloride ionophore complex. The view of a separate development of $GABA_A$ and benzodiazepine receptors is supported by recent findings in the adult cat retina that a certain number of $GABA_A$ receptors are not coupled to the benzodiazepine binding sites (Friedman and Redburn, 1990). Iontophoretic studies revealed that the action of benzodiazepines in the adult cat retina was selective to ON-type ganglion cells, whereas in kittens aged 7–9 weeks both ON- and OFF-type cells could be influenced by benzodiazepines. This suggests that during postnatal development OFF-type ganglion cells lose GABA receptors that are linked to the benzodiazepine receptor (Robins and Ikeda, 1989). From recent molecular biological studies is it known that different combinations of the various GABA receptor subunits with different sensitivities to GABA and benzodiazepines can exist. Therefore, one might speculate that the age-dependent changes in GABA/benzodiazepine receptor with respect to binding and physiological function might be due to changes in the selective expression and/or suppression of distinct combinations of GABA receptor subunits during the postnatal development. A series of *in situ* hybridization experiments

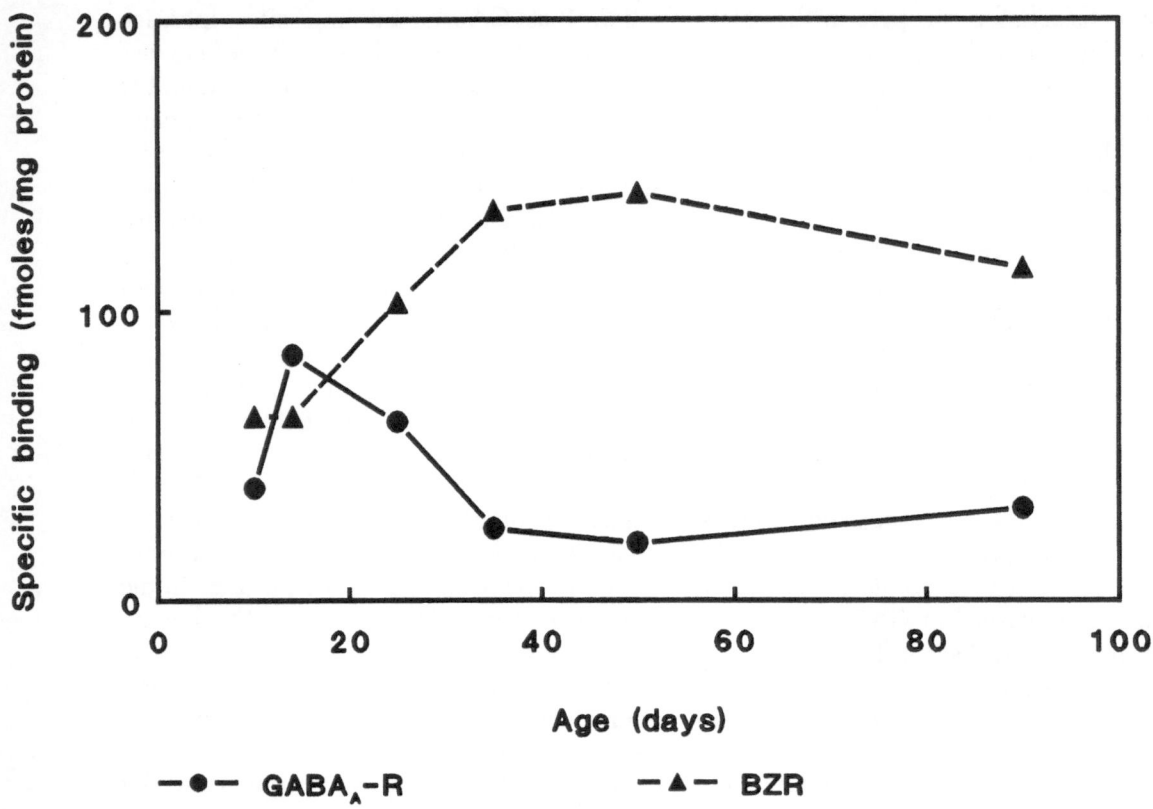

Fig. 7.1. Postnatal development of GABA$_A$ (●) and benzodiazepine (▼) receptor sites in rat retina. Receptor densities were obtained from binding studies in membrane fractions using [^{3}H]muscimol (Schliebs and Rothe, 1988) and [^{3}H]flunitrazepam (Schliebs *et al.*, 1986b) as radioligands. Binding data are expressed as fmol specifically bound radioligand per mg protein content.

performed during the period of early postnatal life should help to elucidate this question.

7.2.2   LATERAL GENICULATE NUCLEUS (LGN)

GABA-mediated inhibition in the LGN has been shown to play a major role in gating, modifying and preserving visual information from the retina. This is emphasized by the high levels of GABA and high activities of GAD found in this region (Ohara *et al.*, 1983; Kanno and Okada, 1988). The dorsal LGN has a laminated structure in many mammals. All layers contain relay cells, local inhibitory interneurons and receive two major excitatory inputs from the retina and the visual cortex

(Somogyi, 1989). Relay neurons in the dorsal LGN receive GABAergic input from axons of the perigeniculate nucleus, and from intrinsic GABAergic interneurons. Electrophysiological experiments demonstrated that GABA$_A$ receptors are involved in such processes like binocular inhibition, center-surround inhibition or spatial frequency tuning difference of X and Y cells.

The postnatal maturation of GABA$_A$ and benzodiazepine receptors in rat LGN is shown in Fig. 7.2. Binding of [^{3}H]muscimol to GABA$_A$ receptors at PD10 is relatively low, increases markedly up to PD30 and then there is a slight decrease until adulthood. In contrast, [^{3}H]flunitrazepam binding at PD10 is relatively high and decreases continuously

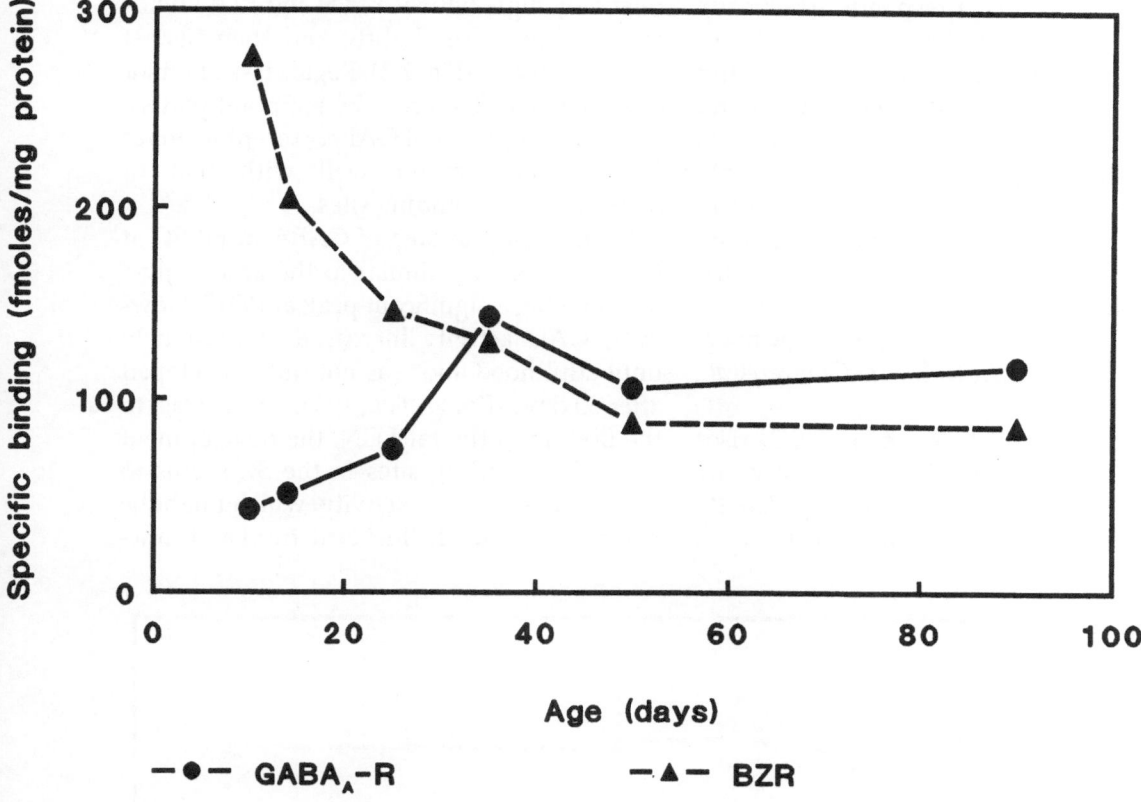

Fig. 7.2. Postnatal development of GABA$_A$ (●) and benzodiazepine (▲) receptor sites in rat lateral geniculate nucleus. Receptor densities were obtained from binding studies in membrane fractions using [^{3}H]muscimol (Schliebs and Rothe, 1988) and [^{3}H]flunitrazepam (Schliebs *et al.*, 1986b) as radioligands. Binding data are expressed as fmol specifically bound radioligand per mg protein content.

until PD90, when the adult binding level is reached (Fig. 7.2).

One problem in describing developmental patterns of small brain areas is the developmental changes in the reference system itself (Shaw *et al.*, 1984). When [^{3}H]flunitrazepam binding is expressed in terms of total binding (binding sites per LGN), then a continuous increase in benzodiazepine binding is observed (Schliebs *et al.*, 1986b). Apparently, the benzodiazepine binding sites in the LGN develop more slowly than the protein content of tissue. In the rat LGN, GAD activity reached the adult level after two weeks of postnatal life. However, the high affinity uptake of GABA, a marker for GABAergic terminals, showed a marked peak on PD15 as

compared to the adult value (Kvale *et al.*, 1983), indicating that the postnatal ontogeny of GABA$_A$ receptors is delayed in comparison to the development of the presynaptic GABAergic parameters. Similar indications for receptor formation preceding the prior formation of functional synapses were also found in ontogenetic studies in other regions (Brooksbank *et al.*, 1981) or other transmitter systems (Bylund, 1979; Coyle and Yamamura, 1976).

7.2.3 SUPERIOR COLLICULUS

The superior colliculus (SC) receives among others projections from the retina and visual cortex. It is also a laminated structure and the upper half of SC layers responds specifically to visual stimulation and has been reported to contain high concentrations of GABA and GAD activity (Fosse *et al.*, 1989; Kanno and Okada, 1988). Intrinsic GABAergic neurons in the superficial layers of SC are probably involved in shaping receptive fields and in determining orientation selectivity of other SC neurons (Stein and Gallagher, 1981). GABA is mainly localized in interneurons (Houser *et al.*, 1983; Okada, 1974) or in GABAergic fibers from substantia nigra to SC (DiChiara *et al.*, 1979).

[^{3}H]muscimol binding in rat SC rises markedly from PD10 until PD37 followed by a slight decrease until adulthood (Fig. 7.3). By PD9, [^{3}H]flunitrazepam binding in the SC is relatively high. Between PD9 and PD14 binding sites increase slightly and then persist until adulthood (Fig. 7.3). Regardless of minor quantitative differences, the temporal pattern of the development of GABA$_A$ receptors in rat SC correlates rather well with that of benzodiazepine binding sites.

High affinity uptake of GABA in rat SC at birth was already similar to the adult value, but showing a significant peak at PD15. However, GAD activity increased continuously until adulthood and was not fully developed after 30 days (Kvale *et al.*, 1983). In contrast to the finding in the rat LGN, the development of GABA$_A$ binding sites in the SC seems to precede that of GAD activity, which might be partly due to the distinct structural and func-

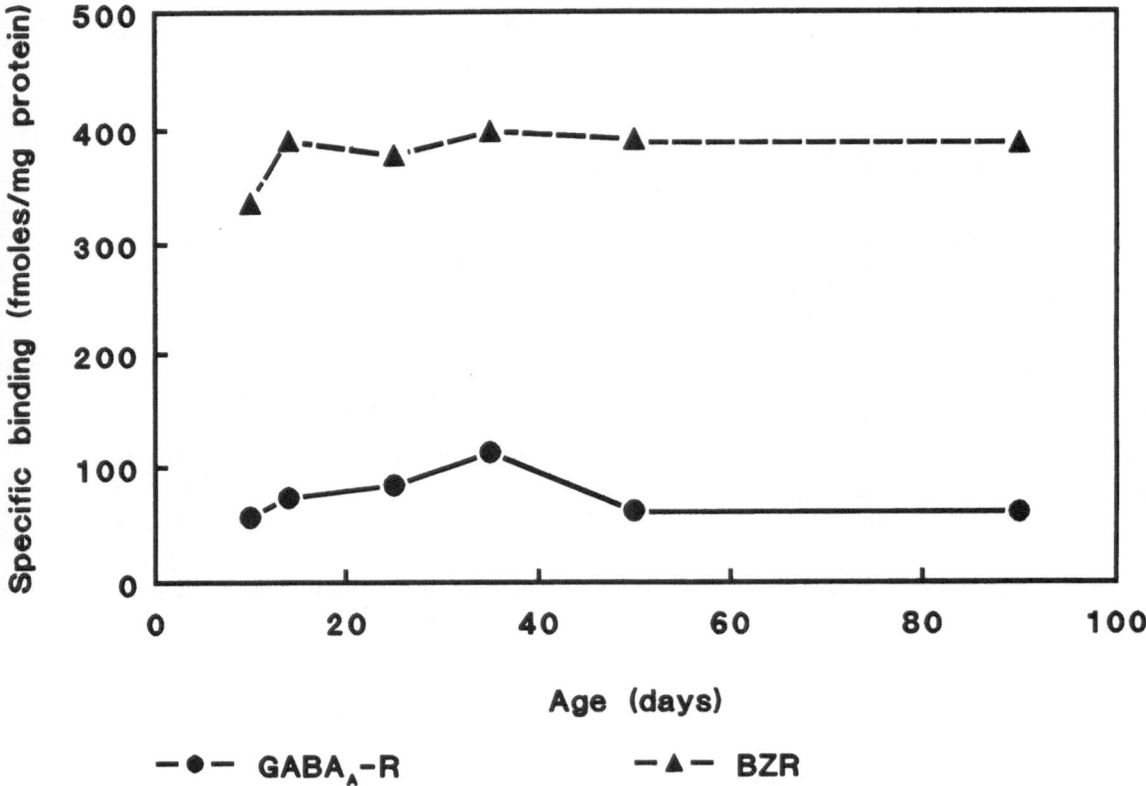

Fig. 7.3. Postnatal development of GABA$_A$ (●) and benzodiazepine (▲) receptor sites in rat superior colliculus. Receptor densities were obtained from binding studies in membrane fractions using [^{3}H]muscimol (Schliebs and Rothe, 1988) and [^{3}H]flunitrazepam (Schliebs *et al.*, 1986b) as radioligands. Binding data are expressed as fmol specifically bound radioligand per mg protein content.

tional development of SC and LGN in the rat (Kvale *et al.*, 1983).

The ontogeny of GABA$_A$ and benzodiazepine receptors in rat SC does not essentially differ from that in other brain regions suggesting that the development of GABA and benzodiazepine receptors in the SC is not correlated with the development of retinal function and the function of the visual system to respond to light stimuli.

7.2.4   VISUAL CORTEX

In the visual cortex GABA has been suggested to play a role in sharpening orientation and direction selectivity and also in determining the ocular dominance of visual cortical neurons (Mower *et al.*, 1988; Jones, 1990). A number of non-pyramidal cells within the visual cortex use GABA as an inhibitory transmitter. Several types of GABAergic neurons, differing in their synaptic connections, are present in all layers of the visual cortex (Somogyi, 1989). GABAergic nerve terminals are also present in all layers but with higher densities in layers IV, I and VI (Lin *et al.*, 1986). This pattern of GABAergic fibers correlates well with the laminar distribution of benzodiazepine receptors in rat visual cortex, but cannot be easily related to the laminar distribution of GABA$_A$ receptors (Rothe and Schliebs, 1989).

Measurements of the maximum number (B_{max}) of [^{3}H]flunitrazepam binding sites in cat visual cortex showed low values at an early age, rising to a peak in receptor density at about PD60, followed by a decline until adulthood (Shaw *et al.*, 1987). At all ages GABA altered K_d but not B_{max} of benzodiazepine binding.

Similarly, at birth, [^{3}H]muscimol binding sites in cat visual cortex are relatively low but increase continuously until postnatal week 13, after which there is a slight decrease until adulthood (Shaw *et al.*, 1984). However, GAD activity exhibited a gradual increase towards adult levels during the first month, and the adult level was reached during postnatal weeks 5–6 (Fosse *et al.*, 1989), thus indicating that the postnatal ontogeny of GABA$_A$ receptors in cat visual cortex is delayed in comparison to that of the presynaptic GABAergic element.

For both receptor populations, K_d values change during postnatal development. Affinity for [^{3}H]flunitrazepam binding was found to decline with age, whereas GABA$_A$ receptors show an initial increase in affinity until the age of 8 weeks, which was followed by a decline until adulthood (Shaw *et al.*, 1986, 1987). This is in contrast to other studies which failed to detect changes in receptor affinity during development (Brooksbank *et al.*, 1981; Guarnieri *et al.*, 1982; Patel *et al.*, 1980; Schliebs and Rothe, 1988; Skerritt and Johnston, 1982).

In order to correlate developmental changes of B_{max} and K_d Shaw *et al.* (1985) introduced the term of receptor sensitivity which is calculated from the ratio of B_{max} and K_d. The ontogeny of benzodiazepine receptors closely parallels that of GABA$_A$ receptors (Shaw *et al.*, 1984), but peak values differ: benzodiazepine binding sites peak near 60 days whereas GABA binding sites peak near PD95. But using the term of receptor sensitivity, peak receptor sensitivity for both receptor populations occurs near 60 days (Shaw *et al.*, 1986, 1987).

It is noteworthy that the peaks in binding sites of both receptor populations are reached during a period of high cortical plasticity (a critical period, which lasts in the cat until the age of about 3 months). Therefore, the peak in [^{3}H]muscimol binding sites just before the end of the critical period during which the cortex is modified may suggest a role for GABA receptors in altering neuronal response properties (Shaw *et al.*, 1984). The observed increase in receptor number during the critical period followed by a decline from the peak into adulthood, closely parallels the development of the synapse to neuron ratios in cat visual cortex, suggesting that receptors are being added or eliminated together with the

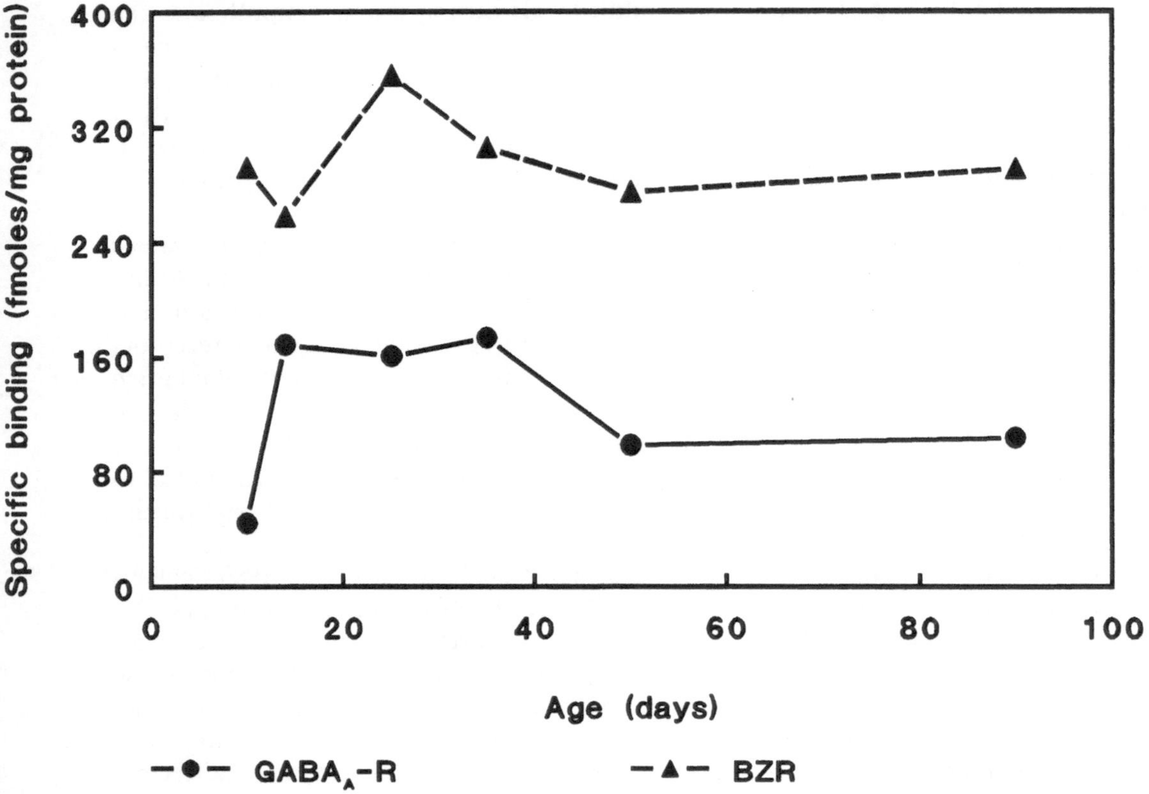

Fig. 7.4. Postnatal development of GABA$_A$ (●) and benzodiazepine (▲) receptor sites in rat visual cortex. Receptor densities were obtained from binding studies in membrane fractions using [^{3}H]muscimol (Schliebs and Rothe, 1988) and [^{3}H]flunitrazepam (Schliebs *et al.*, 1986b) as radioligands. Binding data are expressed as fmol specifically bound radioligand per mg protein content.

synapses during postnatal development (Shaw *et al.*, 1986).

The laminar distribution pattern of both GABA$_A$ and benzodiazepine receptors in cat visual cortex does not change during the entire postnatal development, in contrast to the postnatal ontogeny of muscarinic acetylcholine receptors, which shows age-dependent laminar alterations of receptor binding during an early period of postnatal life (Shaw *et al.*, 1986; Schliebs and Stewart, 1991).

Biochemical and visual behavioral studies showed that during a certain period of early life the visual system in the rat is also highly sensitive to changes of the visual input (Rothblat *et al.*, 1978; Aurich and Bigl, 1988).

In the rat VC, [^{3}H]muscimol binding rises

sharply from PD10 to PD14, remains unchanged until PD37, and then decreases markedly to PD50 when the adult value is reached (Fig. 7.4). This developmental profile is qualitatively similar to that reported previously for the rat cerebral cortex (Aldinio *et al.*, 1980; Patel *et al.*, 1980).

[^{3}H]Flunitrazepam binding to benzodiazepine receptors reached the highest level in the VC on PD25 and then decreased slightly until adulthood (Fig. 7.4). This developmental pattern is very similar to that previously reported in rat cerebral cortex (Candy and Martin, 1979; Mallorga *et al.*, 1980; Regan *et al.*, 1980; Aldinio *et al.*, 1981) and cerebellum (Palacios *et al.*, 1979).

Comparing both receptor populations, the

development of GABA$_A$ receptors precedes that of benzodiazepine receptors, suggesting a distinct maturation of GABA and benzodiazepine receptors in rat VC. This might be partly explained by the finding that in the cortex not all GABA receptors are associated with benzodiazepine receptors. In particular, high affinity GABA$_A$ receptors may not be coupled to benzodiazepine receptors (DeBlas *et al.*, 1988).

As in cat VC, in the rat VC, the peaks in binding sites of both receptor populations are reached during the critical period of development of the rat visual system, which is assumed to last until the age of about 4 weeks (Rothblat *et al.*, 1978).

Regardless of small temporal variations in the developmental profiles, the ontogeny of [^{3}H]muscimol and [^{3}H]flunitrazepam binding sites in rat VC does not essentially differ from that in other non-visual regions. This might indicate that the development of GABA and benzodiazepine binding sites in the visual system is not correlated with the development of retinal function and the functional maturation of the visual system with regard to response to light stimuli. However, it is interesting to note that the development of GABA$_A$ receptors in the rat VC seems to precede that in the subcortical visual regions.

Measurements of GAD activity in individual layers of rat VC indicated a fairly even distribution throughout the cortex, in contrast to the distinct laminar pattern of GABA and benzodiazepine receptors (Rothe and Schliebs, 1989). The activity of GAD in the rat VC increases continuously until PD35 when the adult level is reached, in contrast to the development of high affinity GABA uptake which shows a peak at PD15 (Kvale *et al.*, 1983). This indicates that the ontogeny of GABA$_A$ receptors in rat VC also precedes that of GAD activity, a finding which was also observed for the development of GABA system in rat forebrain (Coyle and Enna, 1976).

In the optic lobe of chicken, a region which corresponds to the VC of mammals, a period of high plasticity was also detected for the development of GABA receptors (Rios *et al.*, 1987). The postnatal development of GABA$_A$ receptor sites in the chick optic lobe shows an increase in binding sites during the first week post-hatching with the highest value at day 6 post-hatching. Between the second and third week GABA$_A$ receptor sites decrease to the adult level (Rios *et al.*, 1987). The descending phase of the developmental course might be explained as a consequence of changes in receptor synthesis or degradation, but it could also result from the elimination of a redundant number of synapses built up during the ascending phase, which is consistent with the hypothesis of selective synaptic stabilization (Changeux and Danchin, 1976). The developmental changes in GABA$_A$ receptor binding in chick optic lobe occur within a period of high cortical plasticity because the effect of dark-rearing on GABA$_A$ receptor binding is maximal at day 6 post-hatching, the time point at which the highest level of GABA$_A$ receptor binding was reached (see also section 7.3).

7.3 GABA AND VISUAL DEPRIVATION

Afferent neuronal activity plays an important role in the development and maturation of the visual system, particularly during a certain period of susceptibility in brain ontogeny. The importance of adequate environmental stimuli for the final tuning of the functional and structural connections during the ontogenetic period of the brain has been strongly emphasized in recent years by a number of electrophysiological, morphological and biochemical studies using the visual system as a model (for reviews, see Singer, 1985; Shaw *et al.*, 1986; Toga, 1987; Hendry and Jones, 1988; Jones, 1990). Light seems to trigger these developmental processes, and once initiated, the completion of the ontogenetic processes can occur without further light input (Mower *et al.*, 1983).

Deprivation of visual stimuli includes such diverse procedures as complete removal of visual input (dark rearing or enucleation) and removal of spatially patterned input by suturing one or both eyelids (monocular or binocular deprivation). Visual deprivation results in abnormalities in visual behavior as well as in electrophysiological and morphological differences in the nuclei of the central visual pathway compared with controls (Singer, 1985).

In cats reared under a normal light/dark cycle, GABA plays a role in determining orientation and direction tuning and ocular dominance. However, in abnormally reared cats GABA seems to play a role in the resultant abnormalities in visual cortical dominance (Burchfiel and Duffy, 1981; Sillito et al., 1981; Mower et al., 1984; Mower and Christen, 1989). In the course of normal development the afferent input is capable of inducing changes in the postsynaptic cell leading to a stabilization of the affected synapses. In this case receptors might play an essential role in mediating neuronal cortical plasticity.

Monocular deprivation of rats from PD10 to PD25 did not affect [^{3}H]muscimol binding in any of the visual regions (retina, LGN, SC, VC) examined (Rothe and Schliebs, 1989). This compares well with similar studies in other species which showed that in cats monocular deprivation also did not affect [^{3}H]muscimol binding in the visual cortex (Mower et al., 1986). These data support the suggestion that physiological light stimulation is not a prerequisite for the correct development of GABA$_A$ receptors in the visual cortex. However, results found in monocularly deprived cats are inconsistent, indicating either no effect (Mower et al., 1986) or an increase in receptor number (Shaw and Cynader, 1988) and changes in GABA$_A$ receptor affinity (Skangiel-Kramska and Kossut, 1984). These conflicting results are assumed to be largely due to differences in preparation and methodology (Shaw and Cynader, 1988). In contrast to the effect of monocular depriva-

tion in cats, a reduction in GABA$_A$ receptor density in the deprived eye columns of layer IV in the visual cortex of monkeys has been detected (Hendry et al., 1990; Jones, 1990) suggesting that GABA$_A$ receptors in adult monkeys undergo a rapid, activity-dependent change in density or structure with monocular deprivation. Thus the effect of monocular deprivation on the development of GABA$_A$ receptors in the visual cortex seems to be species-dependent and might reflect a specific response to the distinct conditions found in area 17 of a particular species (Hendry et al., 1990).

[^{3}H]Flunitrazepam binding to benzodiazepine receptors in the rat LGN, SC, and VC was not affected by monocular deprivation until PD25 (Rothe et al., 1985). Obviously, the development of GABA$_A$ and benzodiazepine receptors in the central visual regions of rats does not need the presence of visual experience, at least until the age of 25 days. This does not compare with previous investigations on the effect of monocular deprivation on alpha- and beta-adrenergic (Aurich et al., 1989), serotoninergic (Aurich et al., 1985), and glutamatergic receptors (Schliebs et al., 1986a), which demonstrated permanent changes in receptor sites particularly in the LGN.

Raising rats in complete darkness from birth until the age of 25 days resulted in decreased [^{3}H]flunitrazepam binding levels in the LGN and in the SC as compared to normally raised control rats, whereas in the retina and VC no effect of dark rearing was detected on benzodiazepine binding (Schiebs et al., 1986b). In the cat VC, however, dark rearing from birth until PD30 resulted in elevated densities of GABA$_A$ and benzodiazepine receptors which were only of transient nature (Shaw et al., 1987; Shaw and Cynader, 1988). In contrast, Mower et al. (1988) could not find any differences between normal and dark-reared cats in the number, affinity, or laminar distribution of GABA$_A$ receptors in cat VC. However, dark rearing of chicks resulted in decreased ^{3}H-GABA binding in the optic lobe which was only transient (Rios *et al.*, 1987). But when comparing different species it must be considered that there might be substantial differences in degree and manner in which the visual input can affect visual development.

Total light suppression from birth until adulthood resulted also in increased [^{3}H]serotonin binding in rat SC and VC (Aurich *et al.*, 1985), in contrast to the lack of effect on the development of alpha- and beta-adrenergic receptors in these regions (Schliebs *et al.*, 1982b; Aurich *et al.*, 1989). In the cholinergic system, total visual deprivation led to transient changes of muscarinic acetylcholine receptor binding in the rat SC and in the retina, which seems to reflect alterations in the time course of receptor development rather than permanent changes (Schliebs *et al.*, 1982a). Obviously, the development of various neurotransmitter receptors is differently affected by the complete lack of visual experience indicating that they are involved in different mechanisms of modulating visual information processing.

7.4 SUMMARY AND CONCLUSIONS

The temporal distinct appearance of GABA$_A$ and benzodiazepine receptors in the visual regions suggests that the ontogeny of the GABAergic transmission is controlled in a complex fashion and cannot be easily related to the functional maturation of the visual system. In nearly all visual regions, both receptor populations declined from a peak to a lower binding level in adulthood. The peaks in binding sites of both receptor populations are reached during a period of high cortical plasticity (critical period) and may suggest a role for GABA receptors in altering neuronal response properties. The descending phase of the developmental course might be the consequence of changes in receptor synthesis or elimination of a redundant number of synapses built up during the ascending phase, which is consistent with the hypothesis

of selective synaptic stabilization (Changeux and Danchin, 1976).

Visual deprivation differently affects the ontogeny of GABA receptors in the visual regions of various species reflecting species-dependent differences in which the visual input can influence the maturation of the visual GABAergic system.

Recently, in the visual system some GABA receptor types have been disclosed which showed unusual characteristics with respect to binding and physiology. But whether they will have special functions for visual information processing needs further investigation.

Molecular biological studies revealed a complex pentameric structure of the GABA/benzodiazepine receptor and a high diversity in each class of subunits. Different combinations of the various GABA receptor subunits with differing sensitivity to GABA and benzodiazepines can exist, but no data on developmental characteristics of the various subunits are available at present. Therefore, one might speculate that the age-dependent changes in GABA/benzodiazepine receptor with respect to binding and physiological function might be due to changes in expression and/or suppression of certain combinations of GABA receptor subunits during postnatal development. *In situ* hybridization experiments performed during the period of early life should help to elucidate this question.

REFERENCES

Aldinio, C., Balzano, M. and Toffano, G. (1980) Ontogenetic development of GABA recognition sites in different brain areas. *Pharmacol. Res. Commun.*, **12**, 495–500.

Aldinio, C., Balzano, M., Savoini, G. *et al.* (1981) Ontogeny of ^{3}H-diazepam binding sites in different rat brain areas. Effect of GABA. *Dev. Neurosci.*, **4**, 461–6.

Aurich, M. and Bigl, V. (1988) A critical period of the development of β-adrenergic receptor binding in the visual system of rat during visual deprivation. *Int. J. Dev. Neurosci.*, **6**, 351–7.

Aurich, M., Schliebs, R. and Bigl, V. (1985) Serotoninergic receptors in the visual system of light-deprived rats. *Int. J. Dev. Neurosci.*, **3**, 285–90.

Aurich, M., Schliebs, R., Stewart, M.G. *et al.* (1989) Adaptive changes in the central noradrenergic system in monocular deprived rats. *Brain Res. Bull.*, **22**, 173–80.

Brandon, C., Lam, D.M.K. and Wu, J.Y. (1979) The γ-amino butyric acid system in the rabbit retina: localization by immunocytochemistry and autoradiography. *Proc. Natl. Acad. Sci. USA*, **76**, 3557–61.

Brooksbank, B.W.L., Atkinson, D.J. and Balasz, R. (1981) Biochemical development of human brain. II. Some parameters of the GABAergic system. *Dev. Neurosci.*, **4**, 188–200.

Burchfiel, J.L. and Duffy, F.H. (1981) Role of intracortical inhibition in deprivation amblyopia: reversal by microiontophoretic bicuculline. *Brain Res.*, **206**, 479–84.

Bylund, D.B. (1979) Regulation of central adrenergic receptors, in *Modulators, Mediators, and Specifiers in Brain Function, Advances in Experimental Medicine and Biology*, Vol. 16 (eds Y.H. Ehrlich, J. Volavka, L.G. Davis and E.G. Brunngraber), Plenum Press, New York, pp. 133–62.

Candy, J.M. and Martin, I.L. (1979) The postnatal development of benzodiazepine receptor in the cerebral cortex and cerebellum of the rat. *J. Neurochem.*, **32**, 655–8.

Changeux, J.P. and Danchin, A. (1976) Selective stabilization of developing synapses as a mechanism for the specification of neuronal networks. *Nature*, **264**, 705–12.

Costa, E. (1988) Polytypic signaling at GABAergic synapses. *Life Sci.*, **42**, 1407–17.

Coyle, J.T. and Enna, S.J. (1976) Neurochemical aspects of the ontogenesis of GABAergic neurons in the rat brain. *Brain Res.*, **111**, 119–33.

Coyle, J.T. and Yamamura, H.I. (1976) Neurochemical aspects of the ontogenesis of cholinergic neurons in the rat brain. *Brain Res.*, **118**, 429–40.

Cutting, G.R., Lu, L., O'Hara, B.F. *et al.* (1991) Cloning of the γ-aminobutyric acid (GABA) p_1 cDNA: a GABA receptor subunit highly expressed in the retina. *Proc. Natl. Acad. Sci. USA*, **88**, 2673–7.

deBlas, A.L., Vitorica, J. and Friedrich, P. (1988) Localization of the $GABA_A$ receptor in the rat brain with a monoclonal antibody to the 57,000 M_r peptide of the $GABA_A$ benzodiazepine receptor/Cl⁻ channel complex. *J. Neurosci.*, **8**, 602–614.

DiChiara, G., Porceddu, M.L., Morelli, M. *et al.* (1979) Evidence for GABAergic projections from the substantia nigra to the ventromedial thalamus and to the superior colliculus of the rat. *Brain Res.*, **176**, 273–84.

Drago, F., Gagliano, C. and Cavaliere, S. (1989) Aging related changes of neurotransmitter in the visual system. *Metab. Pediatr. Syst. Ophthalmol.*, **12**, 21–3.

Fitzpatrick, D., Penny, G.R. and Schmechel, D.E. (1984) Glutamic acid decarboxylase-immunoreactive neurons and terminals in the lateral geniculate nucleus of the cat. *J. Neurosci.*, **4**, 1809–29.

Fosse, V.M., Heggelund, P. and Fonnum, F. (1989) Postnatal development of glutamatergic, GABAergic, and cholinergic neurotransmitter phenotypes in the visual cortex, lateral geniculate nucleus, pulvinar, and superior colliculus in cats. *J. Neurosci.*, **9**, 426–35.

Friedman, D.L. and Redburn, D.A. (1990) Evidence for functionally distinct subclasses of γ-aminobutyric acid receptors in rabbit retina. *J. Neurochem.*, **55**, 1189–99.

Gabbott, P.L.A., Somogyi, J., Stewart, M.G. and Hámori, J. (1986) GABA-immunoreactive neurons in the dorsal lateral geniculate nucleus of the rat: characterization by combined Golgi-impregnation and immunocytochemistry. *Exp. Brain Res.*, **61**, 311–22.

Giolli, R.A., Peterson, C.E., Ribak, C.E. *et al.* (1985) GABAergic neurons comprise a major cell type in rodent visual nuclei: an immunocytochemical study of pretectal and accessory optic nuclei. *Exp. Brain Res.*, **61**, 194–203.

Guarnieri, F., Corda, M.G., Concas, A. *et al.* (1982) Age-related changes of benzodiazepine and GABA binding sites in the rat retina. *Neurobiol. Aging*, **3**, 227–31.

Hendrickson, A.F., Ogren, M.P., Vaughn, J.E. *et al.* (1983) Light and electron microscopic immunocytochemical localization of glutamic acid decarboxylase in monkey geniculate complex: evidence for GABAergic neurons and synapses. *J. Neurosci.*, **3**, 1245–62.

Hendry, S.H.C. and Jones, E.G. (1988) Activity-dependent regulation of GABA expression in the visual cortex of adult monkeys. *Neuron*, **1**, 701–12.

Hendry, S.H.C., Fuchs, J., deBlas, A.L. and Jones, E.G. (1990) Distribution and plasticity of immunocytochemically localized $GABA_A$ receptors in adult monkey visual cortex. *J. Neurosci.*, **10**, 2438–50.

Houser, C.R., Lee, M. and Vaughn, J.E. (1983) Immunocytochemical localization of glutamic acid decarboxylase in normal and deafferented superior colliculus: evidence for reorganization of γ-aminobutyric acid synapses. *J. Neurosci.*, **3**, 2030–42.

Jones, E.G. (1990) The role of afferent activity in the maintenance of primate neocortical function. *J. Exp. Biol.*, **153**, 155–76.

Kanno, S. and Okada, Y. (1988) Laminar distribution of GABA (γ-aminobutyric acid) in the dorsal lateral geniculate nucleus, area 17 and area 18 of the visual cortex, and the superior colliculus of the cat. *Brain Res.*, **451**, 172–8.

Kvale, I., Fosse, V.M. and Fonnum, F. (1983) Development of neurotransmitter parameters in lateral geniculate body, superior colliculus and visual cortex of the albino rat. *Dev. Brain Res.*, **7**, 137–45.

Lin, C.S., Lu, S.M. and Schmechel, D.E. (1986) Glutamic acid decarboxylase and somatostatin immunoreactivities in rat visual cortex. *J. Comp. Neurol.*, **244**, 369–83.

Mallorga, P., Hamburg, M., Tallman, J.F. and Gallager, D.W. (1980) Ontogenetic changes in GABA modulation of brain benzodiazepine binding. *Neuropharmacology*, **19**, 405–8.

McDonald, J.K., Speciale, S.G. and Parnavelas, J.G. (1981) The development of glutamic acid decarboxylase in the visual cortex and the dorsal lateral geniculate nucleus of the rat. *Brain Res.*, **217**, 364–7.

Montero, V. and Singer, W. (1984) Ultrastructure and synaptic relations of neural elements containing glutamic acid decarboxylase (GAD) in the perigeniculate nucleus of the cat. *Exp. Brain Res.*, **56**, 115–25.

Mower, G.D. and Christen, W.G. (1989) Evidence for enhanced role of GABA inhibition in visual cortical dominance of cats reared with abnormal monocular experience. *Dev. Brain Res.*, **45**, 211–18.

Mower, G.D., Christen, W.G. and Caplan, C.J. (1983) Very brief visual experience eliminates plasticity in the cat visual cortex. *Science*, **221**, 178–80.

Mower, G.D., Christen, W.G., Burchfiel, J.L. and Duffy, F.H. (1984) Microiontophoretic bicuculline restores binocular responses to visual cortical neurons in strabismic cats. *Brain Res.*, **309**, 168–72.

Mower, G.D., White, W.F. and Rustad, R. (1986) ^{3}H-muscimol binding of GABA receptors in the visual cortex of normal and monocularly deprived cats. *Brain Res.*, **380**, 253–60.

Mower, G.D., Rustad, R. and White, W.F. (1988) Quantitative comparisons of gamma-aminobutyric neurons and receptors in the visual cortex of normal and dark-reared cats. *J. Comp. Neurol.*, **272**, 293–302.

Murashima, Y.L., Ishikawa, T. and Kato, T. (1990) γ-Aminobutyric acid system in developing and degenerating mouse retina. *J. Neurochem.*, **54**, 893–8.

Nistri, A. and Silvilotti, L. (1985) An unusual effect of γ-aminobutyric acid on synaptic transmission on frog tectal neurones *in vitro*. *Br. J. Pharmacol.*, **85**, 917–22.

Ohara, P.T., Liebermann, A.R., Hunt, S.P. and Wu, J.Y. (1983) Neural elements containing glutamic acid decarboxylase (GAD) in the dorsal lateral geniculate nucleus of the rat: immunocytochemical studies by light and electron microscopy. *Neuroscience*, **8**, 189–211.

Okada, Y. (1974) Distribution of γ-amino butyric acid (GABA) in the layers of superior colliculus of the rabbit. *Brain Res.*, **75**, 362–365.

Palacios, J.M., Niehoff, D.L. and Kuhar, M.J. (1979) Ontogeny of GABA and benzodiazepine receptors: effect of Triton X-100, bromide and muscimol. *Brain Res.*, **179**, 390–5.

Patel, A.J., Smith, R.M., Kingsbury, A.E. *et al.* (1980) Effects of thyroid state on brain development: muscarinic acetylcholine and GABA receptors. *Brain Res.*, **198**, 389–402

Penny, G.R., Conley, M., Schmechel, D.E. and Diamond, I.T. (1984) The distribution of glutamic acid decarboxylase immunoreactivity in the diencephalon of the opposum and rabbit. *J. Comp. Neurol.*, **228**, 38–56.

Regan, J.W., Roeske, W.R. and Yamamura, H.I. (1980) The benzodiazepine receptor: its development and its modulation by γ-amino butyric acid. *J. Pharmacol. Exp. Ther.*, **212**, 137–43.

Ribak, C.E. (1978) Aspinous and sparsely-spinous stellate neurons in the visual cortex of rats contain glutamic acid decarboxylase. *J. Neurocytol.*, **7**, 461–78.

Rios, H., Flores, V. and Fiszer de Plazas, S. (1987) Effects of light- and dark-rearing on the postnatal development of GABA receptor sites in the chick optic lobe. *Int. J. Dev. Neurosci.*, **5**, 319–25.

Robins, J. and Ikeda, H. (1989) Benzodiazepines in the mammalian retina. I. Autoradiographic localisation of receptor sites and the lack of effect on the electroretinogram. *Brain Res.*, **479**, 313–22.

Rothblat, L.A., Schwarts, M.L. and Kasdan, P.M. (1978) Monocular deprivation in the rat: evidence for an age-related defect in visual behavior. *Brain Res.*, **158**, 456–60.

Rothe, T. and Schliebs, R. (1989) Laminar distribution of benzodiazepine receptors in visual cortex of adult rat. *Gen. Physiol. Biophys.*, **8**, 371–80.

Rothe, T., Schliebs, R. and Bigl, V. (1985) Benzodiazepine receptors in the visual structures of monocularly deprived rats. Effect of light and dark adaptation. *Brain Res.*, **329**, 143–50.

Schiller, P.H. (1986) The central visual system. *Vision Res.*, **26**, 1351–86.

Schliebs, R. and Rothe, T. (1988) Development of GABA_A receptors in the central visual structures of rat brain. Effect of visual pattern deprivation. *Gen. Physiol. Biophys.*, **7**, 281–92.

Schliebs, R. and Stewart, M.G. (1991) Laminar postnatal development of muscarinic cholinergic receptors in rat visual cortex and the effect of monocular deprivation. *Neurochem. Int.*, **19**, 143–9.

Schliebs, R., Bigl, V. and Biesold, D. (1982a) Development of muscarinic cholinergic receptor binding in the visual system of monocularly deprived and dark reared rats. *Neurochem. Res.*, **7**, 1181–98.

Schliebs, R., Burgoyne, R.D. and Bigl, V. (1982b) The effect of visual deprivation on β-adrenergic receptors in the visual centres of the rat brain. *J. Neurochem.*, **38**, 1038–43.

Schliebs, R., Kullmann, E. and Bigl, V. (1986a) Development of glutamate binding sites in the visual centres of rat brain. Effect of light deprivation. *Biomed. Biochim. Acta*, **45**, 495–506.

Schliebs, R., Rothe, T. and Bigl, V. (1986b) Dark-rearing affects the development of benzodiazepine receptors in the central visual structures of rat brain. *Dev. Brain Res.*, **24**, 179–85.

Schwartz, J.C., Cortes, C.L., Rose, C. *et al.* (1983) Adaptive changes of neurotransmitter receptor mechanisms in the central nervous system, in *Molecular and Cellular Interactions Underlying Higher Brain Functions, Progress in Brain Research*, Vol. 58 (eds J.–P. Changeux *et al.*), Elsevier, Amsterdam, pp. 117–29.

Shaw, C. and Cynader, M. (1988) Unilateral eyelid suture increases GABA_A receptors in cat visual cortex. *Dev. Brain Res.*, **40**, 148–53.

Shaw, C., Needler, M.C. and Cynader, M. (1984) Ontogenesis of muscimol binding sites in cat visual cortex. *Brain Res. Bull.*, **13**, 331–4.

Shaw, C., Needler, M.C., Wilkinson, M. *et al.* (1985) Modification of neurotransmitter receptor sensitivity in cat visual cortex during the critical period. *Dev. Brain Res.*, **22**, 67–73.

Shaw, C., Wilkinson, M., Cynader, M. *et al.* (1986) The laminar distributions and postnatal development of neurotransmitter and neuromodulator receptors in cat visual cortex. *Brain Res. Bull.*, **16**, 661–71.

Shaw, C., Aoki, C., Wilkinson, M. *et al.* (1987) Benzodiazepine ([³H]flunitrazepam) binding in cat visual cortex: ontogenesis of normal characteristics and the effects of dark rearing. *Dev. Brain Res.*, **37**, 67–76.

Sillito, A.M., Kemp, J.A. and Blakemore, C. (1981) The role of GABAergic inhibition in the cortical effects of monocular deprivations. *Nature (Lond.)*, **291**, 318–20.

Silvilotti, L. and Nistri, A. (1991) GABA receptor mechanisms in the central nervous system. *Progr. Neurobiol.*, **36**, 35–92.

Singer, W. (1985) Central control of developmental plasticity in the mammalian visual cortex. *Vision Res.*, **25**, 389–96.

Skangiel-Kramska, J. and Kossut, M. (1984) Increase of GABA receptor binding activity after short lasting monocular deprivation in kittens. *Acta Neurobiol. Exp.*, **44**, 33–9.

Skerritt, J.H. and Johnston, G.A.R. (1982) Postnatal development of GABA binding sites and their endogenous inhibitors in rat brain. *Dev. Neurosci.*, **5**, 189–97.

Somogyi P. (1989) Synaptic organization of GABAergic neurons and GABA_A receptors in the lateral geniculate nucleus and visual cortex, in *Neural Mechanisms of Visual Perception* (eds D.M.-K. Lam and C.D. Gilbert), Portfolio, Houston, pp. 35–62.

Somogyi, P., Freund, T.F., Wu, J.Y. and Smith, A.D. (1983) The section-Golgi impregnation procedure. Immunocytochemical demonstration of glutamic acid decarboxylase in Golgi impregnated neurons and in their afferent synaptic boutons in the visual cortex in the cat. *Neuroscience*, **10**, 261–94.

Stein, B.E. and Gallagher, H.L. (1981) Maturation of cortical control over superior colliculus cells in cat. *Brain Res.*, **223**, 429–35.

Stephenson, F.A. (1988) Understanding the GABA_A receptor: a chemically gated ion channel. *Biochem. J.*, **249**, 21–32.

Sterling, P. and Davis, T.L. (1980) Neurons in the cat lateral geniculate nucleus that concentrate ³H-γ-amino butyric acid (GABA). *J. Comp. Neurol.*, **192**, 737–49.

Toga, A.W. (1987) The metabolic consequences of visual deprivation in the rat. *Dev. Brain Res.*, **37**, 209–17.

Voaden, M.J., Morjaria, B. and Oraedu, A.C.I. (1980) The localization and metabolism of glutamate, aspartate and GABA in the retina. *Neurochem. Int.*, **1**, 151–65.

RECEPTORS FOR GLUTAMATE AND OTHER EXCITATORY AMINO ACIDS: A CAUSE FOR EXCITEMENT IN NERVOUS SYSTEM DEVELOPMENT

Rebecca M. Pruss

8.1 INTRODUCTION

Excitatory amino acid (EAA) receptors are the major mediators of excitatory transmission in the vertebrate nervous system. The pharmacology and physiology of EAA receptors and EAA associated pathology has received much attention since the discovery in the 1950s of the excitatory and convulsive effects of glutamate (Hayashi, 1954; Curtis *et al.*, 1959). The details of contributions to the understanding of EAA receptors and their roles in the developing and adult nervous system can be found in several excellent reviews (Foster and Fagg, 1984; Mayer and Westbrook, 1987; Monaghan *et al.*, 1989; Wroblewski and Danysz, 1989; Choi and Rothman, 1990; McDonald and Johnston, 1990) and in a series of articles recently compiled by Lodge and Collingridge (1991). The major endogenous activators of EAA receptors are the amino acids, L-glutamate and L-aspartate. However, other sulfur-containing derivatives and analogs of cysteine (cysteic acid, cysteine sulfinic acid, homocysteic acid and homocysteine sulfinic acid) as well as quinolinic acid (a tryptophan metabolite) are also endogenous EAA agonists. Glutamate is the most abundant EAA and fulfills the criteria of a true neurotransmitter: high concentrations of glu-

tamate are stored in vesicles found in nerve terminals; high affinity carriers for acidic amino acids exist on the plasma membranes of neurons and glia and selective high affinity uptake sites for glutamate are found on synaptic vesicles; the enzyme glutaminase which converts glutamine into glutamate is present in nerve terminal mitochondria; exocytotic release of glutamate is Ca^{2+} dependent. In addition, glutamate, although not necessarily the most potent agonist, binds to all EAA receptors. For these reasons EAA receptors are usually simply referred to as glutamate receptors. EAA receptors are found both on neurons and astrocytic glial cells (Barres *et al.*, 1990; Cornell-Bell *et al.*, 1990). EAAs stimulate a variety of responses in these cells through activation of a diverse collection of receptors. Although most EAA receptors are coupled to the activation of cation channels, other EAA receptors are G-protein-coupled receptors. The cation channel-coupled receptors have been termed 'ionotropic' receptors because of their ability to stimulate ion flux whereas the G-protein-coupled receptors have been called 'metabotropic' receptors because they stimulate increases in intracellular second messengers. In these respects, glutamate appears to be similar to the neurotransmitters

Receptors in the Developing Nervous System Vol. 2: Neurotransmitters. Edited by Ian S. Zagon and Patricia J. McLaughlin. Published in 1993 by Chapman & Hall. ISBN 0 412 49400 0. Vols. 1 and 2 (set) ISBN 0 412 54520 9.

acetylcholine (ACh) and serotonin (5-HT) all of which activate a large number of heterogeneous receptors linked both to ion channels and G proteins. Although some of the excitatory roles of EAA receptors on neurons may be clear, the function of EAA receptors on glial cells is still a mystery.

Many of the structural proteins that form functional EAA-activated ion channels have now been identified by molecular cloning. The cDNA for a metabotropic EAA receptor that is coupled to the activation of phospholipase C has also been cloned. Since the number of EAA receptors appears to be quite large and the functional responses triggered by these receptors quite diverse, attempts have been made to identify selective ligands for the various members of the EAA receptor family. The integration of the pharmacology and the physiology with the structural information gained from the cloning and expression studies will eventually complete our understanding of the roles that EAA receptors play in nervous system development, synaptic plasticity, learning and memory and neurodegenerative processes.

8.2 PHARMACOLOGY

Nature has been generous, though cruel, in providing neurotoxins that are selective ligands for EAA receptor subtypes. These ligands are usually constrained analogs of glutamate that are competitive but more potent agonists of EAA receptors. Their extreme potency as EAA agonists is the reason behind their neurotoxicity since over-activation of EAA receptors leads to cell death by a process termed excitotoxicity. Neurotoxins that activate EAA receptors have been isolated from mushrooms (ibotenic acid), chick peas (β-*N*-oxalylamino-L-alanine), cycad plants (β-*N*-methylamino-L-alanine), seaweed (kainic acid), and plankton (domoic acid) as well as from organisms, such as shellfish, that ingest these food sources. By using the structural leads provided by naturally occurring agonists,

organic chemists have designed synthetic agonists and antagonists of EAA receptors that competed with glutamate for its binding site on the receptors. The structure–activity relationships of these and other compounds have been well described by Watkins *et al.* (1990). The availability of so many ligands and their apparent selectivity has helped in the characterization and classification of EAA receptors. Naturally occurring and endogenous inhibitors of EAA receptors also exist, including arthropod neurotoxins found in the venom of some types of wasps and spiders. These toxins paralyze or kill insects by blocking glutamate receptors that mediate excitatory neurotransmission at the insect neuromuscular junction (Jackson and Usherwood, 1988; Lodge and Johnson, 1990). The active components in these venoms include cysteine-rich polypeptides and acylated polyamines. Although these molecules have effects on vertebrate EAA receptors, pharmacological use of these toxins has been limited by the supply of purified toxins as well as by their complexity and the difficulty inherent in their synthesis. In addition to these neurotoxin antagonists, an endogenous metabolite of tryptophan, kynurenic acid, has been found to inhibit EAA receptors. The fact that kynurenic acid, an EAA receptor antagonist, and quinolinic acid, an EAA receptor agonist, are both metabolites of tryptophan, has led to the speculation that an imbalance in tryptophan metabolism may be involved in some chronic neurodegenerative diseases such as Alzheimer's, Parkinson's, Huntington's and amyotrophic lateral sclerosis. Numerous synthetic compounds like the dissociative anesthetic phencyclidine (PCP) also inhibit EAA receptor activation. Many analogs of these naturally occurring and synthetic receptor antagonists have been used to characterize EAA receptors and identify sites other than the glutamate binding site on EAA receptors. These pharmacological tools have been used to develop a simple pharmacological system for classifying EAA receptors. However, the variety of physiological responses elicited by EAA receptor agonists indicates a greater complexity than can be explained by ligand binding assays. A fuller understanding of the complexity and heterogeneity of EAA receptors is emerging now that some of the receptor proteins have been cloned.

8.2.1 AGONISTS USED TO CLASSIFY NMDA AND NON-NMDA RECEPTORS

Whereas glutamate activates all EAA receptors, a synthetic analog of aspartate, *N*-methyl-D-aspartate (NMDA), has been found to be a selective agonist for a subtype of ionotropic EAA receptor. The receptors that bind and can be activated by NMDA are called NMDA receptors. All other EAA receptors are therefore called 'non-NMDA' receptors. This nomenclature can be confusing since it is meant to refer only to a subtype of glutamate or EAA receptors, not to receptors for other hormones or neurotransmitters. Most naturally occurring neurotoxins appear to be selective agonists for non-NMDA receptors. Non-NMDA receptors are divided into two categories based on the apparent selectivity of two of these neurotoxins, kainate and quisqualate. It was subsequently discovered that quisqualate could activate both an ionotropic and a metabotropic EAA receptor. These two quisqualate receptor subtypes can be distinguished by two synthetic glutamate analogs. AMPA (α-amino-3-hydroxy-5-methyl-4-isoxazole propionic acid) selectively activates quisqualate-activated ion channels whereas [1S,3R]1-amino-1,3-cyclopentanedicarboxylic acid (*trans*-ACPD) selectively activates the metabotropic receptor (Schoepp *et al.*, 1990). The pharmacological classification system and how it evolved, together with the cloned receptor subunits (see later) that appear to correspond to these EAA receptors is shown in Fig. 8.1. In addition to these four receptor subtypes (NMDA, kainate, AMPA and *trans*-ACPD), a fifth type of EAA receptor can be identified using L-2-amino-4-phosphonobutanoic acid (AP4). The AP4 receptor appears to be an autoreceptor on nerve terminals whose function is to inhibit glutamate release. This classification system is given for historical perspective. Although AMPA appears to be a selective ligand for one type of non-NMDA receptor, these receptors also bind kainate (although with lower affinity) and quisqualate and kainate both bind to the same subset of ionotropic EAA receptors. These ligands are classified as receptor agonists because they activate ion currents or stimulate phospholipase C. However, the magnitude of the physiological responses seen with most of these agonists is variable and depends on the cell type or brain region being examined. This variability in response to agonists indicates the existence of receptor subtypes within these broad pharmacological categories.

8.2.2 ANTAGONISTS OF NMDA AND NON-NMDA RECEPTORS

Antagonists have been very valuable tools for studying the pharmacology and physiology of EAA receptors. Although no antagonists of AP4 receptors (the inhibitory autoreceptor) or metabotropic EAA receptors have yet been identified, many antagonists have been identified for ionotropic NMDA and non-NMDA receptors. Pharmacologically, NMDA receptors are the best characterized EAA receptors because of the large number of selective antagonists that affect both the NMDA or glutamate binding site on the receptor or allosteric modulatory sites that control ion channel opening. Competitive antagonists of NMDA receptors bind to the site for glutamate or NMDA but do not activate the ion channel. Although both NMDA and non-NMDA receptors bind glutamate and aspartate, many synthetic analogs of acidic amino acids are selective competitive antagonists of NMDA receptors. Such analogs include 2-amino-5-phosphonovaleric acid (AP5), 2-amino-7-phosphonoheptanoic acid (AP7) and

Pharmacology of EAA Receptors

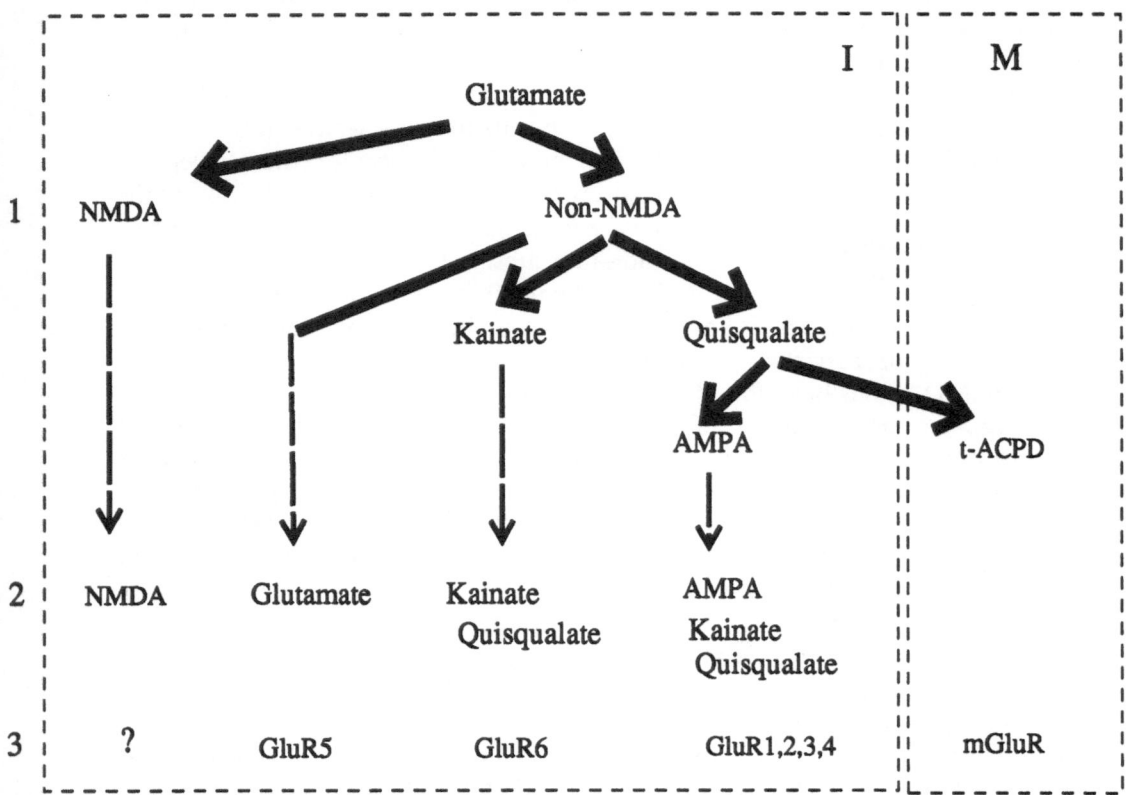

Fig. 8.1. The classification of EAA receptors based on their pharmacology. High affinity ligand binding is indicated in the area of the diagram labeled 1. The ligands these receptor subtypes actually bind (other than glutamate) is indicated in area 2. The EAA receptor proteins that appear to correspond to these pharmacologically defined receptor subtypes is shown in area 3. Ligand-gated ion channels, or ionotropic receptors are grouped in the box labeled I. Metabotropic receptors, those coupled by G proteins to PI turnover are separated by the box labeled M.

3-(2-carboxy-piperazin-4-yl)propyl-1-phosphonic acid (CPP) (Watkins *et al.*, 1990).

The identification of non-competitive antagonists, those that do not displace glutamate or NMDA from the receptor, have been extremely important in the identification of allosteric, modulatory sites on NMDA receptors. In addition they provide a means of identifying NMDA receptor complexes independent of their EAA binding site. Dissociative anesthetics like PCP have been shown to block ion flux through NMDA receptors. Because the blockade of NMDA receptors by PCP is use-

dependent it appears that PCP and its analogs bind to a site within the ion channel of the NMDA receptor complex. MK-801, a high affinity ligand for the PCP binding site on the NMDA receptor, originally synthesized by Merck for use as a neuroprotectant (Wong *et al.*, 1986), has been shown to be a useful tool for characterizing NMDA receptors and responses mediated by NMDA receptors. Divalent cations such as Mg^{2+}, Ni^{2+}, Co^{2+} and Mn^{2+} are also antagonists of NMDA-activated ion channels (Ascher and Nowak, 1988). Their inhibition of Ca^{2+} and Na^+ influx through

NMDA channels can be relieved by depolarization. Since the NMDA receptor is normally blocked by physiological concentrations of Mg^{2+}, activation of NMDA receptors displays voltage dependence, a feature that is unique for a ligand-gated ion channel. Another characteristic of NMDA receptors is their dependence on glycine. Whereas glycine activates receptors coupled to Cl^- channels and, together with γ-aminobutyric acid (GABA), plays an important role as an inhibitory neurotransmitter, the glycine receptor linked to Cl^- channels is blocked by strychnine. By contrast, strychnine has no effect on the ability of glycine to modulate NMDA receptors. Following the discovery that glycine binding is necessary for glutamate to activate NMDA ion channels (Johnson and Ascher, 1987), it was found that kynurenic acid and its more potent analogs antagonize NMDA receptors by displacing glycine from its site on the NMDA receptor complex (Kemp *et al.*, 1988). Although kynurenic acid is also a weak antagonist of non-NMDA receptors, it appears to act as a competitive inhibitor of these receptors since there is no role for glycine in the activation of non-NMDA receptors. Instead, kynurenic acid appears to be a competitive antagonist of non-NMDA receptors. The NMDA receptor is also different from other EAA receptors in its sensitivity to pH (Monaghan and Cotman, 1986; Traynelis and Cull-Candy, 1990) and voltage-independent blockade by Zn^{2+} (Westbrook and Mayer, 1987). In addition, polyamines positively modulate NMDA receptors but not other types of EAA receptors. Although NMDA receptors are generally considered and referred to as a single EAA receptor subtype, there are probably multiple forms of NMDA receptors. Receptor subtypes appear to account for the differential sensitivity of cells with NMDA receptors to the neurotoxic effect of quinolinic acid. In addition, NMDA receptor subtypes can be identified in different brain regions by their preference for agonists or antagonists in ligand binding and receptor autoradiography

studies (Monaghan *et al.*, 1988). A more detailed description of non-competitive NMDA receptor antagonists is given by Lodge and Johnson (1990).

Although NMDA receptors can be antagonized by a variety of different mechanisms, only a few antagonists have been found for non-NMDA receptors and most of these compete for the EAA binding site. Kynurenic acid, as discussed above, is a weak antagonist of non-NMDA as well as NMDA receptors. A class of compounds called quinoxalinediones has been shown to be selective competitive antagonists of kainate, quisqualate and AMPA binding to non-NMDA receptors (Honoré *et al.*, 1988; Sheardown *et al.*, 1990). Some, like 6-nitro,7-cyanoquinoxaline-2,3-dione (CNQX), are also weak inhibitors of NMDA receptors because of their ability to displace glycine from its binding site, others, like 6,7-dinitroquinoxaline-2,3-dione (DNQX) and 2,3-dihydroxy-6-nitro-7-sulfamoyl-benzo(F)quinoxaline (NBQX), have little affinity for the glycine site on NMDA receptors resulting in their increased selectivity for non-NMDA receptors (Watkins *et al.*, 1990). Their potency as non-NMDA receptor antagonists, measured by their ability to displace AMPA and kainate in binding assays, is in the range 10^{-7}–10^{-5} M. Although NBQX appears to have some preference for AMPA binding sites over high affinity kainate binding sites compared to CNQX or DNQX, all three compounds block kainate-, quisqualate-, and AMPA-activated EAA receptors.

8.3 MOLECULAR BIOLOGY

Several strategies have been used to identify and clone cDNAs for EAA receptor proteins. One approach has been to use selective ligands to aid in the purification of EAA receptors or at least their ligand-binding fragments and then use the amino acid sequence of these proteins to design oligonucleotide probes for screening cDNA libraries. By using this strategy and kainate or domoic acid as a ligand two groups have purified and then cloned cDNAs for high affinity kainate binding proteins from frog and chick brain (Gregor *et al.*, 1989; Wada *et al.*, 1989). Both the frog and chick cDNAs code for proteins that are ~450–500 amino acids long. These proteins contain several stretches of hydrophobic amino acids which are predicted to form four transmembrane domains. Although both proteins bind kainate with high affinity and resemble other ligand-gated ion channels, neither the frog nor chick kainate binding protein forms a functional ion channel when they are expressed on their own. Whether these proteins are components of a multimeric complex that is an EAA-activated ion channel remains to be determined.

A second strategy for cloning EAA receptor cDNAs has been to use *Xenopus* oocytes to express mRNA derived from cDNA libraries and assay for the presence of ligand-activated current responses in the oocytes. After successive division of the cDNA library, a single cDNA is isolated that is capable of generating mRNA for a functional receptor. The first cDNA for a protein that forms kainate-activated ion channels and is a member of a large family of ionotropic EAA receptor subunits was cloned from a rat brain cDNA library using this strategy (Hollmann *et al.*, 1989). Oocyte expression was also used successfully to identify and clone cDNAs for a metabotropic EAA receptor (Houamed *et al.*, 1991; Masu *et al.*, 1991).

8.3.1 CLONING OF cDNA FOR SUBUNITS OF IONOTROPIC EAA RECEPTORS

The first subunit for an ionotropic EAA receptor was originally called GluR-K1 because of the large currents elicited in response to kainate compared to other agonists. Using the cDNA sequences corresponding to the protein's transmembrane domains (assumed to be the most highly conserved regions of the

protein) as probes or primers, related proteins have now been identified using low stringency hybridization techniques or the polymerase chain reaction (PCR). Four cDNAs for closely related proteins were isolated: GluR1 (the former GluR-K1), GluR2, GluR3 and GluR4 (also called GluR-A, GluR-B, GluR-C, and GluR-D) (Boulter *et al.*, 1990; Keinänen *et al.*, 1990; Nakanishi *et al.*, 1990). All four of these proteins can be expressed in two different versions, termed flip and flop, depending on the use of alternate exons during mRNA splicing (Sommer *et al.*, 1990). These eight proteins appear to be members of a related family of EAA receptor subunits. All members of this family bind glutamate, kainate, quisqualate and AMPA. Because they bind AMPA with the highest affinity, these receptors have sometimes been referred to as AMPA receptors. Although these subunits have only low affinity for kainate compared to AMPA, kainate is a full agonist for these receptors whereas glutamate, quisqualate and AMPA are variable in their ability to activate the ion channel formed by these subunits.

Additional families of ionotropic EAA receptor subunits have been identified from rat brain cDNA libraries by low stringency hybridization techniques. One receptor subunit called GluR5 was identified based on its 40% sequence identity to GluR4 (GluR1–4 share ~70% sequence identity) (Bettler *et al.*, 1990). Unlike other receptor subunits, GluR5 could only be activated weakly by glutamate and not at all by kainate, quisqualate or AMPA. GluR5 was also found to exist in two forms depending on the presence or absence of a small insert. Recently two additional EAA receptor subunits have been identified that bind glutamate, kainate and quisqualate but not AMPA. One of these subunits, GluR6, was identified by low stringency hybridization using GluR5 as a probe (Egebjerg *et al.*, 1991) whereas the other, KA-1, was identified by PCR using primers based on the sequence of the frog and chick kainate binding proteins (Werner *et al.*, 1991). Both GluR6 and KA-1 have higher affinities for kainate than the other subunits. These receptor subunits may account for some of the high affinity kainate binding sites found in rat brain. When GluR6 was expressed in oocytes, kainate had an EC_{50} of 1 μM for ion current activation compared to an EC_{50} of 35 μM for activation of ion currents when GluR1 was expressed. In ligand binding assays, kainate had a K_d of 5 nM using membranes from mammalian cells expressing the KA-1 subunit. These results are consistent with the receptor binding and autoradiography which indicate that AMPA binding sites and high affinity kainate binding sites are on different proteins and expressed by different cells. However, the results from molecular cloning and expression studies indicate that many, if not all, AMPA binding sites are also low affinity kainate receptors. It remains to be seen whether receptors will be identified that bind AMPA but not kainate. Despite the fact that NMDA receptors are best characterized pharmacologically, none of the EAA receptor subunits that have been identified by molecular cloning have been shown to bind or be activated by NMDA. The presumed association between cloned EAA receptor subunits and pharmacologically characterized EAA receptors is shown in Fig. 8.1.

8.3.2 STRUCTURAL FEATURES OF IONOTROPIC EAA RECEPTOR SUBUNITS

Comparison of the subunits of EAA-activated ion channels with other ligand-gated ion channels such as nicotinic ACh receptors, $GABA_A$ receptors and strychnine-sensitive glycine receptors revealed some structural similarities. The *N*-terminal half of all ligand-gated ion channel proteins, including EAA receptors, is predicted to exist as a large extracellular domain that probably contains the glutamate binding site. The *C*-terminal half of the molecule contains stretches of hydrophobic amino acids that are predicted to form four transmembrane alpha helices followed by a short extracellular tail. The most striking difference between EAA receptors and other types of ligand-gated ion channel proteins is their size. EAA receptor subunits are about twice as large as ACh, $GABA_A$ or glycine receptor subunits. EAA receptor proteins are ~900 amino acids long compared to ~450 amino acids for the other types of receptors. Interestingly, the *C*-terminal half of these receptors appears to be similar to the chick and frog kainate binding proteins discussed earlier and the predicted transmembrane domains of all these proteins are homologous. A comparison of the structural features of EAA receptors with other comparable proteins (other ligand-gated ion channels in the case of the ionotropic EAA receptor) is shown in Fig. 8.2.

From studies using radiolabeled GluR probes to map EAA receptor subunit expression by *in situ* hybridization it appears that individual cells may express several different EAA receptor subunits. Based on the structure and stoichiometry of nicotinic ACh receptors, EAA receptors are predicted to be heteromultimeric complexes of four or five subunits. Therefore, the variations in response to different ligands seen in cells from different parts of the nervous system probably reflect the variation in EAA receptor assembly in different cells. Although it is possible to correlate radioligand binding studies with different GluR subunits, an understanding of the variation in agonist-activated responses will depend on a better understanding of how these receptor subunits combine to form functional glutamate-activated ion channels.

8.3.3 METABOTROPIC EAA RECEPTORS

The cDNA for a metabotropic EAA receptor was cloned independently by two groups using similar oocyte expression strategies (Houamed *et al.*, 1991; Masu *et al.*, 1991). The receptor protein predicted from the cDNA

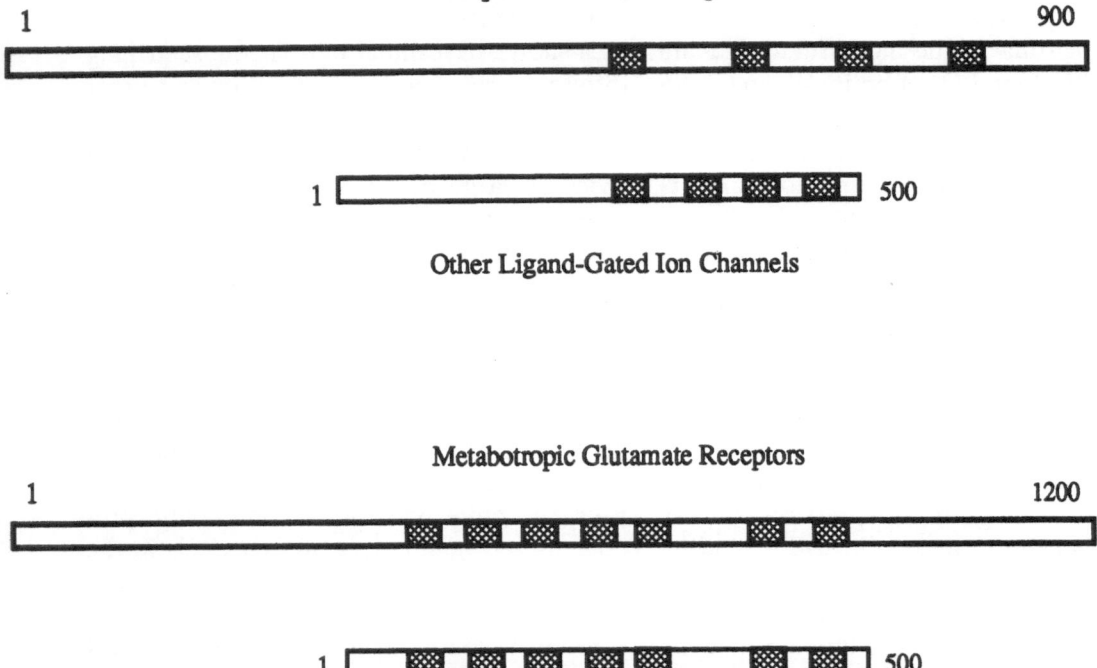

Fig. 8.2.   Illustration of the size and structural organization of ionotropic EAA receptor proteins compared to other ligand-gated ion channels (upper) and a metabotropic EAA receptor protein compared to other G-protein-coupled receptors. Hatched boxes indicate regions of hydrophobic amino acids that are predicted to form transmembrane domains. Numbers refer to the approximate length of each protein in amino acid residues.

sequences has seven putative transmembrane domains typical of other members of the family of G-protein-coupled receptors. However, like the ionotropic EAA receptors, the metabotropic EAA receptor is about twice as large as other comparable proteins. Based on the cDNA sequence, the metabotropic EAA receptor is predicted to be ~1200 amino acids long compared to the usual length of ~400–500 amino acids found for other G-protein-coupled receptors. The increased size of the metabotropic receptor is due to long *N*- and *C*-terminal hydrophilic domains that flank the central hydrophobic region. Sequence comparisons show some homology between the extracellular *N*-terminal domains of the metabotropic and ionotropic EAA receptors and these regions have been postulated to be involved in

ligand binding. If this is true, the metabotropic EAA receptor is different from other G-protein-coupled receptors whose ligand binding pocket is predicted to reside within the hydrophobic transmembrane domains. An illustration of the structural characteristics of metabotropic EAA receptors and other G-protein-coupled receptors is shown in Fig. 8.2.

The metabotropic EAA receptor is coupled to phospholipase C activation by a pertussis toxin-sensitive G protein. Activation of phospholipase C results in the hydrolysis of phosphatidylinositol 4,5-bisphosphate (PI) and the production of two second messengers, inositol 1,4,5-trisphosphate (IP3) and 1,2-diacylglycerol (DAG). IP3 stimulates the release of Ca^{2+} from intracellular storage vesicles associated with the endoplasmic reticulum, sometimes referred to as calciosomes. Activation of this response in oocytes expressing the cloned metabotropic EAA receptor occurs in response to micromolar concentrations of glutamate, quisqualate, ibotenate and *trans*-ACPD whereas millimolar concentrations of aspartate, kainate, and AMPA also elicited weak responses.

Many hormones and neurotransmitters that activate G-protein-coupled receptors have been found to bind to a family of receptors coupled to distinct second messenger pathways. For instance, 5-HT and ACh activate several different G-protein-coupled receptors that can either stimulate phospholipase C or inhibit adenylate cyclase. It is probably only a matter of time before additional G-protein-coupled EAA receptors are identified. It will be interesting to discover whether these receptors are coupled to second messengers other than IP3, DAG and the release of intracellular Ca^{2+}.

The identification of a selective agonist and cDNA that can be used as a probe for these proteins provides two valuable tools for studying the expression of metabotropic EAA receptors in the nervous system. Simply measuring PI hydrolysis does not provide rigorous proof of the existence or involvement of a metabotropic EAA receptor. In neurons, increases in intracellular Ca^{2+} due to glutamate-stimulated Ca^{2+} influx can stimulate phospholipase C, a Ca^{2+}-dependent enzyme. This has led to the conclusion by one group that EAA receptors with the pharmacology of NMDA receptors are capable of stimulating PI hydrolysis (Wroblewski *et al.*, 1987). To rule out this possibility, it is, therefore, necessary to measure IP3 formation or increases in intracellular Ca^{2+} in the absence of external Ca^{2+}.

8.4 PHYSIOLOGY

8.4.1 EAA RECEPTORS INCREASE INTRACELLULAR CALCIUM

Regardless of their type, EAA receptors are responsible for increasing intracellular Ca^{2+}. Ionotropic receptors accomplish this by acti-

vating plasma membrane ion channels. These channels may be permeable to Ca^{2+} or Na^+ or both Ca^{2+} and Na^+. Receptors linked to Ca^{2+}-permeable channels allow the direct entry of Ca^{2+} whereas receptors linked to Na^+-permeable channels result in depolarization and indirectly stimulate Ca^{2+} influx through voltage-operated Ca^{2+} channels. NMDA receptors have been shown to be highly permeable to Ca^{2+} (MacDermott *et al.*, 1986). Non-NMDA receptors on the other hand display a range of cation permeabilities dependent, at least in part, on the ligand that is used to activate the receptor. Although many electrophysiological studies of non-NMDA receptors failed to demonstrate that these receptors are permeable to Ca^{2+} or other divalent cations, more recent studies using cultured hippocampal or retinal bipolar neurons (Iino *et al.*, 1990; Gilbertson *et al.*, 1991) or oocytes expressing cloned GluRs (Hollmann *et al.*, 1991) have found that non-NMDA receptor-activated ion channels are permeable to Ca^{2+}. Because of the limited number of cell types and GluRs examined it is not clear from electrophysiological studies which cell types or what brain regions will express non-NMDA receptors that can directly gate Ca^{2+}.

To get a better picture of the types of cells that express non-NMDA receptors that are permeable to Ca^{2+}, it is possible to use Co^{2+} uptake as a marker of divalent cation-permeability. Co^{2+}, after precipitation with $(NH_4)_2S$, can be detected as a brown–black deposit in the cell bodies and processes of neurons and astrocytic glial cells that express divalent cation permeable non-NMDA receptors (Pruss *et al.*, 1991). Co^{2+} uptake appears to be a reliable indicator of Ca^{2+} influx although Co^{2+} appears to be somewhat less permeable than Ca^{2+}. Co^{2+} uptake can be used to study the functional properties of non-NMDA receptors on cells in different parts of the nervous system. Co^{2+} permeability appears to require complete activation or opening of the non-NMDA ion channel in response to an agonist. Kainate is able to activate Co^{2+} influx through

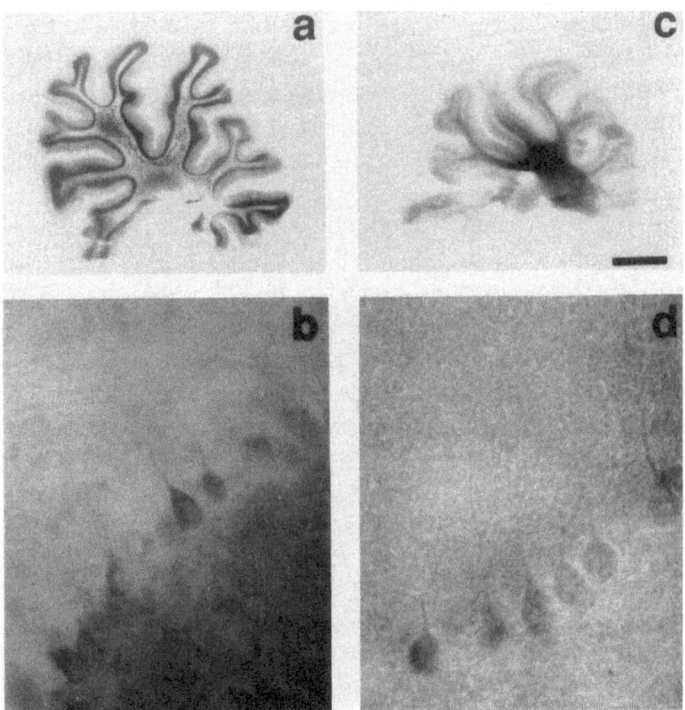

Fig. 8.3. Different patterns of Co²⁺ uptake are found in adult rat cerebellum after stimulation with kainate (a and b) or glutamate (c and d). Glutamate receptor agonists were added to fresh tissue slices in buffer containing 5 mM CoCl₂. After washing and precipitation with (NH₄)₂S, the CoS appears as a dark brown or black deposit in cells that have divalent cation-permeable ion channels coupled to non-NMDA receptors. Quisqualate produces a pattern of labeling identical to that seen with glutamate. Kainate causes Co²⁺ uptake into all cells expressing non-NMDA receptors linked to ion channels whereas glutamate activates only a subpopulation of cells. The Co²⁺ uptake in Purkinje neurons is very clear in both kainate and glutamate-treated slices. Co²⁺ is seen in the granule cell layer only after kainate stimulation, but not in response to glutamate. These labeling patterns indicate that granule neurons express different non-NMDA-activated ion channels than Purkinje neurons. Bar is 1.5 mm in (a) and (c) and 20 μm in (b) and (d). For details see Pruss *et al.* (1991).

all non-NMDA receptors made up of subunits that have been shown to bind kainate. Glutamate and quisqualate, which appear to bind to all kainate receptors, and AMPA, which binds to a subset of these receptors, vary in their ability to activate Co²⁺ influx. The pattern of Co²⁺ uptake stimulated by these agonists in adult rat cerebellum and hippocampus is shown in Figs. 8.3 and 8.4. Non-NMDA receptors appear to be able to stimulate divalent cation influx even early in the development of the nervous system since

the Co²⁺ uptake pattern in the hippocampus of 5-day-old rats is the same as that seen in the adult (Fig. 8.5). Three kainate receptor subtypes (K1, K2 and K3) can be identified based on the pattern of Co²⁺ uptake. K1 is present on cerebellar granule neurons and is activated by kainate but no other EAA receptor agonist. K2 is activated by kainate and glutamate but not by quisqualate or AMPA and generates a response like that seen in the hippocampal dentate gyrus. K3 is found on Purkinje neurons in the cerebellum and all

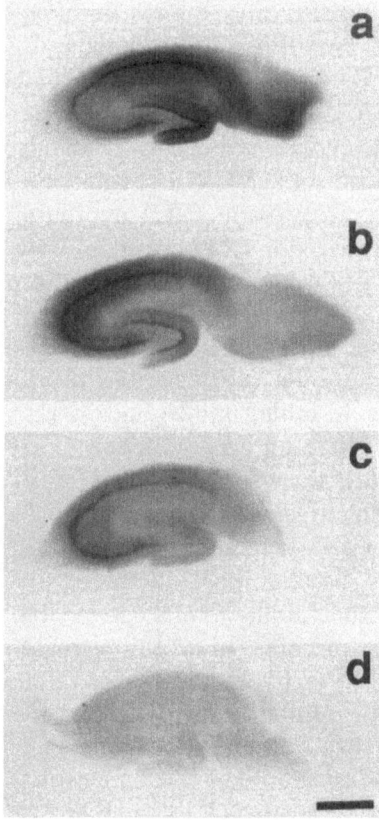

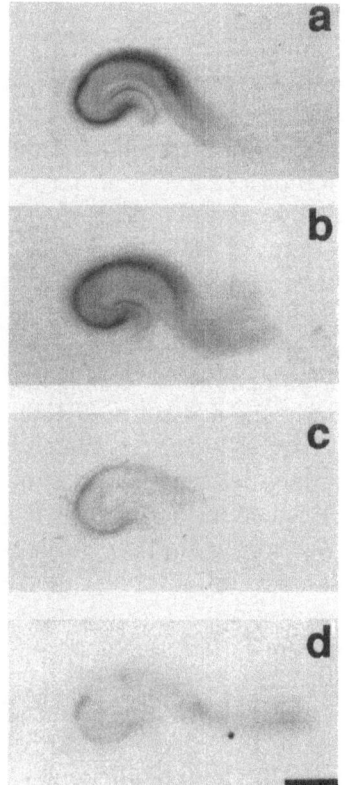

Fig. 8.4. Uptake of Co^{2+} into slices of adult rat hippocampus. Kainate (a), glutamate (b), and quisqualate (c) were added to fresh tissue slices The pyramidal cell layer is labeled by all three agonists whereas the dentate gyrus does not take up Co^{2+} in response to quisqualate indicating the presence of different non-NMDA receptor-activated ion channels on pyramidal neurons and dentate granule neurons. A control slice is shown in (d); bar is 1 mm. For details see Pruss *et al.* (1991).

Fig. 8.5. Stimulation of Co^{2+} uptake by non-NMDA receptor agonists in 5-day-old rat hippocampus. Kainate (a), glutamate (b), and AMPA (c) were used to stimulate fresh tissue slices as in Fig. 8.3. A control slice is shown in (d). The same pattern of agonist-activated Co^{2+} uptake is seen in developing hippocampus as in the adult (AMPA stimulates the same pattern of cobalt uptake as quisqualate) indicating full maturation of divalent cation permeable channels linked to non-NMDA receptors even at this age. Bar is 1 mm.

hippocampal pyramidal neurons and is activated by kainate, glutamate, quisqualate, and AMPA. Since Co^{2+} uptake is a reliable indicator of Ca^{2+} permeability, the results of these studies indicate that most if not all non-NMDA receptors are capable of allowing Ca^{2+} to enter cells directly when activated by the appropriate agonist. Glutamate, the endoge-

nous agonist, may only be able to stimulate divalent cation influx through a subpopulation of non-NMDA receptors. Non-NMDA receptors are mainly recognized for their permeability to monovalent cations. Activation of non-NMDA receptors can indirectly stimulate Ca^{2+} influx due to depolarization caused by Na^+ influx. Depolarization causes Ca^{2+}

entry by stimulating the opening of voltage operated Ca^{2+} channels and by relieving the NMDA receptors from blockade by Mg^{2+}.

The metabotropic EAA receptor also causes increases in intracellular Ca^{2+}. Like other receptors that stimulate IP3 formation, the metabotropic EAA receptor causes an increase in intracellular Ca^{2+} by releasing Ca^{2+} from intracellular stores. In many cells, the formation of IP3 and the release of intracellular Ca^{2+} stores is invariably linked to the activation of Ca^{2+} influx through a non-selective cation channel in the plasma membrane. The exact mechanism involved in the activation of this plasma membrane channel is not completely understood but may be due to inhibition of a plasma membrane K^+ channel that results in membrane depolarization (Charpak *et al.*, 1990). Alternatively IP3 or a subsequent metabolite of IP3 such as inositol 1,3,4,5-tetrakisphosphate may activate a plasma membrane Ca^{2+} channel (Berridge and Irvine, 1989). Although activation of metabotropic EAA receptors has been shown to decrease Ca^{2+} influx through voltage-operated Ca^{2+} channels, this effect may be mediated by DAG and Ca^{2+} activation of protein kinase C. Activation of metabotropic EAA receptors has been shown to increase the firing rate of neurons. This effect may also be due to inhibition of K^+ channels, as discussed earlier, or it may be an indicator of agonist-activated Ca^{2+} entry which acts to facilitate neuron firing.

8.4.2 CONSEQUENCES OF Ca^{2+} INCREASE

The main function of Ca^{2+} inside cells is the regulation of enzyme activity which is accomplished by Ca^{2+} binding directly to a regulatory domain on an enzyme or by binding to calmodulin (CaM), a calcium-binding protein that can associate with and regulate the activity of a variety of intracellular enzymes (Rasmussen and Means, 1989; Heizmann and Hunziker, 1991). Ca^{2+}-activated enzymes include proteases like calpain, lipases such as phospholipase C and phospholipase A_2, and

nucleases. These enzymes participate in catabolic processes but also cause the release of active products such as IP3 or arachidonic acid. Ca^{2+} regulates the level of another second messenger, cyclic AMP, by affecting the activity of adenylate cyclase and cyclic nucleotide phosphodiesterase. Other Ca^{2+}-activated enzymes include nitric oxide (NO) synthase, which forms nitric oxide from arginine. Low concentrations of NO cause vasorelaxation through the activation of guanylate cyclase in vascular smooth muscle cells. However, at high concentrations NO may be neurotoxic due to its ability to inhibit mitochondrial respiration or contribute to the production of reactive free radicals (Dawson *et al.*, 1991). Ca^{2+} also activates protein kinase C and CaM kinases as well as a CaM-dependent phosphatase. These enzymes have important regulatory roles because of their ability to make reversible modifications to proteins. In addition to calmodulin and Ca^{2+}-activated enzymes, Ca^{2+} binds to a large number of proteins whose function may be to sequester intracellular Ca^{2+}. Other Ca^{2+}-binding proteins probably play roles in Ca^{2+}-activated processes, such as cell motility, through their effects on cytoskeleton assembly, and cell and membrane adhesion via annexins and cadherins. Ca^{2+}-binding proteins are also necessary for the fusion of secretory vesicle membranes with the plasma membrane during exocytosis.

Increases in intracellular Ca^{2+} play important roles in the regulation of gene transcription. Increased transcription of mRNA for cellular oncogenes like c-fos and c-jun, and a number of neuropeptides such as enkephalin, vasoactive intestinal peptide, and galanin (Pruss *et al.*, 1985; Pruss and Stauderman, 1988; Rokaeus *et al.*, 1990) occurs in response to increases in intracellular Ca^{2+}. At least two Ca^{2+}-activated protein kinases appear to be able to mediate these transcriptional effects of Ca^{2+}. Protein kinase C activation results in increased formation of c-fos and c-jun heterodimers that act as transcription factors by binding to specific DNA sequences called AP-

1 sites or the TRE (TPA response element) in the promotor regions of genes whose transcription is increased by TPA (12-*O*-tetradecanoylphorbol-13-acetate). CaM kinases have also been shown to mediate Ca^{2+}-activated gene transcription. CaM kinases phosphorylate CREB, a protein that binds to the cyclic AMP response element (CRE). The CRE is a specific sequence in the promoter region of genes that confers inducibility by increased levels of cyclic AMP. Phosphorylation of CREB by either CaM kinases or protein kinase A increases the affinity of CREB for the CRE resulting in a convergence of Ca^{2+} and cAMP signalling pathways (Sheng *et al.*, 1991). The effects of Ca^{2+} on gene transcription are thought to underlie the long-term effects of EAA receptor activation on cell survival and differentiation.

In addition to long-term effects of Ca^{2+} which are dependent on gene transcription, Ca^{2+} triggers acute responses like exocytosis. Trophic or other responses due to EAA receptor activation may be indirectly mediated by the release of growth factors, hormones, neurotransmitters, ATP, ascorbic acid or other factors that influence cell survival, differentiation and function. Whereas the short-term effects of Ca^{2+} on processes like secretion may be triggered by transient elevations in intracellular Ca^{2+}, long-term changes such as increased transcription of genes that code for differentiation-specific products such as neuropeptides appear to require sustained periods of elevated Ca^{2+} (Pruss and Stauderman, 1988). EAA receptor activation may be the means of providing this signal to cells during nervous system development.

In addition to these specific acute and long-term biochemical effects, Ca^{2+} levels play a role in neuronal survival. Although moderate levels of intracellular Ca^{2+} (100–300 nM) appear to enhance neuronal survival, neurotoxicity is associated with high concentrations of intracellular Ca^{2+} (≥ 1 μM) which are sustained for long periods (more than several minutes). This association between excess Ca^{2+} and neurotoxicity has led to the hypothesis that overstimulation of EAA receptors triggers neuronal cell death. Excess release of glutamate can occur as a result of hypoglycemia, hypoxia, uncontrolled seizures, or trauma to the nervous system. Much of the glutamate released during these episodes appears to be due to reversal of the EAA uptake carrier although some is due to Ca^{2+}-dependent release from neurons triggered by EAA receptor activation and/or depolarization (Nicholls and Attwell, 1990). The uncontrolled release of glutamate causes sustained activation of EAA receptors that results in excess Ca^{2+} influx culminating in cell death. This process, termed excitotoxicity, can be provoked by activation of both NMDA and non-NMDA receptors (Meldrum and Garthwaite, 1990). The role of NMDA receptors has received the most attention since these receptors are highly permeable to Ca^{2+} (MacDermott *et al.*, 1986). In models of ischemia and stroke NMDA receptor antagonists have been shown to attenuate neurodegeneration. However, not all cells that are susceptible to excitotoxicity have NMDA receptors. The cerebellar Purkinje cell is one example. Furthermore, quinoxalinediones that are selective antagonists of non-NMDA receptors have been shown to protect against neurodegeneration even when administered up to three hours after an ischemic event. There appear to be at least two forms of neurodegeneration that occur as a result of stroke, ischemia or neurotrauma. First, there is a loss of cells immediately surrounding the site of injury that can be reversed by NMDA receptor antagonists but only within minutes of the injury. Secondarily, there is a loss of cells radiating from the site of injury which is delayed relative to the time of the insult and which can be attenuated by non-NMDA receptor antagonists even hours after the injury. In addition to Ca^{2+}, non-NMDA receptor-activated ion channels are permeable to Na^+. Influx of Na^+, whether alone or in combination with Ca^{2+}, causes cells to swell because of concomitant influx of Cl^- and H_2O. This leads to morphological changes that are different from those seen with Ca^{2+} influx alone. Na^+ influx results in swelling and necrotic cell death whereas Ca^{2+} influx causes the collapse and condensation of the cytoskeleton and other structures inside the cells by a process that, in some cell types, has been called apoptosis. Glutamate and other endogenous EAAs may be able to activate Ca^{2+} entry through NMDA receptors and certain non-NMDA receptor subtypes. Although NMDA receptors are permeable to Ca^{2+}, the passage of Ca^{2+} through the NMDA ion channel may in fact be reduced following ischemia or trauma since the ion channel is blocked by physiological concentrations of Mg^{2+} and acidic pH.

8.5 THE ROLE OF EAA RECEPTORS DURING NERVOUS SYSTEM DEVELOPMENT

8.5.1 EAA RECEPTORS IN THE DEVELOPING NERVOUS SYSTEM

Although most receptor binding and biochemical studies have been done using tissues from rats, other mammalian and non-mammalian species have also been used to study EAA receptors and their roles in the development of the nervous system. Cats and monkeys have been used to study some of the physiological and behavioral effects of EAA receptors. Many cell culture systems and tissue slice preparations come from the mouse as well as rat nervous system, and some studies have been performed measuring EAA receptors in the developing human nervous system. In addition to these mammalian species, EAA receptors and their contribution to nervous system development have been studied using tadpoles and chick embryos and some behavioral studies have been performed using pigeons. In addition to vertebrates, EAA receptors are present, and are important mediators of excitatory neurotransmission in insects. The relative timing of EAA expression and their involvement in vertebrate nervous system

development appears to be similar in all species examined.

Many different methods have been used to measure EAA receptors and their roles in nervous system development. In addition to direct quantitation of EAA receptors using radioligand binding and receptor autoradiography, the physiological response to EAA agonists and antagonists has been measured in whole animals and *in vitro* using tissue slices or cell culture. These physiological approaches include measurements of the effects of EAA receptor agonists and antagonists on electrophysiological responses, neurotransmitter release, second messenger production (IP3 and arachidonic acid), uptake or release of $^{45}Ca^{2+}$, changes in intracellular Ca^{2+} using fluorescent dyes, the response of *Xenopus* oocytes following injection of brain mRNA, transcription of mRNA for c-fos of various neuropeptides, the behavior of animals, ability to cause seizures, changes in synaptic responses, neuronal survival both *in vivo* and *in vitro*, and neurite outgrowth using cultured cells. Recently, with the cloning of cDNAs for EAA receptor subunits, the localization and timing of EAA receptor expression in the developing nervous system has been studied by *in situ* hybridization.

EAA receptors appear to be very important in mediating activity-dependent changes in neuronal survival and synaptic plasticity. The role of EAA receptors, particularly NMDA receptors, has been studied in two models of synaptic plasticity: (1) the development of ocular dominance columns in the visual cortex and (2) the phenomenon of long-term potentiation (LTP) in the hippocampus (Collingridge and Singer, 1990). The development of EAA receptors had been correlated with a critical period in the development of the visual system: the time when neuronal activity plays a role in the establishment of ocular dominance columns in the visual cortex and in the determination of retinotectal connections. Although a peak in NMDA receptors is measured in binding assays during this critical period, this peak may be merely correlative rather than causative since NMDA receptor stimulation alone does not induce the development of ocular dominance columns. Furthermore, the critical period can be delayed or prolonged by dark rearing without affecting the peak in the number of NMDA receptors. Even though NMDA receptor activation may not be sufficient to cause the changes in synaptic connections, NMDA receptors appear to be necessary for the changes produced by neuronal activity to occur. During the critical period NMDA antagonists prevent both the disconnection of deprived visual pathways as well as the recovery of stable synapses once the deprived pathway is reactivated.

Another example of synaptic plasticity that involves EAA receptors is LTP, a stimulus-evoked enhancement in synaptic activity that is believed to be related to learning and memory (Nicoll *et al.*, 1988). In the hippocampus and other brain regions where LTP occurs, brief but high frequency stimulation of excitatory inputs leads to an increase in postsynaptic responsiveness that persists for weeks *in vivo*. Simultaneous depolarization accompanied by NMDA receptor activation and increases in intracellular Ca^{2+} in the postsynaptic cell are necessary to induce most (but not all) forms of LTP. The ability to induce LTP in the CA1 region of the hippocampus appears during the second postnatal week in the rat and is correlated with a transient peak in NMDA binding sites measured in this region. The development of LTP in other regions of the hippocampus is also correlated with increases in NMDA receptors. Although the initiation of LTP requires NMDA receptor activation, the maintenance and expression of LTP, that is the increased postsynaptic responsiveness of the neuron, appears to be mediated by non-NMDA receptors since it can be blocked by CNQX but not by NMDA receptor antagonists. Increased PI turnover and the activation of protein kinase C is also correlated with

LTP formation in the hippocampus suggesting a role for metabotropic EAA receptors in the synaptic changes necessary for the development of LTP.

In addition to these specific examples of synapse formation, remodeling and reinforcement, EAA receptor activation stimulates neurite outgrowth *in vitro* and may play a role in neurite extension and sprouting *in vivo* both during development and neuronal regeneration following injury. EAA receptors are particularly abundant in areas of the nervous system where synapses continue to be replaced or remodeled during the life of an animal. These regions include the olfactory bulb where connections with new olfactory receptor neurons occur throughout life, in the hippocampus where synaptic changes involved in learning and memory are continuously being made, and in the cerebellum where synaptic plasticity is necessary to integrate and adapt to changes in sensory, cortical and motor activity. Even in areas of the adult brain that receive little EAA input, there is a transient peak in the expression of EAA receptors that coincides with periods during development when neuronal connections are being established. In the rat cerebellum, although Purkinje neurons express non-NMDA receptors at all stages of development, NMDA receptors are expressed only transiently between the first and third postnatal week (Dupont *et al.*, 1987). This period is correlated with an increased glutamate binding in the molecular layer and the establishment of connections between granule cell parallel fibers and Purkinje cell dendrites (Garcia-Ladona *et al.*, 1991).

Although EAA receptors appear to be present and be involved in excitatory neurotransmission, synapse reinforcement, and postsynaptic responsiveness, all the functional consequences of EAA receptor activation may not develop at the same time as the receptors themselves. Therefore, using EAA-activated responses as an indicator of EAA receptor expression may not be a valid indicator of their presence or absence. For instance, although EAA receptors are expressed very early during development they do not appear to result in excitotoxicity in very young animals and the pattern of receptor-mediated neurotoxicity changes during development (Garthwaite and Garthwaite, 1986; McDonald *et al.*, 1990). Despite the presence of kainate-activated ion channels, kainate is a relatively weak neurotoxin compared to NMDA in young animals whereas the reverse is true in the adult.

In many cases the use of selective antagonists has been used to classify the EAA receptor subtypes responsible for the observed changes in neuronal connections or responses. In other studies the Ca^{2+}-dependence of the EAA-elicited response has been interpreted to mean that NMDA receptors mediate the response. The realization that other EAA receptor subtypes may also directly gate Ca^{2+} may lead to a re-examination of some of the processes believed to be mediated only by NMDA receptors. The importance of non-NMDA receptors and metabotropic EAA receptors in neuronal development and plasticity as well as neurotoxicity should not be underestimated.

8.5.2   LIGAND BINDING AND AUTORADIOGRAPHY

Although functional responses can be used to evaluate and discriminate EAA receptors during different periods of development, their numbers and distribution in nervous system tissue can be measured directly using ligand binding assays (Young and Fagg, 1990). Several different methods are used to quantitate and localize these receptors using radioactive ligands that are either agonists or antagonists of EAA receptors. Binding assays can be performed using either membranes from different regions of the nervous system or tissue sections prepared from animals at different stages of development. Receptor subtypes can be identified and quantitated by measuring the number of [^{3}H]glutamate binding sites that are

displaced by saturating amounts of unlabeled selective ligands. Alternatively, radiolabeled agonists or antagonists that are selective for EAA receptor subtypes can be used. The metabotropic EAA receptor can be labeled using the selective ligand, [^{3}H]-*trans*-ACPD. NMDA receptors can be measured using [^{3}H]CPP, a competitive NMDA receptor antagonist. And non-NMDA receptors can be measured using [^{3}H]CNQX, a non-selective antagonist of these receptors. [^{3}H]AMPA and [^{3}H]kainate are also used to measure non-NMDA receptor subtypes. Whereas [^{3}H]AMPA identifies high affinity AMPA binding sites some or all of these non-NMDA receptors are low affinity receptors for kainate. [^{3}H]kainate, under the conditions used in most binding assays, measures only high affinity kainate binding sites. These high affinity kainate sites may not all be associated with functional EAA receptors; some may be EAA uptake sites or other kinds of binding sites present on neurons and glia (Somogyi *et al.*, 1990; Ortega *et al.*, 1991). To distinguish EAA receptors from their uptake sites, only Na^+-independent binding is measured since the EAA carrier required is dependent on Na^+ which is co-transported with EAAs.

EAA receptors have usually been measured during postnatal development of the rat nervous system although EAA receptor expression occurs much earlier. The density of binding sites increases up to the second or third week postnatally and then declines to adult levels over the next several weeks. Almost all studies using radioligand binding report an 'overshoot' in the number of EAA receptors relative to adult levels that occurs about one to two weeks after birth with the peak in NMDA receptors preceding the peak in non-NMDA receptors. However, the ontogeny of NMDA receptors measured by radioligand binding is dependent on the ligand being used. The binding site for NMDA precedes the development of the strychnine-insensitive [^{3}H]glycine binding site or the binding site for [^{3}H]TCP, an analog of PCP, associated with the NMDA ion channel. Meanwhile, the ontogenies of the glycine site and the TCP binding site associated with the NMDA receptor complex are nearly identical. The transient peak(s) in EAA receptors corresponds to the occurrence of a critical window of nervous system development when synaptic plasticity results in the formation, modification and stabilization of synaptic contacts. This period also corresponds to the time prior to the onset of naturally occurring neuronal cell death. It is possible that the transient peaks in EAA receptors seen by binding assay may be related to a temporary peak in the number of cells expressing EAA receptors. *In situ* hybridization reveals no apparent increases in mRNA for several EAA receptor subunits that might account for the peak in EAA binding sites.

With the cloning of cDNAs for ionotropic and metabotropic EAA receptors their distribution in the rat nervous system has been mapped by *in situ* hybridization. As described earlier, many non-NMDA receptor subunits have now been cloned although at this time no NMDA selective receptor subunit has been identified. Although there are, as yet, no reports of the distribution of mRNA for the metabotropic EAA receptor during nervous system development, its distribution in adult rat hippocampus and cerebellum was reported at the time the sequence of the mRNA and protein was reported (Masu *et al.*, 1991). In the hippocampus, large amounts of mRNA are seen in CA4-dentate gyrus, moderate amounts in CA2–3 and lower levels in CA1. In the cerebellum the mRNA is particularly abundant in Purkinje cells.

The localization of mRNA for subunits of EAA activated ion channels, including specific examination of flip and flop forms, has been studied in both the adult and developing rat nervous system (Bettler *et al.*, 1990; Pellegrini-Giampietro *et al.*, 1991; Monyer *et al.*, 1991; Werner *et al.*, 1991). These studies show that GluR mRNAs vary in their amount and pattern of expression during the development of the nervous system. Some of these subunits have been mapped as early as embryonic day (ED) 10 in the rat where mRNA for GluRs 4 and 5 were found in postmitotic neurons. The mRNA for all EAA receptor subunits was detected at every stage of development in the rat. The distribution of different GluRs changed during development as did the expression of the flip and flop versions of GluR1–4. The flip versions were almost exclusively expressed before postnatal day (PD) 8 and even after that the flop versions of some GluRs were found only in very restricted patterns only in selected regions of the nervous system. Flip versions of GluRs give greater current responses than the flop versions when the receptor subunits are expressed in mammalian cells (Sommer *et al.*, 1990). The appearance of flop versions occurs about the time of the peak in synapse formation in the rat forebrain (Aghajanian and Bloom, 1967). A switch in EAA receptors to forms that allow less current flow may be important for protecting cells from the neurotoxic consequences of high local concentration of EAAs following synapse formation. The relative abundance of GluRs changes during development: at ED14, GluR2 > GluR4 >GluR1 = GluR3, at PD1 GluR1 > GluR2 > GluR3 >> GluR4, at PD8 GluR1 = GluR3 > GluR2 >> GluR4, and at PD14 GluR1 = GluR2 (due to profoundly increased expression of the flop version of GluR2 in the cerebellum) = GluR3 > GluR4 (where both flip and flop versions are expressed predominantly in cerebellum) (Monyer *et al.*, 1991). In addition to brain, GluR mRNA can be detected in spinal cord, retina, and dorsal root ganglia. For instance, GluR4 was detected in relatively large amounts in retinal ganglion cells and amacrine cells whereas GluR5 mRNA was particularly abundant in dorsal root ganglion neurons. Most GluRs are expressed in large amounts in cerebellum, hippocampus and olfactory bulb. GluR5 is conspicuous by its very restricted pattern of expression in the brain: it is very low in or absent from the hippocampus (while other GluR mRNAs are high) and in the cerebellum GluR5 mRNA is found only in Purkinje neurons. The pattern of EAA subunit expression has yet to provide a clear explanation or understanding of the radio-ligand binding patterns seen with AMPA and kainate or account for changing patterns of susceptibility to EAA-mediated neurotoxicity. Since agonist and antagonist binding as well as ion channel activation will be determined by the combination of subunits expressed, a complete understanding of ligand binding and EAA receptor-mediated functional responses awaits the development of techniques to get stable expression of the receptor subunits in various combinations. Although the possible number of receptors that could be formed from these subunits is quite large, the actual number of combinations will be limited by rules of assembly which are as yet unknown. It will be some time before the receptor subunit combinations can be assigned to functional responses but we are beginning to get a picture of their structure– function relationships.

8.6 SUMMARY

In conclusion, the pharmacology and physiology of EAA receptors has suggested that there are many different receptor subtypes. However, they can be classified in a systematic way on the basis of ligand binding. Cloning and expression studies have just begun to provide information about the structural basis for the sometimes contradictory pharmacology, biochemistry, ligand binding and electrophysiology seen with EAA receptors. The knowledge of the true structural and functional diversity of EAA receptors and their differential distribution in the nervous system will lead to a better understanding of their contribution to nervous system development and function. This information should make it possible to identify and design selective agonists or antagonists of EAA receptors that will be important therapeutically. Since excitatory neurotransmission is of fundamental importance to nervous system function, nature has probably designed multiple receptors to mediate excitatory amino acid-mediated transmission for specialized purposes. Increases in intracellular Ca^{2+} appear to be the primary

consequence of EAA receptor activation. The identification of the processes used by neural cells to interpret EAA receptor activation and translate those signals will be the future goal of neurobiologists. The integration of the molecular cloning information with the pharmacology and physiology of EAA receptors will be an exciting challenge that holds the key to our full understanding of the roles of these receptors in nervous system function and dysfunction.

GLUTAMATE RECEPTOR UPDATE

During 1991 and 1992 there was a real explosion in the number of EAA receptors identified, the cDNA for one providing the means to hunt for others like it. At last, several cDNAs for the much sought after NMDA receptor were cloned (Moriyoshi *et al.*, 1991; Meguro *et al.*, 1992; Monyer *et al.*, 1992; Yamazaki *et al.*, 1992b). The four NMDA receptor subunits identified so far can be divided into two families based on their structural similarity: NMDAR1 is only ~20% similar to NMDAR2A, NMDAR2B, and NMDAR2C which are between 55 and 70% similar to each other. Although the mRNAs for these subunits have only been localized in adult rat brain they show interesting differences in their distribution. NMDAR1 and NMDAR2A are distributed throughout the CNS although NMDAR1 mRNA appears to be relatively more abundant than the mRNA for NMDAR2A. NMDAR2B is found mainly in the hippocampus and appears to be absent from cerebellum while NMDAR2C is found only in the cerebellum (Monyer *et al.*, 1992). Two more kainate-selective ion channel subunits have now been described (Bettler *et al.*, 1992; Herb *et al.*, 1992; Morita *et al.*, 1992; Sakimura *et al.*, 1992). Although members of the kainate receptor family do not form functional ion channels on their own, they do when expressed in certain combinations and can even be activated by AMPA (Herb *et al.*, 1992). Mishina and co-workers, who have been cloning cDNAs for EAA receptors expressed in mouse brain

(Sakimura *et al.*, 1990), have also described a cDNA which is structurally related but distinct from previously described NMDA, AMPA or kainate receptor subunits (Yamazaki *et al.*, 1992a). More members of the metabotropic EAA receptor family have also been reported along with descriptions of their coupling to multiple second messenger systems (Aramori and Nakanishi, 1992; Tanabe *et al.*, 1992). Antibodies, as well as *in situ* hybridization, have now been used to map the distribution of the largest metabotropic EAA receptor, mGluRlα, in adult rat brain (Martin *et al.*, 1992). However, no papers have appeared describing the distribution or functional roles of metabotropic EAA receptors in the developing nervous system.

Beyond the identification of additional members of the EAA receptor families by molecular cloning strategies, functional studies have revealed structural elements that determine divalent cation permeability of EAA-activated ion channels. A glutamine residue near the external face of the second putative transmembrane domain of AMPA and kainate receptor subunits appears to play a key role in determining calcium permeability of the ion channel (Hume *et al.*, 1991; Verdoorn *et al.*, 1991). In NMDA receptors an asparagine residue appears to play the same role as glutamine making NMDA channels highly permeable to calcium. An unusual finding is that this strategic glutamine is replaced in some AMPA and kainate receptor subunits by an arginine through the unconventional process of mRNA editing, switching the receptor from a calcium permeable to calcium impermeable ion channel (Sommer *et al.*, 1991). Although the gene for most, if not all, AMPA and kainate receptors codes for a glutamine in this position, (CAG), the adenosine is modified to an inosine resulting in the insertion of an arginine instead of a glutamine residue during protein synthesis. But only in the case of GluR-B, GluR5 and GluR6. And editing appears to be developmentally regulated. Although 100% of the GluR-B mRNA in adult rat brain appears to be edited to the arginine or calcium impermeable form, analysis of rat brain at ED14 and PD0 indicates that ~1% of the total GluR-B message is in the unedited or calcium permeable form (Burnashev *et al.*, 1992). For other EAA subunits that undergo editing the process is not as thorough. In adult rat brain only 40% of the GluR5 mRNA and 75% of the GluR6 mRNA appears in the edited, arginine-containing, calcium impermeable form. It remains to be seen if the editing process shows regional differences either in the adult or developing CNS.

Developmental differences in EAA receptors appear to determine the sensitivity of a cell to ambient levels of glutamate or other endogenous agonists. Variations in splicing and the remarkable process of mRNA editing regulate the magnitude of total current flux and the amount of calcium entry into neurons. EAA receptors expressed early in development, when neurons are dividing, sprouting, migrating, differentiating and establishing synaptic connections, activate greater levels and/or more sustained depolarization and probably allow more calcium entry than adult forms of these receptors. Both the increases in ion flux along with their ability to trigger release of neurotransmitters, neuropeptides and growth factors onto neighboring cells are probably important factors guiding CNS development. However, once synaptic connections have been established, conversion of EAA receptors to lower conductance, less calcium permeable adult forms may be a necessary requirement to prevent excitotoxicity due to increasing ambient levels of glutamate. It will be interesting to see how these developmental changes are triggered and whether a delay or failure in the alteration of EAA receptor subunits plays any role in developmental regulated neuronal cell death in pathological neurodegeneration.

REFERENCES

Aghajanian, G.K. and Bloom, F.E. (1967) The formation of synaptic junctions in developing rat brain: a quantitative electron microscopic study. *Brain Res.*, **6**, 716–27.

Aramori, I. and Nakanishi, S. (1992) Signal transduction and pharmacological characterization of a metabotropic glutamate receptor, mGluR1, in transfected CHO cells. *Neuron*, **8**, 757–65.

Ascher, P. and Nowak, L. (1988) The role of divalent cations in the N-methyl-D-aspartate responses of mouse central neurons in culture. *J. Physiol. (Lond.)*, **399**, 247–66.

Barres, B.A., Chun, L.L.Y. and Corey, D.P. (1990) Ion channels in vertebrate glia. *Annu. Rev. Neurosci.*, **13**, 441–74.

Berridge, M.J. and Irvine, R.F. (1989) Inositol phosphates and cell signalling. *Nature*, **341**, 197–205.

Bettler, B., Boulter, J., Hermans-Borgmeyer, I. *et al.* (1990) Cloning of a novel glutamate receptor subunit GluR5: expression in the nervous system during development. *Neuron*, **5**, 583–95.

Bettler, B., Egebjerg, J., Sharma, G. *et al.* (1992) Cloning of a putative glutamate receptor: a low affinity kainate binding subunit. *Neuron*, **8**, 257–65.

Boulter, J., Hollmann, M., O'Shea-Greenfield, A. *et al.* (1990) Molecular cloning and functional expression of glutamate receptor subunit genes. *Science*, **249**, 1033–7.

Burnashev, N., Monyer, H., Seeburg, P.H. and Sakmann, B. (1992) Divalent ion permeability of AMPA receptor channels is dominated by the edited form of a single subunit. *Neuron*, **8**, 189–98.

Charpak, S., Gähwiler, B.H., Do, K.Q. and Knöpfel, T. (1990) Potassium conductances in hippocampal neurons blocked by excitatory amino-acid transmitters. *Nature*, **347**, 765–7.

Choi, D.W. and Rothman, S.M. (1990) The role of glutamate neurotoxicity in hypoxic ischemic neuronal death. *Annu. Rev. Neurosci.*, **13**, 171–82.

Collingridge, G. and Singer, W. (1990) Excitatory amino acid receptors and synaptic plasticity. *Trends Pharmacol. Sci.*, **11**, 290–6.

Cornell-Bell, A.H., Finkbeiner, S.M., Cooper, M.S. and Smith, S.J. (1990) Glutamate induces calcium waves in cultured astrocytes: long-range glial signalling. *Science*, **247**, 470–3.

Curtis, D.R., Phillis, J.W. and Watkins, J.C. (1959) Chemical excitation of spinal neurons. *Nature*, **183**, 611.

Dawson, V.L., Dawson, T.M., London, E. *et al.* (1991) Nitric oxide mediates glutamate neurotoxicity in primary cortical cultures. *Proc. Natl. Acad. Sci. USA*, **88**, 6368–71.

Dupont, J.-L., Gardette, R. and Crepel, F. (1987) Postnatal development of the chemosensitivity of rat cerebellar Purkinje cells to excitatory amino acids. An *in vitro* study. *Dev. Brain Res.*, **34**, 59–68.

Egebjerg, J., Bettler, B., Hermans-Borgmeyer, I. and Heinemann, S. (1991) Cloning of cDNA for a glutamate receptor subunit activated by kainate but not AMPA. *Nature*, **351**, 745–8.

Foster, A.C. and Fagg, G.E. (1984) Acidic amino acid binding sites in mammalian neuronal membranes: their characteristics and relationship to synaptic receptors. *Brain Res. Rev.*, **7**, 103–64.

Garcia-Ladona, F.J., Palacios, J.M., Girard, C. and Gombos, G. (1991) Autoradiographic characterization of [^{3}H]-glutamate binding sites in developing mouse cerebellar cortex. *Neuroscience*, **41**, 243–55.

Garthwaite, G. and Garthwaite, J. (1986) *In vitro* neurotoxicity of excitatory acid analogues during cerebellar development. *Neuroscience*, **17**, 755–67.

Gilbertson, T.A., Scobey, R. and Wilson, M. (1991) Permeation of calcium ions through non-NMDA glutamate channels in retinal bipolar cells. *Science*, **251**, 1613–15.

Gregor, P., Mano, I, Maoz, I. *et al.* (1989) Molecular structure of the chick cerebellar kainate-binding subunit of a putative glutamate receptor. *Nature*, **342**, 689–92.

Hayashi, T. (1954) Effects of sodium glutamate on the nervous system. *Keio J. Med.*, **3**, 183–92.

Heizmann, C.W. and Hunziker, W. (1991) Intracellular calcium-binding proteins: more sites than insights. *Trends Biochem. Sci.*, **16**, 98–103.

Herb, A., Burnashev, N., Werner, P. *et al.* (1992) The KA-2 subunit of excitory amino acid receptors shows widespread expression in brain and forms ion channels with distinctly related subunits. *Neuron*, **8**, 775–85.

Hollmann, M., O'Shea-Greenfield, A., Rogers, S.W. and Heinemann, S. (1989) Cloning by functional expression of a member of the glutamate receptor family. *Nature*, **342**, 643–8.

Hollmann, M., Hartley, M. and Heinemann, S. (1991) Calcium permeability of KA-AMPA-gated glutamate receptor channels: dependence on subunit composition. *Science*, **252**, 851–3.

Honoré, T., Davies, S.N., Drejer, J. *et al.* (1988) Quinoxalinediones: potent competitive non-NMDA glutamate receptor antagonists. *Science*, **241**, 701–3.

Houamed, K.M., Kuijper, J.L., Gilbert, T.L. *et al.* (1991) Cloning, expression and gene structure of a G protein-coupled glutamate receptor from rat brain. *Science*, **252**, 1318–21.

Hume, R.J., Dingledine, R. and Heinemann, S.F. (1991) Identification of a site in glutamate receptor subunits that controls calcium permeability. *Science*, **253**, 1028–31.

Iino, M., Ozowa, S. and Tsuzuki, K. (1990) Permeation of calcium through excitatory amino acid receptor channels in cultured rat hippocampal neurones. *J. Physiol. (Lond.)*, **424**, 151–65.

Jackson, H. and Usherwood, P.N.R. (1988) Spider toxins as tools for dissecting elements of excitatory amino acid transmission. *Trends Neurosci.*, **6**, 278–83.

Johnson, J.W. and Ascher, P. (1987) Glycine potentiates the NMDA response in cultured mouse brain neurons. *Nature*, **325**, 529–31.

Keinänen, K., Wisden, W., Sommer, B. *et al.* (1990) A family of AMPA-selective glutamate receptors. *Science*, **249**, 556–60.

Kemp, J.A., Foster, A.C., Leeson, P.D. *et al.* (1988) 7-Chlorokynurenic acid is a selective antagonist at the glycine modulatory site of the *N*-methyl-D-aspartate receptor complex. *Proc. Natl. Acad. Sci. USA*, **85**, 6547–50.

Lodge, D.W. and Collingridge, G. (eds) (1991) *The Pharmacology of Excitatory Amino Acids*, Elsevier, Cambridge.

Lodge, D. and Johnson, K.M. (1990) Noncompetitive excitatory amino acid receptor antagonists. *Trends Pharmacol. Sci.*, **11**, 81–6.

MacDermott, A.B., Mayer, M.L., Westbrook, G.L. *et al.* (1986) NMDA receptor-activation increases cytoplasmic calcium concentration in cultured spinal cord neurons. *Nature*, **321**, 519–22.

Maguro, H., Mori, H., Araki, K. *et al.* (1992) Functional characterization of a heteromeric NMDA receptor channel expressed from cloned cDNAs. *Nature*, **357**, 70–4.

Martin, L.J., Blackstone, C.D., Huganir, R.L. *et al.* (1992) Cellular localization of a metabotropic glutamate receptor in rat brain. *Neuron*, **9**, 259–70.

Masu, M., Tanabe, Y., Tsuchida, K. *et al.* (1991) Sequence and expression of a metabotropic glutamate receptor. *Nature*, **349**, 760–5.

Mayer, M.L. and Westbrook, G.L. (1987) Permeation and block of *N*-methyl-D-aspartic acid receptor channels by divalent cations in mouse cultured central neurons. *J. Physiol. (Lond.)*, **394**, 501–27.

McDonald, J.W. and Johnston, M.V. (1990) Physiological and pathophysiological roles of excitatory amino acids during central nervous system development. *Brain Res. Rev.*, **15**, 41–70.

McDonald, J.W., Trescher, W.H. and Johnston, M.V. (1990) The selective ionotropic-type quisqualate receptor agonist AMPA is a potent neurotoxin in immature rat brain. *Brain Res.*, **526**, 165–8.

Meguro, H., Mori, H., Araki, K. *et al.* (1992) Functional characterization of a heteromeric NMDA receptor channel expressed from cloned cDNAs. *Nature*, **357**, 70–4.

Meldrum, B. and Garthwaite, J. (1990) Excitatory amino acid neurotoxicity and neurodegenerative disease. *Trends Pharmacol. Sci.*, **11**, 379–87.

Monaghan, D.T. and Cotman, C.W. (1986) Identification and properties of *N*-methyl-D-aspartate receptors in rat brain synaptic plasma membranes. *Proc. Natl. Acad. Sci. USA*, **83**, 7532–5.

Monaghan, D.T., Olvermann, H.J., Nguyen, L. *et al.* (1988) Two classes of *N*-methyl-D-aspartate recognition sites: differential distribution and differential regulation by glycine. *Proc. Natl. Acad. Sci. USA*, **85**, 9836–40.

Monaghan, D.T., Bridges, R.J. and Cotman, C.W. (1989) The excitatory amino acid receptors: their classes, pharmacology, and distinct properties in the function of the central nervous system. *Annu. Rev. Pharmacol. Toxicol.*, **29**, 365–402.

Monyer, H., Seeburg, P.H. and Wisden, W. (1991) Glutamate-operated channels: developmentally early and mature forms arise by alternative splicing. *Neuron*, **6**, 799–810.

Monyer, H., Sprengel, R., Schoepfor, R. *et al.* (1992) Heteromeric NMDA receptors: Molecular and functional distinction of subtypes. *Science*, **256**, 1217–21.

Morita, T., Sakimura, K., Kushiya, E. *et al.* (1992) Cloning and functional expression of a cDNA encoding the mouse β2 subunit of the kainate-selective glutamate receptor channel. *Mol. Brain Res.*, **14**, 143–6.

Moriyoshi, K., Masu, M., Ishii, T. *et al.* (1991) Molecular cloning and characterization of the rat NMDA receptor. *Nature*, **354**, 31–7.

Nakanishi, N., Shneider, N.A. and Axel, R. (1990) A family of glutamate receptor genes: evidence for the formation of heteromultimeric receptors with distinct channel properties. *Neuron*, **5**, 569–81.

Nicholls, D. and Attwell, D. (1990) The release and uptake of excitatory amino acids. *Trends Pharmacol. Sci.*, **11**, 462–8.

Nicoll, R.A., Kauer, J.A. and Malenka, R.C. (1988) The current excitement in long-term potentiation. *Neuron*, **1**, 97–103.

Ortega, A., Eshhar, N. and Teichberg, V.I. (1991) Properties of kainate receptor/channels on cultured Bergman glia. *Neuroscience*, **41**, 335–49.

Pellegrini-Giampietro, D.E., Bennett, M.V.L. and Zukin, R.S. (1991) Differential expression of three glutamate receptor genes in developing rat brain: an *in situ* hybridization study. *Proc. Natl. Acad. Sci. USA*, **88**, 4157–61.

Pruss, R.M. and Stauderman, K.A. (1988) Voltage regulated calcium channels involved in the regulation of enkephalin synthesis are blocked by phorbol ester. *J. Biol. Chem.*, **263**, 13173–8.

Pruss, R.M., Moskal, J.R., Eiden, L.E. and Beinfeld, M. (1985) Specific regulation of VIP biosynthesis by phorbol ester in bovine chromaffin cells. *Endocrinology*, **117**, 1020–6.

Pruss, R.M., Akeson, R.L., Racke, M.M. and Wilburn, J.L. (1991) Agonist-activated cobalt uptake identifies divalent cation-permeable kainate receptors on neurons and glial cells. *Neuron*, **7**, 1–10.

Rasmussen, C.D. and Means, A.R. (1989) Calmodulin, cell growth and gene expression. *Trends Neurosci.*, **11**, 433–8.

Rokaeus, Å., Pruss, R.M. and Eiden, L.E. (1990) Regulation of galanin expression by calcium, protein kinase A and protein kinase C messenger systems in chromaffin cells. *Endocrinology*, **127**, 3096–102.

Sakimura, K., Bujo, H., Kushiya, E. *et al.* (1990) Functional expression from cloned cDNAs of glutamate receptor species responsive to kainate and quisqualate. *FEBS Lett.*, **272**, 73–80.

Sakimura, K., Morita, T., Kushiya, E. and Mishina, M. (1992) Primary structure and expression of the τ2 subunit of the glutamate receptor channel selective for kainate. *Neuron*, **8**, 267–374.

Schoepp, D., Bockaert, J. and Sladeczek, F. (1990) Pharmacological and functional characteristics of metabotropic excitatory amino acid receptors. *Trends Pharmacol. Sci.*, **11**, 508–14.

Sheardown, M.J., Nielsen, E.O., Hansen, A.J. *et al.* (1990) 2,3-Dihydroxy-6-nitro-7-sulfamoyl-benzo-(F)quinoxaline: a neuroprotectant for cerebral ischaemia. *Science*, **247**, 571–4.

Sheng, M., Thompson, M.A. and Greenberg, M.E. (1991) CREB: a Ca^{2+}-regulated transcription factor phosphorylated by calmodulin-dependent kinases. *Science*, **252**, 1427–30.

Sommer, B., Keinänen, K., Verdoorn, T.A. *et al.* (1990) Flip and flop: a cell-specific functional switch in glutamate-operated channels of the CNS. *Science*, **249**, 1580–5.

Sommer, B., Köhler, M., Sprengel, R. and Seeburg, P.H. (1991) RNA editing in brain controls a determinant of ion flow in glutamate-gated channels. *Cell*, **67**, 11–19.

Somogyi, P., Eshhar, N., Teichberg, V.I. and Roberts, J.D.B. (1990) Subcellular localization of a putative kainate receptor in Bergmann glial cells using a monoclonal antibody in the chick and fish cerebellar cortex. *Neuroscience*, **35**, 9–30.

Tanabe, Y., Masu, M., Ishii, T. *et al.* (1992) A family of metabotropic glutamate receptors. *Neuron*, **8**, 169–79.

Traynelis, S.F. and Cull-Candy, S. (1990) Proton inhibition of N-methyl-D-aspartate receptors in cerebellar neurons. *Nature*, **345**, 347–50.

Verdoorn, T.A., Burnashev, N., Monyer, H. *et al.* (1991) Structural determinants of ion flow through recombinant glutamate receptor channels. *Science*, **252**, 1715–18.

Wada, K., Dechesne, C.J., Shimaski, S. *et al.* (1989) Sequence and expression of a frog brain complementary DNA encoding a kainate-binding protein. *Nature*, **342**, 684–9.

Watkins, J.C., Krogsgaard-Larsen, P. and Honoré, T. (1990) Structure–activity relationships in the development of excitatory amino acid receptor agonists and competitive antagonists. *Trends Pharmacol. Sci.*, **11**, 25–33.

Werner, P., Voigt, M., Keinänen, K. *et al.* (1991) Cloning of a putative high-affinity kainate receptor expressed predominantly in hippocampal CA3 cells. *Nature*, **351**, 741–4.

Westbrook, G.L. and Mayer, M.L. (1987) Micromolar concentrations of Zn^{2+} antagonize NMDA and GABA responses of hippocampal neurons. *Nature*, **328**, 640–3.

Wong, E.H., Kemp, J.A., Priestley, T. *et al.* (1986) The anti-convulsant MK-801 is a potent N-methyl-D-aspartate antagonist. *Proc. Natl. Acad. Sci. USA*, **83**, 7104–8.

Wroblewski, J.T., Nicoletti, F., Fadda, E. and Costa, E. (1987) Phencyclidine is a negative allosteric modulator of signal transduction at two subclasses of excitatory amino acid receptors. *Proc. Natl. Acad. Sci. USA*, **84**, 5068–72.

Wroblewski, J.T. and Danysz, W. (1989) Modulation of glutamate receptors: molecular mechanisms and functional implications. *Annu. Rev. Pharmacol. Toxicol.*, **29**, 441–74.

Yamazaki, M., Araki, K., Shibate, A. and Mishima, M. (1992a) Molecular cloning of a cDNA encoding a novel member of the mouse glutamate receptor channel family. *Biochem. Biophys. Res. Commun.*, **183**, 886–92.

Yamazaki, M., Mori, H., Araki, K. *et al.* (1992b) Cloning, expression and modulation of a mouse NMDA receptor subunit. *FEBS Lett.*, **300**, 39–45.

Young, A.B. and Fagg, G.E. (1990) Excitatory amino acid receptors in the brain: membrane binding and receptor autoradiographic approaches. *Trends Pharmacol. Sci.*, **11**, 126–33.

F. Javier Garcia-Ladona, Guadalupe Mengod and José M. Palacios

During the last two decades increasing evidence has accumulated supporting a role for neuropeptides as classical neurotransmitters or neuromodulators at synapses of the central and peripheral nervous systems (see Polak, 1989 for a review). Although their involvement in brain function is well documented, other possible roles, for example that of neurotrophic factor, have also been proposed based on their mitogenic activity (Hanley, 1985). Studies on neuropeptide receptor changes during brain development are suggestive of such a role. Neurotensin is one of those peptides whose receptors present developmentally regulated patterns in brain. Because of the availability of experimental tools for the analysis of its pre- and post-synaptic components, both at the mRNA and protein levels, the neurotensinergic synapse offers an interesting model to analyse transmitter receptor interactions during development. This chapter reviews the available information on changes in neurotensin receptor (NTR) binding and NTR mRNA levels and their correlation with neurotensin (NT) mRNA during the development of the rat central nervous system (CNS) the species for which most of the available data have been generated.

9.1 NEUROTENSIN AS A NEUROTRANSMITTER

Neurotensin, a 13 amino acid peptide (Fig. 9.1), was first isolated from bovine hypothalamus and later localized in several other regions of the central and peripheral nervous system (Kahn *et al.*, 1980; Jennes *et al.*, 1982). When injected into different brain regions, NT induces several physiological and behavioral effects such as decrease of locomotor activity (Kalivas *et al.*, 1983), hypothermia and antinociception (Kalivas *et al.*, 1982). Radioimmunoassay and immunohistochemical studies demonstrated its synaptosomal localization and uneven regional localization in the brain (Jennes *et al.*, 1982; Kalivas *et al.*, 1982; Inagaki *et al.*, 1982; Roberts *et al.*, 1984). Although the precise molecular mechanisms mediating the actions of the peptide are not fully understood, it has been shown that NT can modulate second messenger pathways (Goedert *et al.*, 1984a; Gilbert and Richelson, 1984; Bozou *et al.*, 1986), inducing increases of cytosolic Ca^{2+} (Sato *et al.*, 1991) and protein phosphorylation via Ca^{2+}/calmodulin dependent protein kinases (Kasckow *et al.*, 1991).

The rat gene coding for a NT precursor has been cloned and sequenced (Kislauskis *et al.*,

Receptors in the Developing Nervous System Vol. 2: Neurotransmitters. Edited by Ian S. Zagon and Patricia J. McLaughlin. Published in 1993 by Chapman & Hall. ISBN 0 412 49400 0. Vols. 1 and 2 (set) ISBN 0 412 54520 9.

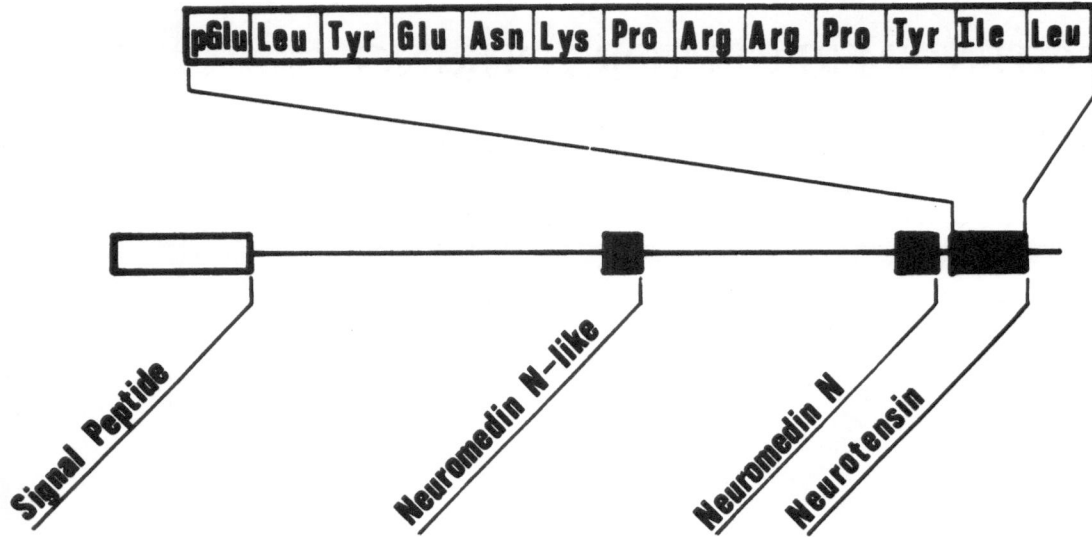

Fig. 9.1. Neurotensin/neuromedin N gene and neurotensin amino acid sequence.

1988). This gene contains, in addition, sequences coding for a NT-like peptide, neuromedin N (Fig. 9.1). The expression of the NT gene can be induced by several agents such as nerve growth factor (NGF), activators of adenylate cyclase, protein kinase C inhibitors such as staurosporine and lithium (Dobner *et al.*, 1988; Tischler *et al.*, 1991). Nerve cells expressing NT mRNA (Alexander *et al.*, 1989) are detectable only in discrete areas of the adult brain such as hippocampus, striatum, cingulate cortex and hypothalamus (see later).

9.2   NEUROTENSIN BINDING SITES IN THE CNS: CHARACTERIZATION AND DISTRIBUTION

NT actions are mediated by its interaction with specific membrane receptors. Ligand binding studies using either ^{125}I-labeled NT or [3H]NT have allowed the characterization of NT binding sites in brain (for review see Uhl, 1990).

NT binding sites have been detected in brain membrane fractions from several mammalian species. Different authors have reported the presence of at least two types of NT-binding site exhibiting different affinities for the peptide. NT-binding sites (K_d=3 nM) are present in high density in the frontal cortex, hypothalamus, midbrain, striatum and thalamus, and are less abundant in the hippocampus, olfactory bulb and cerebellum (Uhl and Snyder, 1977; Goedert *et al.*, 1984). Agonists for these high affinity sites are NT itself and xenopsin (an NT analog from *Xenopus*). Displacement data have shown that the partial sequence NT(8–13) is the more potent NT-analog, indicating that the C-terminal part of NT is the active region. Recently, a protein (mol.wt. 72 kDa) containing NT-binding sites (K_d=5 nM) was purified from bovine brain (Mills *et al.*, 1988).

Binding of NT to synaptic membranes has revealed the presence of high and low affinity NT-binding sites (Mazella *et al.*, 1983, 1988). In these studies, the high affinity component is located in a protein of 100 kDa and displays a K_d of 0.10 nM whereas the low affinity component has a K_d around 4 nM. These values are similar to those reported in studies describing only one type of NT binding site. Moreover, the low affinity NT binding site displayed the special characteristic of being sensitive to the antihistaminic agent levocastine (Mazella *et al.*, 1988). However, a low affinity site insensitive to this agent has also been characterized (Kitabgi *et al.*, 1987) and purified (Mills *et al.*, 1988).

Ligand binding autoradiography has been used to determine the precise anatomical distribution of NT binding sites in CNS. The pharmacological characteristics of NT binding sites detected by autoradiography *in vitro* agree in general with previous results obtained with membrane preparations: the K_d is around 9 nM and NT(8–13) is the more potent fragment of the native peptide inhibiting the binding of radiolabeled NT (Moyse *et al.*, 1987). Although NT binding sites are widely distributed in brain, the highest densities are localized to discrete brain areas.

In adult animals, NT binding sites are specially abundant in parts of the nigrostriatal and mesolimbic systems. High densities of NT binding sites are present in the substantia nigra pars compacta and ventral tegmental area (Young and Kuhar, 1981; Moyse *et al.*, 1987). Several studies indicate that NTR are localized to dopaminergic cell bodies (Palacios and Kuhar, 1981) and can be presynaptically located on dopaminergic terminals in caudate-putamen, olfactory tubercle and nucleus accumbens (Quirion *et al.*, 1985). High levels of NT binding are also detected in the cingulate cortex, olfactory bulb and olfactory tubercle, Islands of Calleja, vertical limb of diagonal band of Broca and nucleus basalis, amygdala, pre- and postsubiculum, ventral part of dentate gyrus, anterior dorsal thalamic nucleus, suprachiasmatic nucleus, and zona incerta (Young and Kuhar, 1981; Goedert *et al.*, 1984b; Kohler *et al.*, 1985; Moyse *et al.*, 1987).

9.3 CHARACTERISTICS OF NEUROTENSIN RECEPTOR GENE

The isolation and characterization of a gene coding for a rat NTR has been reported by Tanaka *et al.* (1990). This gene, isolated from a rat brain cDNA library by expression cloning in an oocyte system encodes for a 424 amino acid protein. The analysis of its sequence confirms that this NTR belongs to the G-protein-coupled receptor family. It has seven hydrophobic putative transmembrane domains. Northern blot analyses with poly $(A)^+$ mRNA purified from several tissues show that NTR mRNA is relatively abundant in brain whereas the levels in peripheral tissues are only 15% of those observed in the brain.

In oocytes injected with *in vitro* transcribed NTR mRNA, NT elicits a response which is mediated by an endogenous Ca^{2+}-dependent Cl^- channel and results from the activation of G protein coupled to phosphoinositide-Ca^{2+} second messenger systems. This result confirms NT activation of different second messenger pathways in the cell (Goedert *et al.*, 1984a; Gilbert and Richelson, 1984; Bozou *et al.*, 1986; Sato *et al.*, 1991; Kasckow *et al.*, 1991).

The transfection of COS cells, a monkey kidney cell line, with the NTR gene allowed the pharmacological characterization of this receptor. [³H]NT binds specifically and in a saturable fashion to membrane fractions from these transfected cells. Scatchard plots revealed a single site of binding with a K_d of

Table 9.1. IC$_{50}$ Values (M) for the displacement of [^{125}I]NT binding to membranes of cDNA NTR transfected cells (a) and [³H]NT binding to rat brain membranes (b) by different peptide analogs of NT. Data from (a) Tanaka *et al.* (1990) and (b) Goedert *et al.* (1984a)

	IC$_{50}$ (M)	
	(a)	(b)
NT(8–13)	2.5×10^{-10}	27×10^{-10}
NT	7.4×10^{-10}	31×10^{-10}
Xenopsin	8.6×10^{-10}	34×10^{-10}
Neuromedin N	26×10^{-10}	—
NT(9–13)	240×10^{-10}	603×10^{-9}

0.16 nM, close to the value observed in rat brain membranes. Displacement experiments revealed that among different NT analogs the NT(8–13) was the more potent and effective inhibiting ^{125}I-labeled NT binding, followed by xenopsin that was equipotent to NT. Neuromedin N and NT(9–13) were less effective than NT, and NT(1–8) displayed no activity (Table 9.1), thus confirming that the C-terminal part of NT is more important for the interaction with NTR (Mazella *et al.*, 1989). The antihistaminic levocastine acting on the low affinity binding sites was ineffective in the cloned NTR. Therefore, the cloned NTR is a NT receptor exhibiting the high affinity binding site described earlier. At present, the sequence of NTR reported by Tanaka *et al.* (1990) is the only one available. However, the possibility that there are other subtypes of NTR, with different molecular or pharmacological characteristics, cannot be ruled out. Multiplicity of receptor subtypes for a given neurotransmitter appears to be the rule rather than the exception.

9.4 NEUROTENSIN RECEPTORS DURING THE DEVELOPMENT OF THE CNS

9.4.1 NEUROTENSIN BINDING SITES DURING RAT BRAIN ONTOGENY

The evolution of NTR levels during ontogeny of the rat brain has been studied using membrane binding assays and autoradiographic techniques (Kiyama *et al.*, 1987; Schotte and Laduron, 1987; Palacios *et al.*, 1988).

At least two different patterns of NTR development have been observed in the rat CNS. A 'classical' one, where NTR density progressively increases during embryonic life to reach the adult patterns in the 2nd/3rd postnatal week, for example in the substantia nigra (Fig. 9.2b) and a second one in which

NT binding sites are present at very high levels around the second week of postnatal life (Fig. 9.2a) and then decrease to adult values, for example in the neocortex.

NT binding sites can be detected early in the prenatal period. At gestational day 14 (GD14) they are mainly present in the spinal cord. The density of NTR increases between GD16 and GD18, especially in the developing neocortex where two bands of high density were observed: one in the molecular layer and parts of the cortical plate and the other localized in the more internal part corresponding to layer VI (Figs 9.3 and 9.5). High densities of binding sites were also observed in the developing hypothalamus and substantia nigra. Low levels were found in most of the brainstem nuclei with the exception of the developing nucleus tractus solitarius where a medium density of binding sites was present.

After birth, the pattern of transient expression was particularly dramatic in the neocortex (Kiyama *et al.*, 1987; Palacios *et al.*, 1988) (Figs 9.3 and 9.4). A double band of labeling was seen, with highest densities in the more external (layer I) and more internal layers (layers V and VI) respectively (Fig. 9.5). In other brain areas the distribution of NT binding sites was comparable to the adult pattern. High densities of NT binding sites were also observed in the septum, Islands of Calleja, olfactory bulb, medial habenula and dentate gyrus. At postnatal day (PD)21 the distribution of NT binding sites was that of the adult brain (Fig. 9.6).

9.4.2 THE DEVELOPMENT OF NEUROTENSIN SYSTEM IN OTHER SPECIES

An early and transient expression of NT binding sites has also been reported in human brain. Mailleux *et al.* (1990) have observed high levels of NT binding transiently expressed in the embryonic human inferior

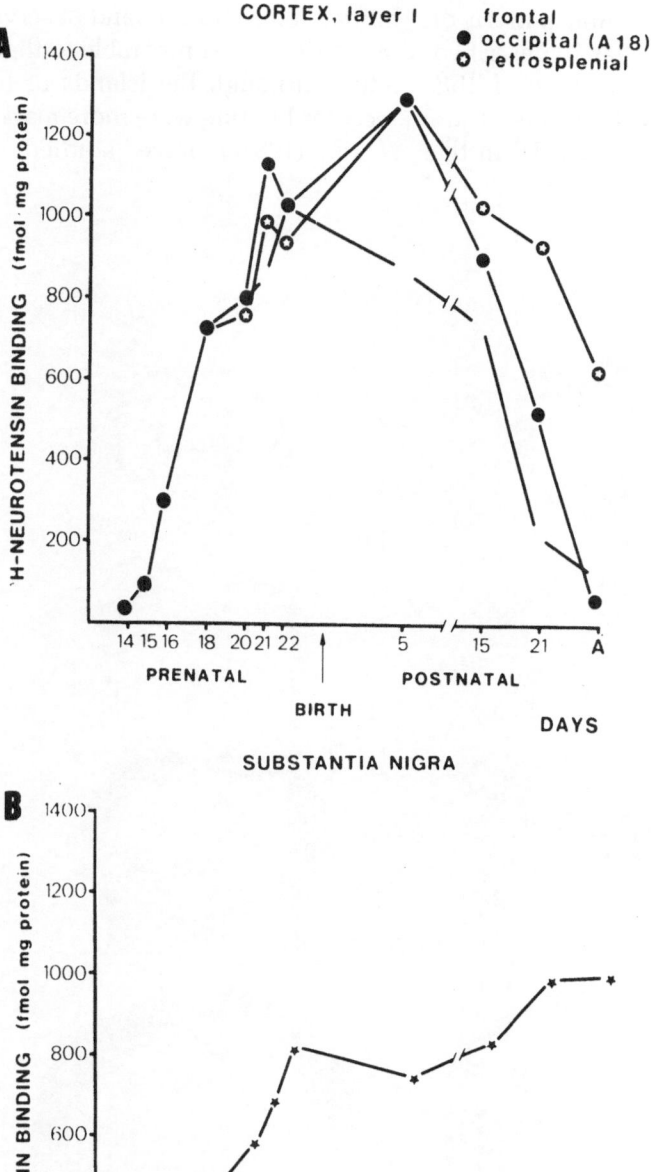

Fig. 9.2. Microdensitometric quantification of [³H]neurotensin binding in (a) layer I of different cortical areas and (b) substantia nigra, during ontogeny.

olive. A different developmental pattern of NT immunoreactivity was described (Mailleux and Vanderhaeghen, 1988) in this nucleus. In our laboratory (Palacios *et al.*, unpublished) we have examined NTR in the human fetus striatum and observed a pattern of distribution comparable to that seen in the adult, although the islands of high density receptor binding were more marked. Goedert *et al.* (1984c) have studied the neuro-

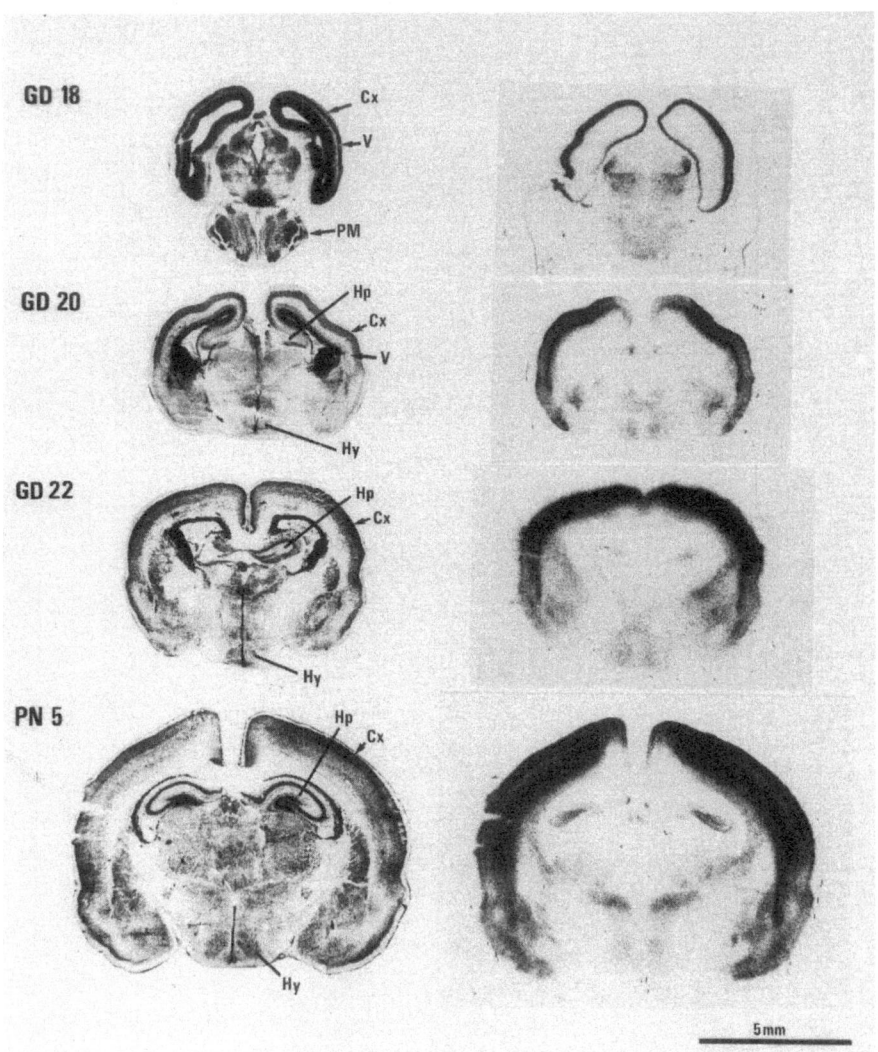

Fig. 9.3.   Photomicrographs of coronal sections of rat brain at different ages during the development of the animal (GD18, GD20, GD22, PN5). On the left side sections stained for cresyl violet are shown, on the right side the autoradiograms of sections incubated with [³H] neurotensin. Reproduced from Palacios *et al.* (1988) with permission. Cx, neocortex; Hp, hippocampus; Hy, hypothalamus; PM, pons medulla.

tensinergic system in cat striatum, showing that no postnatal modifications occur in this area as suggested by the developmental pattern of NT immunoreactivity and [³H]NT binding sites. Transient expression of NT binding sites has also been reported in the cat inferior olive (Mailleux *et al.*, 1989). Thus, as in the rat, in other species different regions of the brain present differential patterns of ontogenic changes.

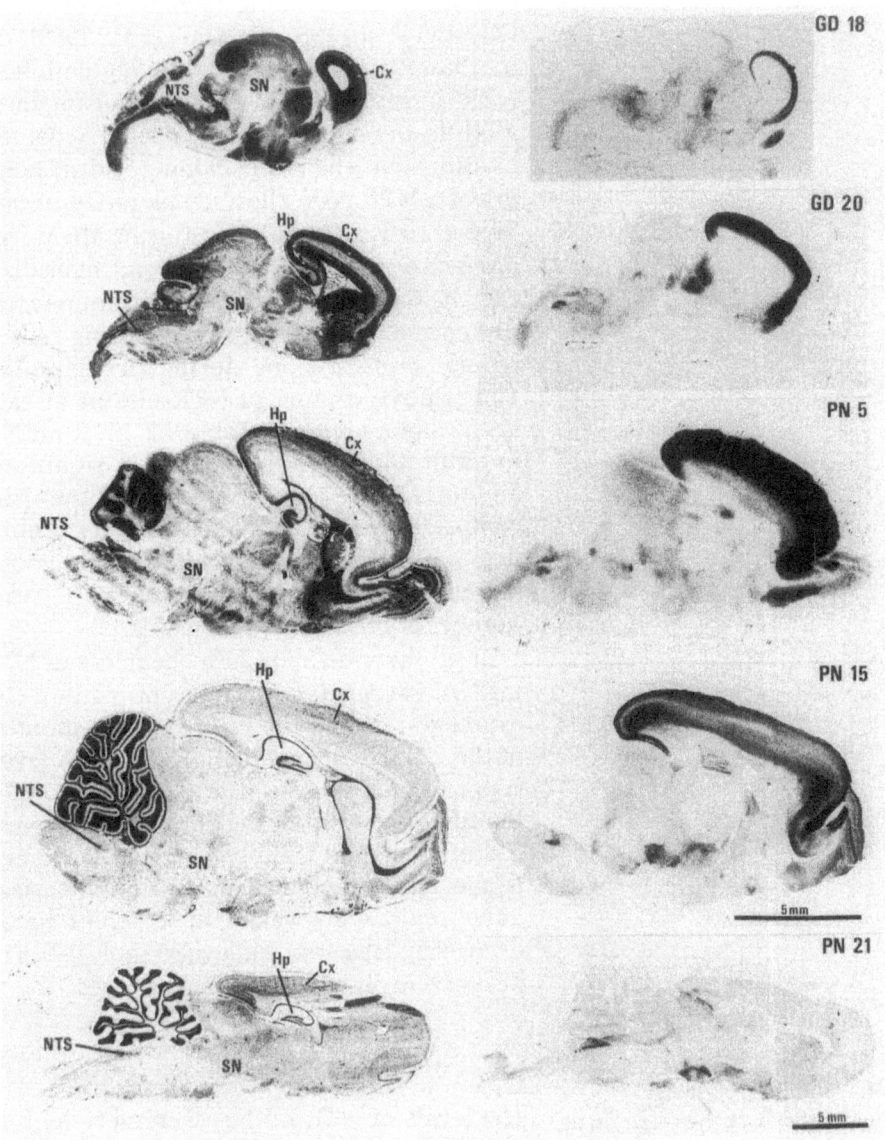

Fig. 9.4. Photomicrographs of sagittal sections of rat brain and cervical spinal cord at different ages during gestation (GD18, GD20) and early postnatal stages (PN5, PN15, PN21). See legend to Fig. 9.3 for details. NTS, nucleus tractus solitarius; SN, substantia nigra.

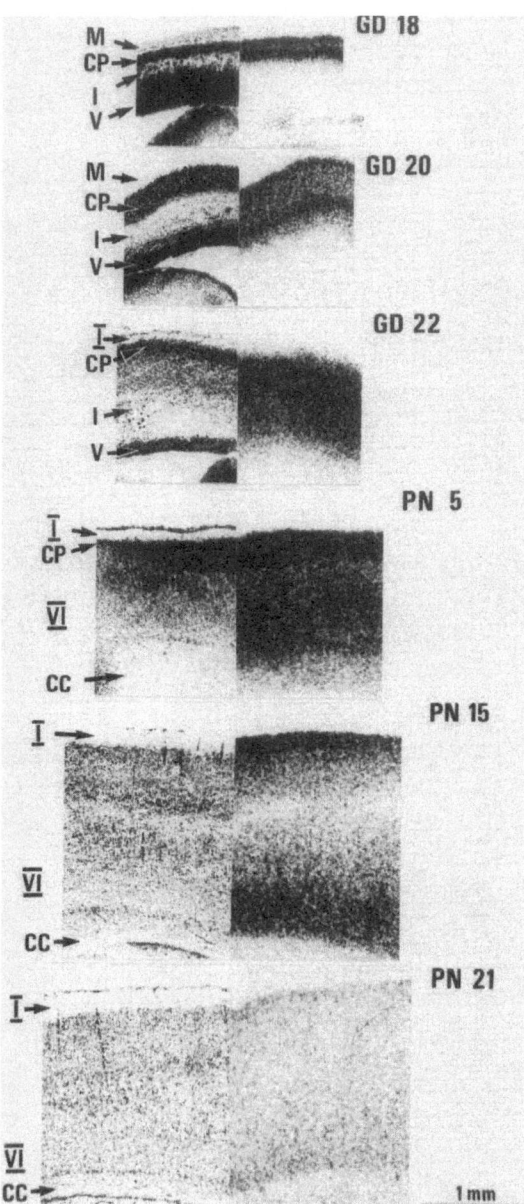

Fig. 9.5. [³H] Neurotensin binding sites in coronal sections of the neocortex of the rat brain at different pre- (GD18, GD20, GD22) and postnatal (PN5, PN15, PN21) ages. See legend to Fig. 9.3 for details. CC, corpus callosum; CP, cortical plate; I, subplate layer; V, VI, layers V and VI.

9.4.3 NEUROTENSIN RECEPTOR mRNA DISTRIBUTION IN DEVELOPING BRAIN: CORRELATION WITH NEUROTENSIN BINDING SITES

Although developmental changes in the density of NT binding sites were observed (Kiyama *et al.*, 1987; Palacios *et al.*, 1988), the mechanisms involved in the regulation of NTR densities which may account for these modifications have not yet been extensively investigated. The recent cloning and sequencing of a NTR gene allows exploration of new aspects of NTR regulation during the development of the CNS. Using *in situ* hybridization histochemistry we have examined the presence of NTR mRNA synthesizing cells in several brain regions during development. The relative content of NTR mRNA at each age is summarized in Table 9.2. NTR mRNA in adult animals was detected in significant amounts only in substantia nigra pars compacta, ventral tegmental area, retrorubral field and medial lemniscus whereas in young animals NTR mRNA was seen in a larger number of brain areas (Table 9.2).

In situ hybridization histochemistry of NTR mRNA revealed two different patterns of expression during the postnatal development of the CNS. In some areas NTR mRNA levels progressively increased after birth, reaching the adult pattern after the third postnatal week (i.e. in substantia nigra pars compacta, ventral tegmental area and nucleus basalis) whereas in other regions it peaked at different times, decreasing later to adult levels (Table 9.2). The peaks were observed at different times.

(a) Cortical areas

The levels of NTR mRNA were very high in cortex of young animals but followed a distinct complex pattern depending on both age and cortical area. The motor area of frontal cortex expressed NTR mRNA at higher levels in the internal layers than in the sensory area, where it was expressed in both

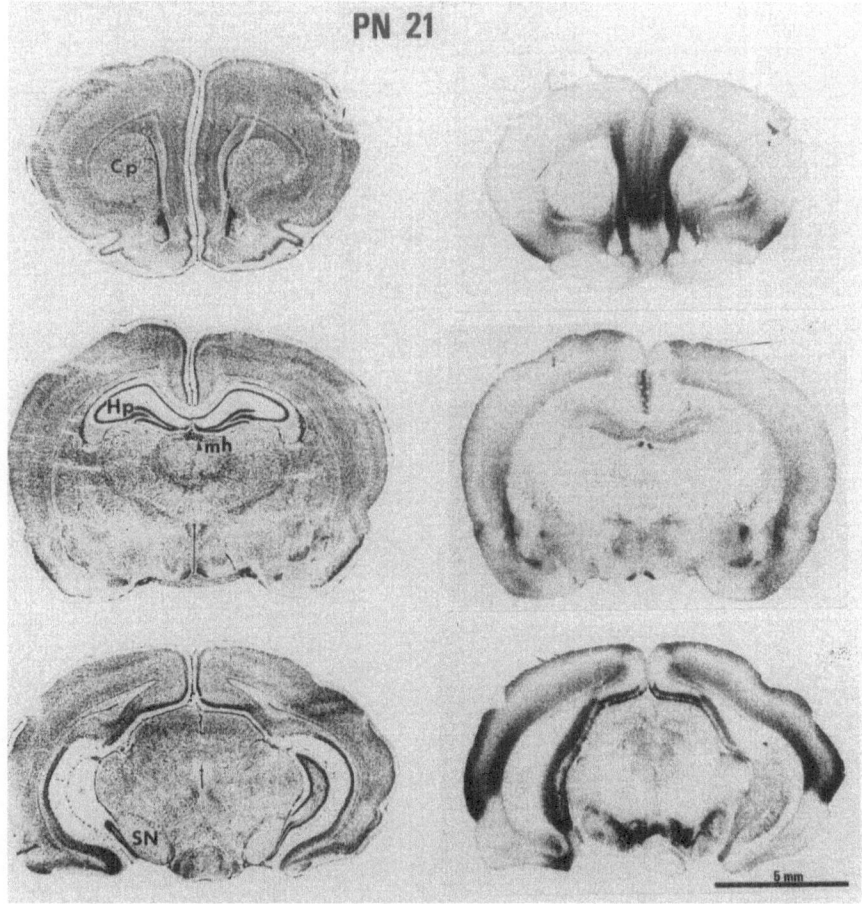

Fig. 9.6. Coronal sections of the rat brain at PN15. See Fig. 9.3 legend for details. Cp, caudate putamen; mh, medial habenula.

the external and the internal layers (Table 9.2). By PD5, in the anterior part of frontal cortex motor area, the differences in the hybridization signal between the external and the internal layers were more evident, being stronger in the internal layer (Fig. 9.7D,F). However, in the posterior part of the brain the mRNA levels were high in the external layer (Fig. 9.7H,J). In contrast, at this time in the sensory area the highest mRNA levels were always in the internal layers (Fig. 9.7J,L). At PD1 all layers of retrosplenial cortex express NTR mRNA. During development the internal layers dis-

play very low levels of mRNA and by PD10 it is only detectable in the more external layers (Table 9.2). At PD15 moderate levels of mRNA expression are seen (Fig. 9.10F) becoming undetectable by PD20 (Table 9.2).

(b) Nigrostriatal system and basal forebrain

The dopaminergic efferents are the more important projections originating in substantia nigra and innervating the caudate-putamen, amygdala and globus pallidus. Palacios and

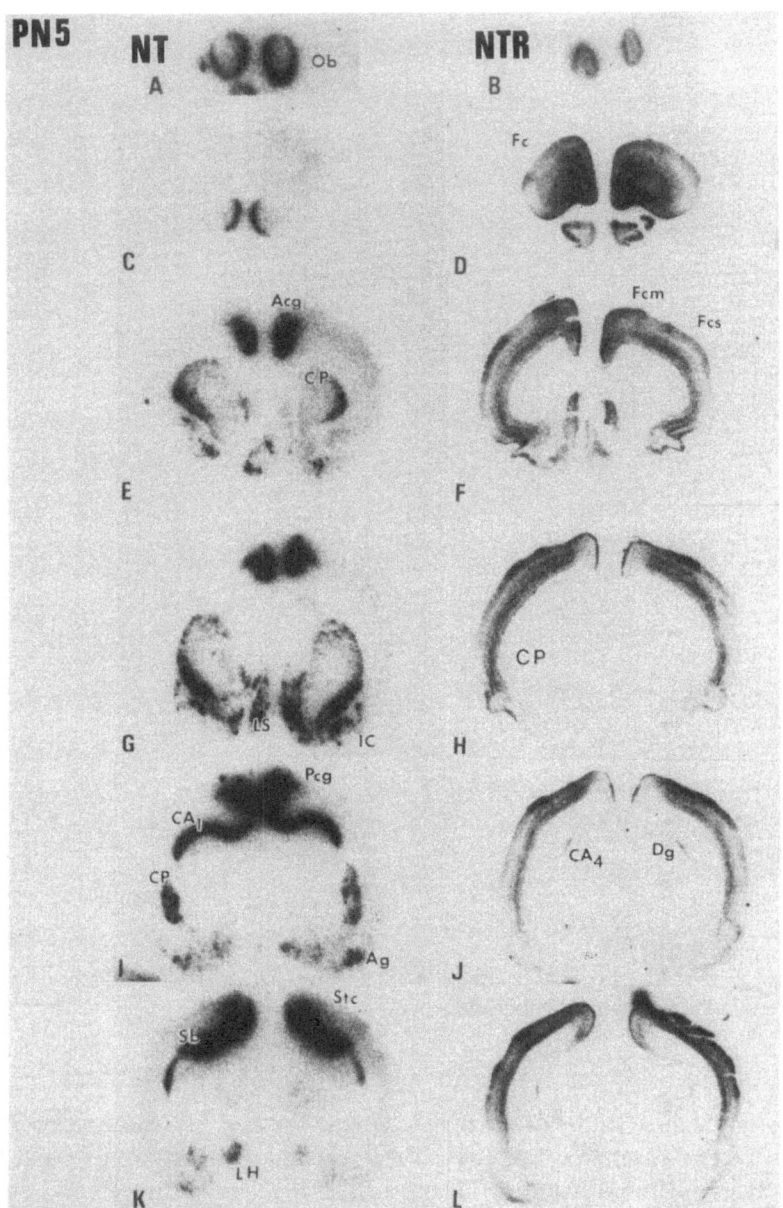

Fig. 9.7.　Autoradiographic visualization of mRNA coding for neurotensin (left column) and neurotensin receptor (right column) in the rat brain at PN5. The images are photographs from autoradiograms generated by hybridization of tissue sections with oligonucleotide probes complementary to selected regions of the NT and NTR genes. Dark areas are rich in hybridization signal. Abbreviations: Acg, anterior cingulate cortex; Ag, amygdaloid nuclei; CA1, CA4, fields of Ammon's Horn; CP, cortical plate; Dg, dentate gyrus; Fcm, Fcs, frontal cortex motor and sensory areas; GB, globus pallidus; Hb, habenular nuclei; HG, hypoglossus nucleus; IC, inferior colliculus; LH, lateral hypothalamic area; LS, lateral septum; Ltdg, laterodorsal tegmental nucleus; Ms, medial septum; NB, nucleus basalis; Ob, olfactory bulb; Pcg, posterior cingulate cortex; Po, primary olfactory cortex; Pst, principal sensory trigeminal nucleus; RSS, retrosplenial cortex; SNC, substantia nigra compacta; Sb, subiculum; Stc, striate cortex; VDB, diagonal band of Broca's vertical limb; VMH, ventromedial hypothalamic nucleus; VTA, ventral tegmental area. Reproduced from García-Ladona *et al.* (1991), with permission.

Table 9.2.   Relative levels of NTR mRNA in different rat brain regions during postnatal development

Regions	PD1	PD5	PD10	PD15	PD20	Adult
Nucleus basalis	-	-	+++	++++	+++++	+++++
Substantia nigra	-	-	++	+++	++++	++++
Ventral tegmental area	-	-	++	+++	++++	++++
Olfactory bulb	++	+++	++++	+++++	+++	+
Ammon's Horn	++	++	++	++	-	-
Dentate gyrus granule cells	++	++	++	++	-	-
Hilus	++	++++	+++	++	-	-
Habenular nucleus	-	-	+++	+++	-	-
Nucleus geniculate lateral	+++	+++	+	-	-	-
Neocortex						
Frontal cortex somatosensory internal	+++++	++++++	++	-	-	-
Frontal cortex somatosensory external	+++++	++++	++	-	-	-
Frontal cortex motor internal	++++++	++++++	++	-	-	-
Frontal cortex motor external	++++	+++	++	-	-	-
Anterior cingulate internal	+++	+++	++	-	-	-
Anterior cingulate medium	+++	++++	+++	-	-	-
Anterior cingulate external	+++	+++	+	-	-	-
Retrosplenial internal	+++++	++++	-	-	-	-
Retrosplenial external	+++++	++++	++++	++++	+	-
Striate internal	+++++	+++++	+++	++	-	-
Striate medium	+++++	+++++	+++	++	-	-
Striate external	+++++	+++++	+++	-	-	-

-, no significant amounts were detected.

Kuhar (1981) have shown that NTR binding sites are localized in the dopaminergic cells of substantia nigra, further adding to the wealth of data on NT interactions with the dopaminergic system (Nemeroff, 1980; Govoni *et al.*, 1980; Hökfelt *et al.*, 1984; Shi and Bunney, 1991; Levant *et al.*, 1991). There is increasing evidence for a general presence of NT binding sites in the nigrostriatal and other dopaminergic systems (Hokfelt *et al.*, 1984; Govoni *et al.*, 1980; Levant *et al.*, 1991; Nemeroff *et al.*, 1980; Shi and Bunney, 1991). Autoradiographic studies (Young and Kuhar, 1981; Quirion *et al.*, 1982; Moyse *et al.*, 1987) reported the presence of NT binding sites in basal ganglia and related zones (i.e. caudate-putamen, islands of Calleja, olfactory tubercle, substantia nigra, and ventral tegmental area). However, in adult animals NTR mRNA expression was found mainly in substantia nigra and ventral tegmental area (Fig. 9.11A), whereas in caudate-putamen and islands of Calleja no significant levels were observed in these animals.

During development, significant levels of NTR mRNA were seen in the substantia nigra pars compacta and ventral tegmental area as early as PD10 and increased thereafter (Table 9.2). Although a postnatal increase in [^{3}H]NT binding sites was described at the same period, high levels of binding were found even at younger ages (Palacios *et al.*, 1988) (Figs 9.2B and 9.4).

(c)   Hippocampal formation

Intermediate levels of NTR mRNA were observed at PD1 in the Ammon's horn, hilus and granule cells of dentate gyrus (Table 9.2). Between PD5 and PD10 high mRNA levels were observed in the hilus but decreased by PD15 (Fig. 9.10B,C). NT mRNA content decreased during development in the Ammon's horn fields, thus in these areas no detectable amount of NTR mRNA was present in adult animals.

The pattern of expression of NTR mRNA in the hippocampal formation during development agrees with previous data on NT binding sites (Palacios *et al.*, 1988). NTR is mainly expressed until PD15 in the CA4 field of Ammon's horn and infragranular and granular layers of dentate gyrus. During the postnatal period, the dentate gyrus is still developing and, at the time we detected NTR mRNA and NT binding sites (Palacios *et al.*, 1988), granule cells migrate from the hilus towards the granular layer (Bayer, 1980), thus supporting the hypothesis that NT plays a role in the maturation of the dentate gyrus and its related connections.

Autoradiographic studies have shown the presence of high levels of NT binding sites in the subiculum of adult hippocampus (Moyse *et al.*, 1987) which disagrees with *in situ* hybridization data. This mismatch possibly indicates that NT binding sites present in this area are presynaptic receptors in the afferent fibers.

(d)   Olfactory system

NTR mRNA was expressed in the first postnatal week (Fig. 9.8B) in the internal granule cell layer and at lower levels in the external plexiform cell layer. It increased until PD15 (Fig. 9.9B) decreasing thereafter (Table 9.2). NTR binding was detected in the plexiform and internal granular layers of adult olfactory system (Young and Kuhar, 1981; Moyse *et al.*,

1987), indicating the existence of NT circuits in the olfactory bulb.

The transient expression of NTR could indicate a neurotrophic effect of NT during the maturation of different CNS regions. This effect could be important for migrating cells in cortical structures as well as granule cells in dentate gyrus. These hypotheses were also suggested by Gonzalez *et al.* (1991) on somatostatin receptors, showing also a changing developmental pattern of somatostatin binding in cortex. Transient expression of NTR in cortex is possibly related to cortical cell death. In fact, such a phenomenon seems to be the cause of observed transient expression of other neuropeptidergic systems in brain cortex (Parnavelas and Cavanaugh, 1988; Fitzpatrick-McElligott *et al.*, 1991).

Another possible explanation for the transient presence of NTR mRNA could be a developmental regulation of NTR while synaptic contacts are taking place in target areas (being NTR presynaptic components) which could also explain the mismatch between the mRNA and the binding site. However, because of the different levels of sensitivity of the techniques used in these studies, caution has to be exerted in interpreting these results. Finally, the possibility of a neuropeptide–receptor interaction during maturation cannot be ruled out as it was proposed for cortical structures (Cowan *et al.*, 1984). For instance, the transient expression of NTR mRNA in hippocampus could be related to neurotensinergic efferent connections from other lymbic structures such as septum and diagonal band given that during development a high expression of NT mRNA was observed in these areas peaking around the second postnatal weak (see below).

Data from lesion studies led to the hypothesis that levels of the agonist may regulate the presence of neurotransmitter receptors in target areas. Surgical or chemical denervation paradigms demonstrated an up-regulation of receptor density in several neurotransmitter systems such as adrenergic receptors in lateral geniculate body (Menkes *et al.*, 1983) and

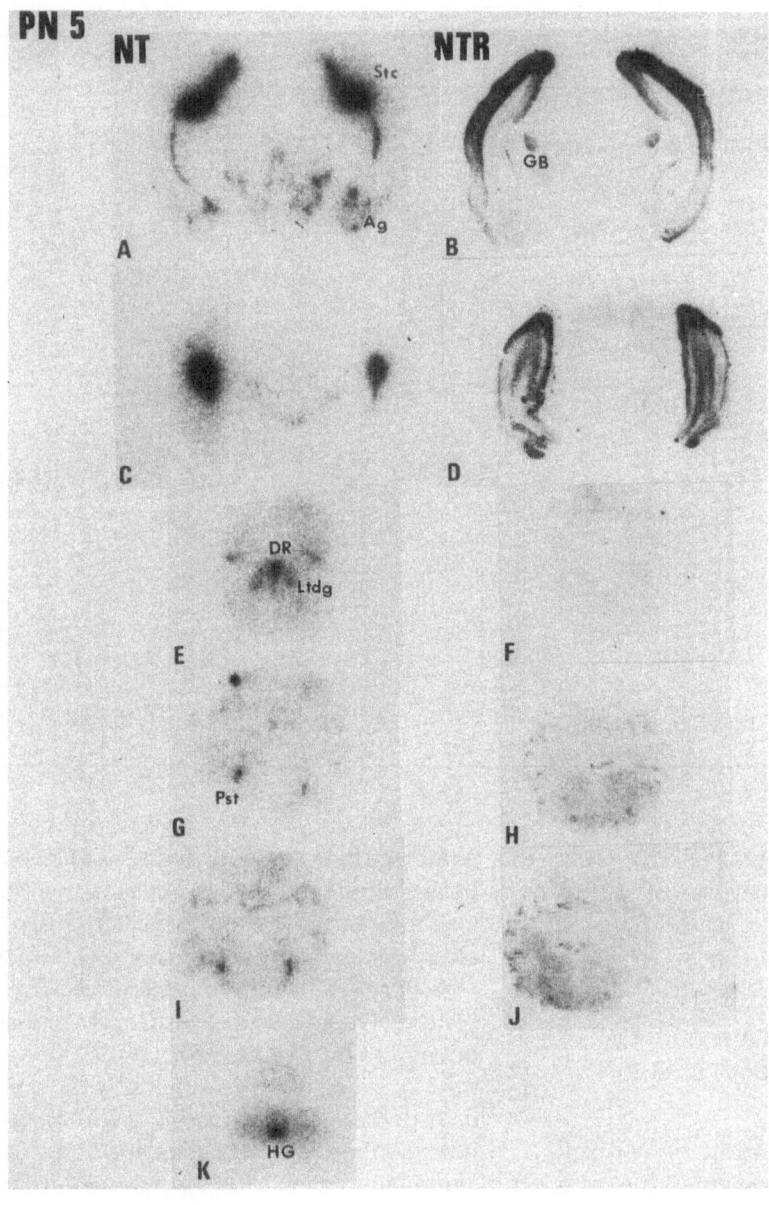

Table 9.3.   Relative levels of NT mRNA in different rat brain areas during postnatal development

Regions	PD1	PD5	PD10	PD15	PD20	Adult
Caudate putamen	+++++	+++++	++++	+++	+++	+++
Islands of Calleja	+++++	+++++	+++	+++	+++	+++
Olfactory bulb	++++++	++++	+++	++	++	+
Subiculum	++++++	+++++	++++	++++	++++	++++
Ammon's Horn	+++	++	++	++	-	-
Dentate gyrus	+++	++	++	++	-	-
Amygdala	++++	++++	+++	++	+	+
Septum	++	++++	+++	+++	++	++
Diagonal band	+	+++	++++	+++	+++	+++
Lateral hypothalamus	+++++	++++	++++	+++	-	-
Inferior olive	-	-	++++	++++	++	++
Dorsal raphe	++++	+++	+++	++	++	+
Lateral tegmental nucleus	+++	++	+	+	+	+
Dorsal parabrachialis nucleus	+++	+++	+++	++	-	-
Ventral parabrachialis nucleus	+++	+++	+++	++	-	-
Dorsal lateral lemniscus	-	+++	++	+	+	+
Dorsal cochlear	-	++++	++++	++++	++++	++++
Lateral vestibular	-	++++	++++	++++	++++	++++
Parvocellular nucleus	-	++++	++++	++++	++++	++++
Solitarius nucleus	-	++++	++++	++++	++++	++++
Spinal trigeminal nucleus	-	-	++++	++++	+++	+++

-, no significant levels were detected.

expression of NTR and NT can help to understand how the presence of the transmitter may control the presence of the receptor during ontogeny.

9.4.4   DEVELOPMENTAL PATTERN OF NEUROTENSIN mRNA LEVELS IN RAT BRAIN

The levels and distribution of NT mRNA during postnatal development in the rat brain have been examined using *in situ* hybridization histochemistry (Sato *et al.*, 1990; Kiyama *et al.*, 1991a,b; our unpublished results). In adult animals, NT mRNA was detected in more regions that NTR mRNA: dorsal and ventral areas of the striatum, nucleus accumbens, Calleja islands, olfactory tubercle, inferior olive, nucleus hypoglossus, nucleus spinalis trigemini, dorsal raphe, cerebellar peduncle (interal tegmental nucleus),

parabranchial nucleus, dorsal cochlear nucleus, vestibular nucleus, nucleus tractus solitarius. During development NT mRNA was also detected in other CNS regions (Table 9.3).

As for NTR, different patterns of NT mRNA content were observed during the postnatal development (Table 9.3). High levels of NT mRNA were observed in the first postnatal days decreasing progressively with time in the intermediate layers of anterior and posterior cyngulate cortex, caudate-putamen, islands of Calleja, olfactory bulb, CA1 field of Ammon's horn, dentate gyrus, amygdala, lateral hypothalamic area (Figs 9.7A–K; 9.9A–I; 9.10A–G), dorsal raphe, lateral tegmental nucleus, dorsoparabranchial nucleus and ventroparabranchial nucleus and inferior olive (Fig. 9.8A–K). Relatively high levels of NT mRNA were found through the developmental period mainly in subiculum, dorsal cochlear nucleus, lateral vestibular nucleus,

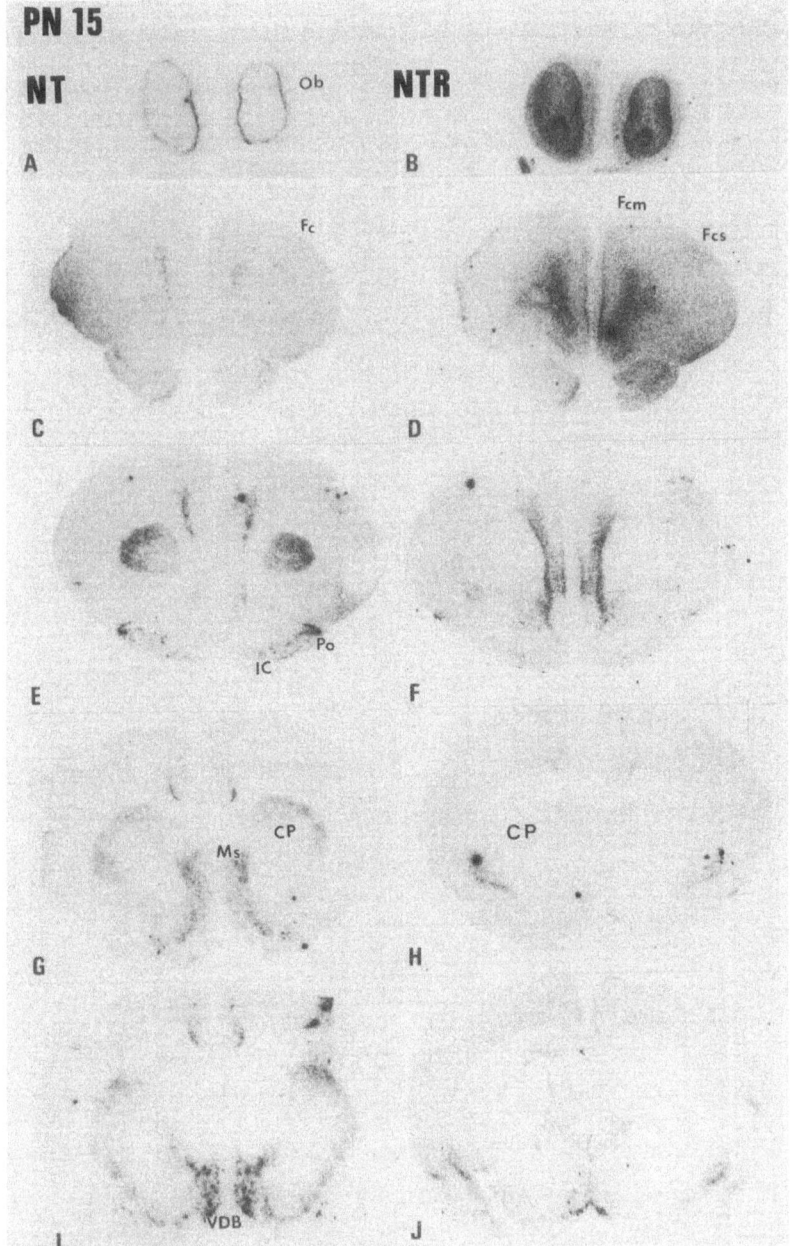

Fig. 9.9. Postnatal day 15. See legend to Fig. 9.7 for details.

parvocellular nucleus, nucleus solitarius and spinal trigeminal nucleus, although these brainstem nuclei displayed detectable levels of NT mRNA only after PD5. Finally, NT mRNA content peaked at different ages of postnatal development in septum, diagonal band and inferior olive.

In general, hybridization data for NT

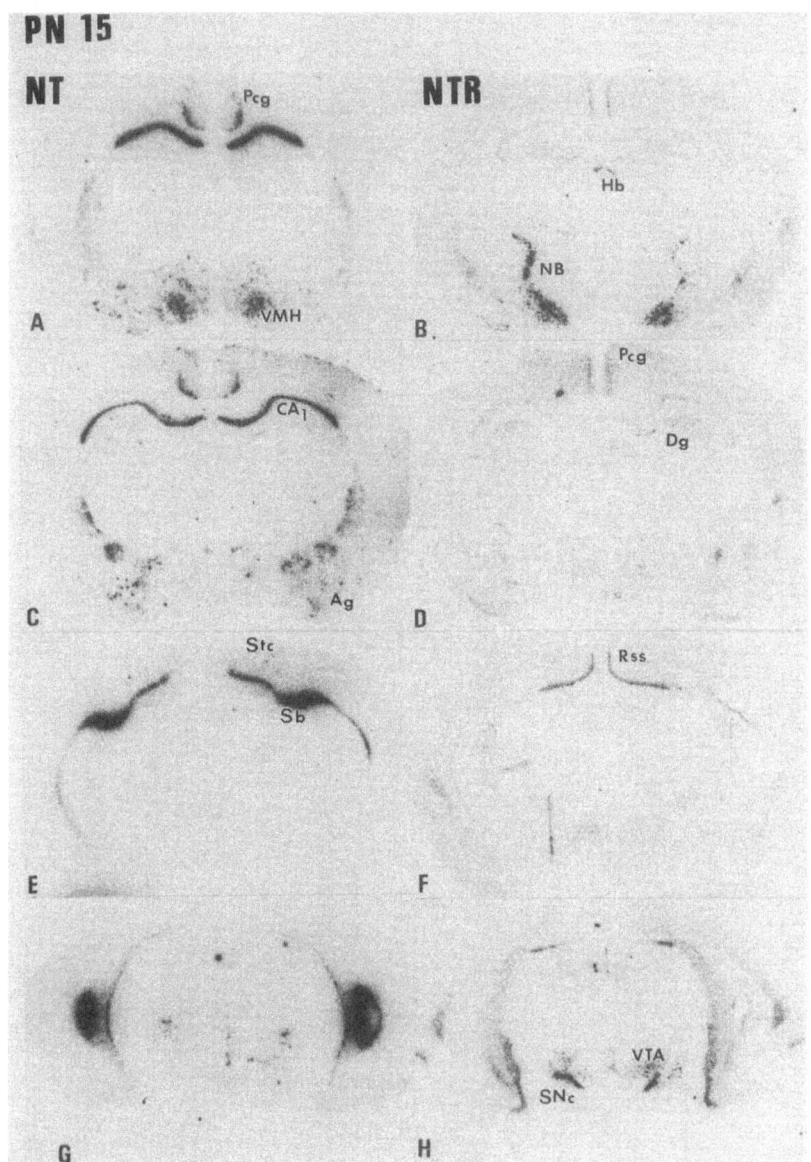

Fig. 9.10. Postnatal day 15. See legend to Fig. 9.7 for details.

mRNA and immunohistochemical results are in good agreement in adult and developing brain (Hara *et al.*, 1982; Minagawa *et al.*, 1983). However, mismatches between the levels of peptide and mRNA can be observed. Several brain areas displayed high levels of NT mRNA through development although immunopositive cells or fibers have not been described (subiculum, CA1 field of the hippocampus) or are seen only in the early postnatal period (brain stem nuclei). These mismatches between NT and NT mRNA can

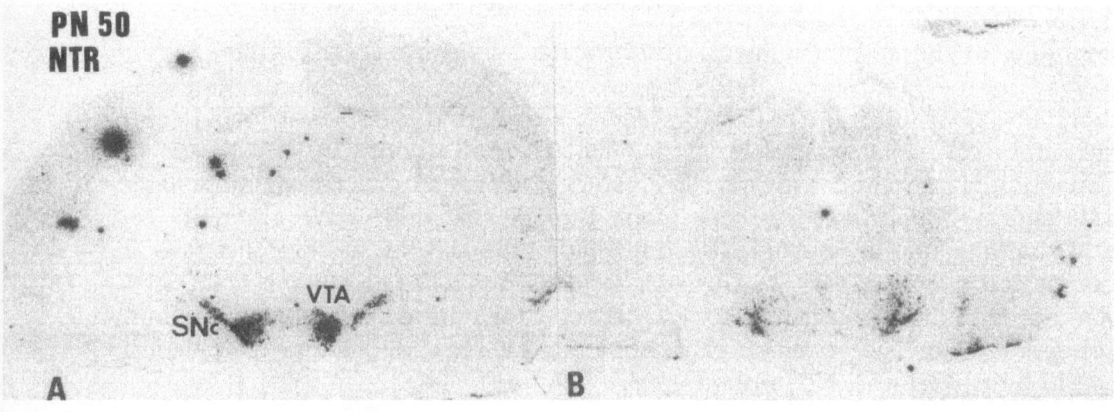

Fig. 9.11. Visualization of NTR mRNA at postnatal day 50. Patterns at PN50 are similar to adult.

be explained by different mechanisms involving NT gene expression and processing in different steps (Uhl and Nishimori, 1990).

The NT gene can be actively expressed, translated and the peptide transported to distant target areas or turnover rates of both NT and NT mRNA are very slow. As mentioned above, the NT gene expression is regulated by second messenger pathways (Dobner *et al.*, 1988; Tischler *et al.*, 1991) and this could be specially important in the perinatal period. We cannot rule out the possibility that the processing of the peptide is attenuated in the earlier steps of development because the machinery to process the peptide is still immature, as suggested for other systems (Rehfeld, 1990). Thus, although the NT mRNA is present, NT is not easily detectable by immunohistochemistry.

Differential processing of neuropeptide genes is a common characteristic of several peptidergic systems (Uhl and Nishimori, 1990). It was demonstrated that NT gene contains a NT-like sequence as well as the neuromedin N sequences (Dobner *et al.*, 1987; Kislauskis *et al.*, 1988). During development a different processing of the gene coding for NT and neuromedin can be activated at different developmental steps, thus, in adults the expression of NT-containing gene does not necessarily imply NT

production. This mechanism of differential processing has been suggested for NT in a recent study demonstrating different ratios of NT/neuromedin gene products in several brain regions (Kitabgi *et al.*, 1991).

The comparative analysis of the immunohistochemical and autoradiographic studies revealed the existence of several mismatches between the presence of fibers and/or NT-immunopositive cells and NTR. To explain these mismatches between neurotransmitters and receptors, different hypotheses have been proposed (for extensive review see Herkenham, 1987). In the neurotensinergic system, the most striking mismatch is possibly that found in developing neocortex. High levels of NTR mRNA and NT binding sites are observed during early development whereas either NT-immunoreactive fibers or cells, or NT mRNA levels were not detected except in some specific cases. In this case, NT would act as a paracrine peptide by diffusion from distant areas. A direct regulation of NTR by the presence of NT can be expected for instance in the septohippocampal and inferior olive system. In these areas, NT mRNA is present in increasing amounts during the first postnatal week whereas NTR mRNA and NT binding progressively decrease in the same period (Tables 9.2 and 9.3).

9.5 CONCLUDING REMARKS

The understanding of the molecular mechanisms involved in the maturation of the neurotensinergic system during postnatal development of the CNS is still incomplete.

Receptor autoradiography studies (Palacios *et al.*, 1988) have revealed the transient presence of NT binding sites in several brain regions even at the prenatal period. This phenomenon has also been described for other neuropeptides (Gonzalez *et al.*, 1991; Tribollet *et al.*, 1991). Although transient expression for some of the classical neurotransmitters can be the result of simple synaptic modifications occurring during postnatal development, for the case of NT and other neuropetides it may also imply a direct neurotrophic role (Hanley, 1985) during this period of maturation. It is evident that there are mismatches between the presence of the NT and mRNAs (Uhl and Nishimori, 1990) and/or by the transport of the molecule far from the area of synthesis. Mismatches between the presence of NT-immunopositive structures and binding sites or NTR and their respective mRNAs can be explained by the different rate of turnover of both protein and mRNAs.

Whereas in some brain regions the NT and NTR mRNA contents may account for a down-regulation induced by increasing concentration of the neuropeptide (i.e. septo-hippocampal formation), in others, such regulation is not so clear as NT mRNA decreases at the same time as NTR mRNA is up-regulated (striatonigral system). However, further studies should be carried out to verify if lack of detection of NT immunopositive cells or fibers is due to a rapid turnover or transport of the peptide. Also, the obtaining of antibodies raised against synthetic peptides from the NTR protein sequence may help in the elucidation of real mismatches of neurotensinergic system as well as the mechanisms of synaptic maturation during development.

ACKNOWLEDGEMENTS

Part of the studies reviewed in this chapter were carried out at Preclinical Research, Sandoz Pharma Ltd, Basel (Switzerland), and in collaboration with the Institute of Pathology, University of Basel, particularly with Professor A. Probst. We gratefully acknowledge the help of all our colleagues at these institutions. We also thank Dr R.G.W. Gristwood for critical reading of the manuscript, and Ms M.D. Moliné for expert clerical help.

REFERENCES

Alexander, M.J., Miller, M.A., Dorsa, D.M. *et al.* (1989) Distribution of neurotensin/neuromedin N mRNA in rat forebrain: unexpected abundance in hippocampus and subiculum. *Proc. Natl. Acad. Sci. USA*, **86**, 5202–6.

Bayer, S.A. (1980) Development of the hippocampal region in the rat. I. Neurogenesis examined with ^{3}H-thymidine autoradiography. *J. Comp. Neurol.*, **190**, 87–114.

Bozou, J.C., Amar, S., Vincent, J.P. and Kitabgi, P. (1986) Neurotensin mediated inhibition of cyclic AMP formation in neuroblastoma N1E-115 cells: involvement of the inhibitory GTP-binding component of adenylate cyclase. *Mol. Pharmacol.*, **29**, 489–96.

Cowan, W.M., Fawcett, J.W., O'Leary, D.D.M. and Stanfield, B.B. (1984) Regressive events in neurogenesis. *Science*, **225**, 1258–65.

Dobner, P.R., Barber, D.L., Villa-Komaroff, L. and McKiernan, C. (1987) Cloning and sequence analysis of cDNA for the canine neurotensin/neuromedin N precursor. *Proc. Natl. Acad. Sci. USA*, **84**, 3516–20.

Dobner, P.R., Tischler, A.S., Lee, Y.C. *et al.* (1988) Lithium dramatically potentiates neurotensin/neuromedin N gene expression. *J. Biol. Chem.*, **263**, 13983–6.

Fitzpatrick-McElligott, S., Card, J.P., O'Kane, T.M. and Baldino, F. (1991) Ontogeny of somatostatin mRNA-containing perikarya in the rat central nervous system. *Synapse*, **7**, 123–34.

Garcia-Ladona, F.J., Palacios, J.M. and Mengod, G. (1992) Neurotransmitter/receptor mismatches during ontogeny: an in situ hybridization correlative study of the ontogenic expression of neurotensin and neurotensin receptor mRNAs in rat brain. (Submitted.)

Gilbert, J.A. and Richelson, E. (1984) Neurotensin stimulates formation of cyclic GMP in murine neuroblastoma clone N1E-115. *Eur. J. Pharmacol.*, **99**, 245–6.

Goedert, M., Pinnock, R.D, Downes, C.P. *et al.* (1984a) Neurotensin stimulates inositol phospholipid hydrolysis in rat brain slices. *Brain Res.*, **323**, 193–7.

Goedert, M., Pittaway, K., Williams, B.J. and Emson P.C. (1984b) Specific binding of tritiated neurotensin to rat brain membranes: characterization and regional distribution. *Brain Res.*, **304**, 71–81.

Goedert, M., Mantyh, P.W., Emson, P.C. and Hunt, S.P. (1984c) Inverse relationship between neurotensin receptors and neurotensin-like immunoreactivity in the cat striatum. *Nature*, **307**, 543–6.

Gonzalez, B.J., Leroux, P., Bodenant, C. and Vaudry, H. (1991) Ontogeny of somatostatin receptors in the rat somatosensory cortex. *J. Comp. Neurol.*, **305**, 177–88.

Govoni, S., Hong, J.S., Yang, H.Y.T. and Costa, E. (1980) Increase of neurotensin content elicited by neuroleptics in nucleus accumbens. *J. Pharmacol. Exp. Ther.*, **215**, 413–17.

Hanley, M.R. (1985) Neuropeptides as mitogens. *Nature*, **315**, 14–15.

Hara, Y., Shiosaka, S., Senba, E. *et al.* (1982) Ontogeny of the neurotensin-containing neuron system of the rat: immunohistochemical analysis. I. Forebrain and diencephalon. *J. Comp. Neurol.*, **208**, 177–95.

Herkenham, M. (1987) Mismatches between neurotransmitter and receptor localizations in brain: observations and implications. *Neuroscience*, **23**, 1–38.

Hökfelt, T., Everitt, B.J., Theodorson-Norheim, E. and Goldstein, M. (1984) Occurrence of neurotensin-like immunoreactivity in subpopulations of hypothalamic, mesencephalic, and medullary catecholamine neurons. *J. Comp. Neurol.*, **222**, 543–59.

Hunt, S.P. (1988) The development of neurotransmitter receptors, in *The Making of the Nervous System* (eds J.G. Parnavelas, C.D. Stern and R.V. Stirling), Oxford University Press, Oxford, pp. 454–72.

Inagaki, S., Shinoda, K., Kubota, Y. *et al.* (1983) Evidence for the existence of a neurotensin-containing pathway from the endopiriform nucleus and the adjacent prepiriform cortex to the ante-rior olfactory nucleus and nucleus of diagonal band (Broca) of the rat. *Neuroscience*, **8**, 487–93.

Jennes, L., Stumpf, W.E. and Kalivas, P.W. (1982) Neurotensin: topographical distribution in rat brain by immunohistochemistry. *J. Comp. Neurol.*, **210**, 211–24.

Kahn, D., Abrams, G.M., Zimmerman, E.A. *et al.* (1980) Neurotensin neurons in the rat hypothalamus: an immunocytochemical study. *Endocrinology*, **107**, 47–54.

Kalivas, P.W., Jennes, L., Nemeroff, C.B. and Prange Jr, A.J. (1982) Neurotensin: topographical distribution of brain sites involved in hypothermia and antinociception. *J. Comp. Neurol.*, **210**, 225–38.

Kalivas, P.W., Burgess, S.K., Nemeroff, C.B. and Prange Jr., A.J. (1983) Behavioural and neurochemical effects of neurotensin microinjection into the ventral tegmental area of the rat. *Neuroscience*, **8**, 495–505.

Kasckow, J., Cain, S.T. and Nemeroff, C.B. (1991) Neurotensin effects on calcium/calmodulin-dependent protein phosphorylation in rat neostriatal slices. *Brain Res.*, **545**, 343–6.

Kislauskis, E., Bullock, B., McNeil, S. and Dobner, P.R. (1988) The rat gene encoding neurotensin and neuromedin N. *J. Biol. Chem.*, **263**, 4963–8.

Kitabgi, P., Rostene, W., Dussaollant, M. *et al.* (1987) Two populations of neurotensin binding sites in murine brain: discrimination by the antihistaminic levocastine reveals markedly different radioautographic distribution. *Eur. J. Pharmacol.*, **140**, 285–93.

Kitabgi, P., Masuo, Y., Nicot, A. *et al.* (1991) Marked variations of the relative distributions of neurotensin and neuromedin N in micropunched rat brain areas suggest differential processing of their common precursor. *Neurosci. Lett.*, **124**, 9–12.

Kiyama, H., Inagaki, S., Kito, S. and Tohyama, M. (1987) Ontogeny of [^{3}H]neurotensin binding sites in the rat cerebral cortex: autoradiographic study. *Dev. Brain Res.*, **31**, 303–6.

Kiyama, H., Emson, P.C., Sato, M. and Tohyama, M. (1991a) The transient appearance of proneurotensin mRNA in the rat hypoglossal nucleus during development. *Dev. Brain Res.* **58**, 293–6.

Kiyama, H., Sato, M., Emson, P.C. and Tohyama, M. (1991b) Transient expression of neurotensin mRNA in the mitral cells of rat olfactory bulb during development. *Neurosci. Lett.*, **128**, 85–9.

Kiyama, H., Shiosaka, S., Sakamoto, N. *et al.* (1986) A neurotensin immunoreactive pathway from the subiculum to the mammillary body in the rat. *Brain Res.*, **375**, 357–9.

Kohler, C., Radesater, A.C., Hall, H. and Winblad, B. (1985) Autoradiographic localization of [³H]neurotensin-binding sites in the hippocampal region of the rat and primate brain. *Neuroscience*, **16**, 577–87.

Levant, B., Bissette, G., Widerlov, E. and Nemeroff, C.B. (1991) Alterations in regional brain neurotensin concentrations produced by atypical antipsychotic drugs. *Regul. Pept.*, **32**, 193–201.

Mailleux, P. and Vanderhaeghen, J.-J. (1988) Transient neurotensin in the human inferior olive during development. *Brain Res.*, **456**, 199–203.

Mailleux, P., Schiffmann, S.N., Halleux, P. *et al.* (1989) Transient neurotensin in the cat inferior olive during development. *Neurochem. Int.*, **14**, 159–61.

Mailleux, P., Pelaprat, D. and Vanderhaeghen, J.-J. (1990) Transient neurotensin high-affinity binding sites in the human inferior olive during development. *Brain Res.*, **508**, 345–8.

Mazella, J., Poustis, C., Labbe, C. *et al.* (1983) Monoiodo-[Trp11]neurotensin, a highly radioactive ligand of neurotensin receptors. *J. Biol. Chem.*, **258**, 3476–81.

Mazella, J., Chabry, J., Kitabgi, P. and Vincent, J.P. (1988) Solubilization and characterization of active neurotensin receptors from mouse brain. *J. Biol. Chem.*, **263**, 144–9.

Mazella, J., Chabry, J. and Vincent, J.P. (1989) Purification of the neurotensin receptor from mouse brain by affinity chromatography. *J. Biol. Chem.*, **264**, 5559–63.

Menkes, D.B., Gallager, D.W., Reinhard, J.F. and Aghajanian, G.K. (1983) Alpha-adrenoceptor denervation supersensitivity in brain: physiological and receptor binding studies. *Brain Res.*, **272**, 1–12.

Mills, A., Demoliou-Mason, C.D. and Barnard, E. (1988) Purification of the neurotensin receptor from bovine brain. *J. Biol. Chem.*, **263**, 13–16.

Minagawa, H., Shiosaka, S., Inagaki, S. *et al.* (1983) Ontogeny of neurotensin-containing neuron system of the rat: immunohistochemical analysis II. Lower brain stem. *Neuroscience*, **8**, 467–86.

Minneman, K.P., Pittman, R.N. and Molinoff, P.B. (1981) β-adrenergic receptor subtypes: properties, distribution and regulation. *Annu. Rev. Neurosci.*, **4**, 419–61.

Moyse, E., Rostène, W., Vial, M. *et al.* (1987) Distribution of neurotensin binding sites in rat brain: a light microscopic radioautographic study using monoiodo[¹²⁵I]Tyr3-neurotensin. *Neuroscience*, **22**, 525–36.

Nemeroff, C.B. (1980) Neurotensin: perchance an endogenous neuroleptic? *Biol. Psychiatry*, **15**, 283–302.

Palacios, J.M. and Kuhar, M.J. (1981) Neurotensin receptors are located on dopamine-containing neurones in rat midbrain. *Nature*, **294**, 587–9.

Palacios, J.M., Pazos, A., Dietl, M.M. *et al.* (1988) The ontogeny of brain neurotensin receptors studied by autoradiography. *Neuroscience*, **25**, 307–17.

Parnavelas, J.G. and Cavanagh, M.E. (1988) Transient expression of neurotransmitters in developing neocortex. *Trends Pharmacol. Sci.*, **11**, 92–3.

Penney, J.B., Pan, H.S., Young, A.B. *et al.* (1981) Quantitative autoradiography of [³H]muscimol binding in the rat brain. *Science*, **214**, 1036–8.

Polak, J.M. (ed.) (1989) *Regulatory Peptides*, Birkhäuser Verlag, Basel.

Quirion, R., Gaudreau, P., St Pierre, S. *et al.* (1982) Autoradiographic distribution of [³H]neurotensin receptors in rat brain: visualization by tritium sensitive film. *Peptides*, **3**, 757–63.

Quirion, R., Chiueh, C.C., Everist, H.D. and Pert, A. (1985) Comparative localization of neurotensin receptors on nigrostriatal and mesolimbic dopaminergic terminals. *Brain Res.*, **327**, 385–9.

Rehfeld, J.F. (1990) Posttranslational attenuation of peptide gene expression. *FEBS Lett.*, **268**, 1–4.

Roberts, G.W., Woodhams, P.L., Polak, J.M. and Crow, T.J. (1984) Distribution of neuropeptides in the limbic system of the rat: the hippocampus. *Neuroscience*, **11**, 35–77.

Sato, M., Lee, Y., Zhang, J.H. *et al.* (1990) Different ontogenic profiles of cells expressing preproneurotensin/neuromedin N mRNA in the rat posterior cingulate cortex and the hippocampal formation. *Dev. Brain Res.*, **54**, 249–55.

Sato, M., Shiosaka, S. and Tohyama, M. (1991) Neurotensin and neuromedin N elevate the cytosolic calcium concentration via transiently appearing neurotensin binding sites in cultured rat cortex cells. *Dev. Brain Res.*, **58**, 97–103.

Schotte, A. and Laduron, P.M. (1987) Different postnatal ontogeny of two [³H]neurotensin binding sites in rat brain. *Brain Res.*, **408**, 326–8.

Shi, W.X. and Bunney, B.S. (1991) Neurotensin modulates autoreceptor mediated dopamine effects on midbrain dopamine cell activity. *Brain Res.*, **543**, 315–21.

Tanaka, K., Masu, M. and Nakanishi, S. (1990)

Structure and functional expression of the cloned rat neurotensin receptor. *Neuron*, **4**, 847–54.

Tischler, A.S., Ruzicka, L.A. and Dobner, P.R. (1991) A protein kinase inhibitor, staurosporine, mimics nerve growth factor induction of neurotensin/neuromedin N gene expression. *J. Biol. Chem.*, **266**, 1141–6.

Tribollet, E., Goumaz, M., Raggenbass, M. *et al.* (1991) Early appearance and transient expression of vasopressin receptors in the brain of rat fetus and infant. An autoradiographical and electrophysiological study. *Dev. Brain Res.*, **58**, 13–24.

Uhl, G.R. (1990) Neurotensin receptors, in *Handbook of Chemical Neuroanatomy*, Vol.9. (eds A. Bjorklund, T. Hökfelt and M.J. Kuhar), Elsevier Science, Amsterdam, pp. 443–53.

Uhl, G.R. and Kuhar, M. (1984) Chronic neuroleptic treatment enhances neurotensin receptor binding in human and rat substantia nigra. *Nature*, **309**, 350–2.

Uhl, G.R. and Nishimori, T. (1990) Neuropeptide gene expression and neural activity: assessing a working hypothesis in nucleus caudalis and dorsal horn neurons expressing preproenkephalin and preprodynorphin. *Neurobiology*, **10**, 73–99.

Uhl, G.R. and Snyder, S.H. (1977) Neurotensin receptor binding, regional and subcellular distributions favor transmitter role. *Eur. J. Pharmacol.*, **41**, 89–91.

Young, W.S. and Kuhar, M.J. (1981) Neurotensin receptor localization by light microscopic autoradiography in rat brain. *Brain Res.*, **206**, 273–85.

OPIOID RECEPTORS AND THE DEVELOPING NERVOUS SYSTEM

10

Sandra E. Loughlin and Frances M. Leslie

10.1 INTRODUCTION

Three endogenous opioid peptide families have been identified, including those derived from the proenkephalin, prodynorphin and proopiomelanocortin genes (Imura *et al.*, 1985, for review). Peptides derived from these genes interact with three major classes of opioid receptor, mu, delta and kappa (Leslie, 1987). These receptors are widely distributed throughout the central nervous system and mediate the biological effects of endogenous opioid peptides, as well as of certain non-peptide analgesic drugs such as morphine and ketocyclazocine. In addition, epsilon (Schulz *et al.*, 1979; Chang *et al.*, 1984) and zeta (Zagon *et al.*, 1989) receptor types have been described, which have high affinity and selectivity for the opioid peptides beta-endorphin and enkephalin, respectively.

Both opioid peptides and their receptors appear very early in development, are expressed transiently in a number of brain regions, and reach their mature distributions relatively late in ontogeny (Leslie and Loughlin, 1992). It has, therefore, been suggested that the endogenous opioid systems may mediate basic developmental processes such as cell division and differentiation (Kent *et al.*, 1982). A large body of literature supports this hypothesis (see Zagon and McLaughlin, 1983; McDowell and Kitchen, 1987; Leslie and Loughlin, 1992, for reviews). Clinical data on the subtle, but long-lasting effects of maternal opioid exposure are consistent with a large number of studies demonstrating influences of *in vivo* and *in vitro* administration of opioid agonists and antagonists on developmental processes. Such effects are distinct from those of opioids in the adult nervous system (Olson *et al.*, 1991, for recent review).

Whereas manipulation of opioid systems may modulate developmental processes, such actions may be either direct or indirect and it is not yet clear whether opioids serve as neurotrophic factors. Neurotrophic factors may act by regulating cell division or cell survival, or have effects on process outgrowth and/or synthesis of proteins necessary to differentiated functions of neurons (Hefti *et al.*, 1992). In order to demonstrate that a compound has a neurotrophic role, several criteria must be met, including the presence of the endogenous compound and the receptor to which it binds at the appropriate time and in the appropriate neuronal elements (Fallon and Loughlin, 1992). This chapter describes the ontogeny of endogenous peptide and receptor expression and briefly reviews the evidence for a role of endogenous systems in basic developmental processes.

10.2 ENDOGENOUS OPIOID PEPTIDES

Endogenous opioid peptides are synthesized in the nervous system as large precursors and

Receptors in the Developing Nervous System Vol. 2: Neurotransmitters. Edited by Ian S. Zagon and Patricia J. McLaughlin. Published in 1993 by Chapman & Hall. ISBN 0 412 49400 0. Vols. 1 and 2 (set) ISBN 0 412 54520 9.

are cleaved enzymatically to yield a variety of peptide fragments, at least 17 of which have opioid activity (Douglass *et al.*, 1984; Imura *et al.*, 1985; Hollt, 1986). Three genes encode the opioid peptide precursors, proopiomelanocortin (POMC), proenkephalin and prodynorphin (see Hollt, 1991, for review). They are differentially expressed in the nervous system, as well as in a number of peripheral tissues, including endocrine glands. The peripheral expression of opioid peptides is of particular importance during developmental time periods before the blood–brain barrier is effective, since central nervous system ontogeny may be under the influence of circulating opioids.

Immunocytochemical and radioimmunoassay studies have provided a detailed description of the developmental localization of opioid peptide-containing neuronal elements in brain. Since the peptide products of the three genes exhibit many structural similarities, detailed anatomical studies on the distributions of these systems have relied on the use of multiple antisera directed at specific amino acid sequences. More recently, molecular biological techniques have allowed the study of the developmental expression of mRNA for each of the genes. However, both approaches are limited by the sensitivity of assays for detection of low levels of products which may be expressed transiently or in restricted regions.

10.2.1 ENDOGENOUS OPIOID PEPTIDES IN THE ADULT

While recent *in situ* hybridization studies have confirmed the neurochemical identity of many cell groups previously localized by immunocytochemistry, the precise distribution of opioid pathways and terminals remains somewhat controversial. Combined immunocytochemical and retrograde tracing studies have elucidated many of the discrete opioid pathways. A detailed, critical review of the distribution of endogenous opioids is, however, outside the scope of this chapter. The adult systems are described briefly here and the subsequent discussion concentrates on what is known of the developmental appearance of each system.

(a) Proopiomelanocortin

The POMC gene (Uhler and Herbert, 1983) is expressed in the brain, the pituitary, and a number of peripheral tissues. Expression of the gene is regulated by steroids and neurotransmitters which modulate c-AMP and intracellular calcium (Hollt, 1991). The POMC precursor is the source of the opioid peptide beta-endorphin, as well as beta-lipotropin, adrenocorticotropic hormone and melanocyte-stimulating hormones (Mains *et al.*, 1977).

In the central nervous system, beta-endorphin and other POMC products are present in neurons of the hypothalamus and in the brainstem nucleus tractus solitarius. POMC neurons in the nucleus tractus solitarius innervate the caudal brainstem, perhaps including the lateral reticular nucleus (Schwartzberg and Nakane, 1982). Processes of POMC cells in the arcuate extend for long distances to innervate many brain regions (Khachaturian *et al.*, 1985a). Many periventricular regions, including septum, bed nucleus of the stria terminalis, periventricular thalamus and periaqueductal gray are innervated, as well as brainstem reticular and autonomic nuclei. A marked absence of POMC fibers is observed in striatum, hippocampus and cortex.

(b) Proenkephalin

The proenkephalin gene (Rosen *et al.*, 1984) is expressed in the pituitary, the adrenal medulla and a number of other peripheral tissues, as well as the central nervous system. Intracellular calcium and factors which modulate adenylate cyclase or protein kinase C regulate gene expression (Hollt, 1991). The opioid peptides met- and leu-enkephalin, and other enkephalin-containing peptides such as met-enkephalin-arg-gly-leu, met-enkephalin-arg-phe, peptide E, peptide F and BAM-22P, are derived from preproenkephalin (Imura *et al.*, 1985).

In contrast to the fairly limited distribution of POMC-containing neurons, the population of cells in the central nervous system expressing proenkephalin is relatively ubiquitous (Khachaturian *et al.*, 1985a; Petrusz *et al.*, 1985; Merchenthaler *et al.*, 1986; Harlan *et al.*, 1987; Fallon and Ciofi, 1990). *In situ* hybridization studies have generally found larger numbers of cell bodies than immunocytochemical studies, probably due to low concentrations of peptide in the normal soma (Schafer *et al.*, 1991). Enkephalinergic systems consist of both long- and short-projecting neurons. Cell bodies are located in most regions of the telencephalon, including the cerebral cortex, olfactory tubercle, amygdala, hippocampus, striatum, septum, bed nucleus of the stria terminalis and preoptic area. Enkephalinergic cells are also present in most hypothalamic nuclei, certain thalamic nuclei, a number of brainstem cell groups including colliculi, periventricular gray, interpeduncular nucleus, reticular and autonomic nuclei, Golgi cells of the cerebellum and spinal cord. Fibers and terminals are observed in a large number of brain structures. Cortical regions are sparsely innervated, with scattered fibers in entorhinal and cingulate cortices, whereas the hippocampus is more densely innervated. Basal telencephalic structures such as the nucleus accumbens, bed nucleus of the stria terminalis, caudate putamen and amygdala contain many enkephalinergic fibers. A dense projection from the caudate putamen to the globus pallidus, substantia nigra and other pallidal structures exists. Most hypothalamic nuclei are moderately innervated, whereas only certain thalamic nuclei, including the medial and lateral geniculate bodies and paraventricular nuclei contain enkephalinergic fibers. Fibers are also observed in the colliculi, central gray, ventral tegmental area, interpe-duncular nuclei, raphe nuclei, locus coeruleus and brainstem motor, autonomic and reticular nuclei.

(c) Prodynorphin

Expression of the prodynorphin, also known as proenkephalin B, gene (Civelli *et al.*, 1985) is greatest in the brain, especially in the striatum, hypothalamus and hippocampus. Little is known of the regulation of expression of prodynorphin, but, like the other opioid genes, it is positively regulated by c-AMP (Hollt, 1991). Prodynorphin is the source of the opioid peptide, dynorphin A, as well as leu-enkephalin and other leu-enkephalin containing peptides including alpha- and beta-neoendorphin (Hollt, 1986).

The majority of the brain regions which contain enkephalin cells and terminals also contain dynorphin (Khachaturian *et al.*, 1983; Morris *et al.*, 1986; Schafer *et al.*, 1988; Fallon and Ciofi, 1990). However, there are significant differences in the distribution of the two opioid peptide systems. In cortex, there is a large number of dynorphin-containing cells and processes, especially in entorhinal and cingulate cortices. Hippocampal dynorphin terminals are of much greater density than enkephalin and are distributed to all hippocampal subfields except CA1. In the basal telencephalon, dynorphin is distributed more sparsely in the bed nucleus of the stria terminalis. Dynorphin cells and fibers are densely localized to the ventral pallidum and largely avoid the enkephalin-rich globus pallidus. In the hypothalamus, dynorphin cells and fibers are found in most of the nuclei which contain enkephalin, with minor differences in distribution. The distribution in thalamus is restricted to the same nuclei as enkephalin, including geniculate bodies and paraventricular nuclei. In the substantia nigra, dynorphin and enkephalin are differentially distributed such that the substantia nigra pars reticulata contains dynorphin, whereas enkephalin is localized to the substantia nigra pars compacta. Dynorphin is not present in the interpeduncular nucleus, but is distributed similarly to enkephalin in brainstem and spinal cord.

(d) Co-localization

The localization of the three endogenous opioid peptide families is, thus, quite extensive. The beta-endorphin system is organized such that a limited number of cell bodies, located in discrete nuclei, have long projections to large portions of the basal forebrain, diencephalon and brainstem, particularly periventricular regions. In contrast, the enkephalin and dynorphin systems are characterized by a diffuse distribution of cell bodies throughout the neuraxis, with projections to the majority of brain systems. Whereas enkephalin and dynorphin are co-localized in a few structures, they are largely present in separate populations of neurons and fibers which are co-distributed (Fallon and Ciofi, 1990). It is likely that the enkephalin and dynorphin inputs may converge on individual target cells to modulate their output. In some of the regions where all three peptide systems are co-distributed, it is possible that neurons may be bathed in a complex mixture of opioid peptides.

(e) Central nervous system opioids in nonneuronal cells

In addition to its neuronal expression, there is evidence suggesting that proenkephalin is expressed in glial cells in the central nervous system (CNS). Proenkephalin mRNA has been localized to astrocytes in culture (Vilijn *et al.*, 1989; Hauser *et al.*, 1990; Stiene-Martin *et al.*, 1991b) and cultured astrocytes have been shown to release enkephalin (Batter *et al.*, 1991). Although enkephalin immunoreactivity within glia of the adult CNS has not been observed, immunoreactive glia have been detected in developing cerebellum (Zagon and McLaughlin, 1990).

10.2.2 ONTOGENY OF ENDOGENOUS OPIOID PEPTIDES

All three opioid genes are expressed early in embryonic development (Rosen and Polakiewicz, 1989), as are their peptide products (see Khachaturian *et al.*, 1985a; McDowell and Kitchen, 1987, for reviews). In addition to the CNS expression of these opioid peptides, the prenatal brain may be exposed to circulating endogenous opioids from peripheral sources. These include POMC products from the pituitary (Schwartzberg and Nakane, 1982; Khachaturian *et al.*, 1983) and ovaries (Shaha *et al.*, 1984), proenkephalin products from the adrenal medulla (Keshet *et al.*, 1989; Coulter *et al.*, 1990) and prodynorphin products from the posterior pituitary derived from hypothalamic magnocellular neurons (Watson *et al.*, 1982).

By embryonic day 17 (ED17), the rat brain contains 20% of adult levels of a 'morphine-like' substance (Garcin and Coyle, 1976). The opioid peptides met-enkephalin, dynorphin and beta-endorphin are detectable in mouse brain by ED11.5, before any opioid binding sites are detectable (Rius *et al.*, 1991a). In early postnatal ontogeny, the relative distributions of the opioid peptides and receptors suggest that no simple relationships can be defined between any one peptide and any one receptor type (Loughlin *et al.*, 1985). Therefore, the developmental appearance of each peptide family will be discussed and subsequently the ontogeny of each receptor type will be reviewed.

(a) Proopiomelanocortin

POMC mRNA has been detected in the rat central nervous system by ED10.5 (Elkabes *et al.*, 1989). Although levels are still low at birth and during the first postnatal week, they are higher at postnatal day 8 (PD8) than during the next two weeks. In the fourth postnatal week, there is a rapid increase to adult levels (Rosen and Polakiewicz, 1989). Studies on the distribution of beta-endorphin immunoreactivity have shown a similar biphasic appearance of the peptide (Bayon *et al.*, 1979; Bloom *et al.*, 1980). Beta-endorphin is present by ED16 in most brain regions at approximately 2% of adult levels. The concentration of peptide relative to brain protein increases little after ED18 (Ng *et al.*, 1984). Within some brain regions, peptide content remains level or decreases slightly until PD14, then increases to adult levels (Kapcala, 1983). Immunoreactive cells and fibers are visible in hypothalamus in the second embryonic week (Schwartzberg and Nakane, 1982; Khachaturian *et al.*, 1983; Elkabes *et al.*, 1989; Rius *et al.*, 1991a), with earliest visualization of hypothalamic cells by ED10.5 and hypothalamic fibers by ED11.5 (Elkabes *et al.*, 1989). Immunoreactive fibers are present in mesencephalon by ED11.5, in spinal cord by ED12.5 and the adult pattern of terminal arborization is recognizable by ED13.5–15.5 (Elkabes *et al.*, 1989).

Transient expression of POMC peptides has been reported in spinal cord from ED 14–PD28 (Haynes *et al.*, 1982) and in the early postnatal forebrain germinal zone (Kent *et al.*, 1982; Loughlin *et al.*, 1991). This transient POMC expression may be in both neuronal and non-neuronal cells. POMC peptide expression in adult peripheral nervous system may reappear following injury (Berry and Haynes, 1989). It has been suggested that POMC peptides may modulate cellular proliferation or guide migration of postmitotic cells (Loughlin *et al.*, 1991) or may regulate the transition from the mitotic cycle to the onset of differentiation (Berry and Haynes, 1989).

(b) Proenkephalin

Proenkephalin mRNA is detectable in rat brain by ED11–12 (Keshet *et al.*, 1989), with enkephalin immunoreactivity visible by ED 11.5 (Rius *et al.*, 1991a). By ED13, both met- and leu-enkephalin like immunoreactivity are present in multiple forms (Dahl *et al.*, 1982).

Enkephalin immunoreactive fibers are present in hindbrain by ED15 and throughout the neuraxis by ED17–18 (Palmer *et al.*, 1982). Although enkephalin is present in many brain regions by ED16, it is present in much lower concentrations than beta-endorphin (Bayon *et al.*, 1979). Unlike beta-endorphin, the developmental distribution of enkephalin does not resemble that of the adult. Met- and leu-enkephalin levels increase independently in postnatal ontogeny, suggesting that differential processing occurs or that leu-enkephalin is a significant product of prodynorphin (Patey *et al.*, 1980).

The ontogeny of striatal enkephalin has been examined by a number of investigators. In fetal pig, hippocampal mRNA increases linearly, whereas striatal mRNA peaks in midgestation, then declines (Pittius *et al.*, 1987). Proenkephalin mRNA is detectable in rat striatum by ED16, followed by the appearance of immunoreactivity by ED17–18. This appearance follows the gradient of cell generation (Tecott *et al.*, 1989). Schwartz and Simantov (1988) have reported that striatal enkephalin mRNA peaks at PD2, then decreases to embryonic levels by PD7, increasing in postnatal weeks 2–4 to adult levels. This developmental profile is similar to that reported for enkephalin immunoreactivity (Bayon *et al.*, 1979). In contrast, Rosen and Polakiewicz (1989) have reported a peak at PD8–10, followed by a 50-fold decrease at PD28.

Hippocampal enkephalin immunoreactivity contrasts with the early appearance of striatal enkephalin in that fibers are first seen at postnatal stages (Gall *et al.*, 1984). At ED 16, enkephalin immunoreactivity is detectable in hippocampal extracts, but at very low levels (Bayon *et al.*, 1979). It should be noted that the hippocampus is a late developing structure, with cell division continuing into postnatal stages.

A transient appearance of enkephalin immunoreactivity has been reported in the cerebellar germinal zone (Zagon *et al.*, 1985). While enkephalin mRNA was not detected in the neurons of the external granular layer, Purkinje cells and some Golgi cells express enkephalin mRNA (Osborne *et al.*, 1991). Levels of expression have been shown to follow the maturational gradient of these cells. In the retina, enkephalin mRNA is present during development, but not in the adult (Isayama and Zagon, 1991). Enkephalin mRNA is also observed in mesodermally derived cells before differentiation (Keshet *et al.*, 1989). It has, thus, been suggested that enkephalin may be an important regulator of cell division or differentiation (Zagon and McLaughlin, 1993).

(c) Prodynorphin

Less is known of the developmental appearance of prodynorphin in the central nervous system. As the distribution parallels that of enkephalin, many similarities exist in the developmental appearance of enkephalin and dynorphin systems in brain, especially in striatum (Pittius *et al.*, 1987; Rosen and Polakiewicz, 1989). Thus, in the fetal pig, prodynorphin mRNA increases linearly in hippocampus, whereas in the striatum it peaks at midgestation, then declines (Pittius *et al.*, 1987). In mouse brain, dynorphin immunoreactivity is first seen at ED11.5 (Rius *et al.*, 1991b) and prodynorphin products have been detected in rat brain at ED20 (Zamir *et al.*, 1985). Prodynorphin mRNA is first detectable in rat striatum at ED15 (Alvarez-Bolado *et al.*, 1990). By the day of birth, striatal dynorphin immunoreactivity has a patchy appearance (Loughlin *et al.*, 1985). Striatal prodynorphin mRNA levels have been reported to peak at PD8–10, then decrease 50-fold by PD28 (Rosen and Polakiewicz, 1989). Hypothalamic and medullary prodynorphin mRNA levels show similar postnatal increases followed by decreases. The postnatal appearance of prodynorphin mRNA has also been seen in cortical regions with iso- and allocortices expressing this gene in the first postnatal week and the hippocampus at PD14 (Alvarez-Bolado *et*

al., 1990; Sato *et al.*, 1991). Dynorphin immunoreactivity, in contrast, is seen at PD6 (Gall 1984).

(d) Opioid peptide processing

As has been noted, the interpretation of studies on the developmental appearance of opioid peptides is complicated by cross-reactivity caused by similarities in peptide structure. In addition, synthesis, degradation and post-translational processing of the peptides might affect the results obtained. Post-translational processing of propeptide precursors giving rise to active peptide products has been shown to occur early in ontogeny (Seizinger *et al.*, 1984a,b; McMillen *et al.*, 1988; Coulter *et al.*, 1990; Sei and Dores, 1990; Hylka and Thommes, 1991; Quinlan and Alessi, 1991; Rius *et al.*, 1991b). Opioid precursors may be differentially processed at different developmental stages, such that the ratio of long- to short-chain peptides decreases as maturation occurs. The degradative enzyme, neutral endopeptidase, which is also known as enkephalinase, has been detected in fetal brain (Patey *et al.*, 1980). Thus, mechanisms for peptide processing, release and degradation are apparent early in development.

10.3 OPIOID RECEPTORS

The receptors to which both exogenous and endogenous opioid compounds bind have been intensively studied. The pharmacology of the opioid receptor types is reviewed here followed by their localization in adult brain. Subsequent sections discuss the developmental appearance of each type and review the literature suggesting that early receptors may differ in their pharmacological characteristics from those of the adult.

10.3.1 PHARMACOLOGY

Three major classes of opioid receptor, mu, delta and kappa have been identified within the mammalian CNS and periphery (Leslie, 1987). An epsilon receptor, with high affinity and selectivity for beta-endorphin, has also putatively been identified from functional (Schulz *et al.*, 1981) and radioligand binding (Nock *et al.*, 1990) studies, although this is somewhat controversial (reviewed by Leslie, 1987). Most recently, a zeta receptor, with high affinity and selectivity for enkephalin peptides, has been identified in immature tissues (Zagon *et al.*, 1990, 1991). The properties and possible functional roles of this receptor are described by Zagon and McLaughlin (1993) and will not be addressed further at this point.

The ontogeny of opioid receptors has been examined primarily by radioligand binding. Although membrane binding and autoradiographic studies have yielded generally similar findings (McDowell and Kitchen, 1987; Leslie and Loughlin, 1992), the use of quantitative autoradiography has provided a higher degree of anatomical resolution such that regional and temporal changes within a single brain structure can be studied in great detail. Although early reports of opioid receptor ontogeny involved the use of radioligand binding conditions which were not highly selective for a given receptor type (Clendennin *et al.*, 1976; Coyle and Pert, 1976; Kirby, 1981), later studies have discriminated the developmental appearance of multiple receptor types (reviewed by Leslie and Loughlin, 1992).

Recent functional and radioligand binding data suggest that mu, delta and kappa opioid receptors may consist of multiple subtypes. Early evidence of mu receptor heterogeneity was provided by the work of Pasternak *et al.* (1980), who demonstrated a differential appearance of high and low affinity morphine binding sites in developing brain. The ontogeny of these two affinity states correlated with the functional expression of morphine analgesia and respiratory depression, respectively. Although later studies have provided additional support for the concept of inde-

pendent mu_1 and mu_2 receptors (Pasternak and Wood, 1986), the differential developmental appearance of these sites has not been examined in further detail and will not be discussed here. In most recent studies described in this review, mu receptor ontogeny has been examined using [^{3}H]DAGOL, a ligand with equal affinity for mu_1 and mu_2 sites (Clark *et al.*, 1988).

An increasing body of evidence has indicated heterogeneity of kappa receptors within mammalian CNS (reviewed by Unterwald *et al.*, 1991). Although some investigators have reported the existence of as many as four receptor subtypes (Rothman *et al.*, 1990; Kinouchi and Pasternak, 1991), there is presently strongest evidence for the subclassification of $kappa_1$ and $kappa_2$ sites (Zukin *et al.*, 1988). Whereas most early studies of kappa receptor ontogeny used binding conditions which were somewhat selective for the $kappa_1$ site, we have recently compared the ontogeny of $kappa_1$ and $kappa_2$ subtypes in rat brain (Leslie and Smith, 1992) (section 10.3.3).

Most recently, functional and radioligand binding studies have suggested the possible existence of more than one type of delta receptor (Negri *et al.*, 1991; Wild *et al.*, 1991). As yet, however, radioligand binding conditions have not been described which clearly discriminate these putative subtypes. In the present review, delta receptors will therefore be discussed as a single class.

10.3.2 OPIOID RECEPTOR DISTRIBUTION IN THE ADULT

Quantitative autoradiography studies have shown that mu, delta and kappa opioid binding sites are present in most regions of the CNS (Khachaturian *et al.*, 1985a; Leslie and Loughlin, 1992, for review). They are heterogeneously distributed such that each type has a distinct localization. Whether or not these sites represent functional receptors remains to be determined, but a large number of studies have indicated that opioids exert receptor-mediated effects within many systems of adult and developing brain (DeVries *et al.*, 1991; Olson *et al.*, 1991; also see below).

We review here the distribution of mu, delta and kappa receptors in the brain of the adult rat, the most studied species. Descriptions for forebrain areas are drawn from a number of recent mapping studies which report largely similar distributions (Kornblum *et al.*, 1987a; Mansour *et al.*, 1987, 1991; Sharif and Hughes, 1989).

(a) Mu binding sites

High densities of mu receptors are present in a number of telencephalic structures. Binding in cerebral cortex is moderately dense, but exhibits regional and laminar heterogeneity. In many cerebral cortical regions, binding is greater in layers I, III and IV or V. Within the hippocampus, mu receptors are especially dense in the dentate granule cell layer and are also dense in the pyramidal cell layer and stratum lacunosum moleculare. Olfactory bulb, accessory olfactory bulb and accessory olfactory nucleus exhibit high levels of binding which correspond to specific cell layers. The distribution of mu binding sites in olfactory tubercle is patchy and probably corresponds to a dense accumulation of receptors over striatal compartments within sparse background labeling. This patchy distribution of mu binding is also present in the lateral nucleus accumbens, with more homogeneous labeling in medial nucleus accumbens. The patchy distribution of mu binding in the caudate putamen is superimposed on a moderately dense matrix. The patches of dense labeling correspond to the striosomes, which are an important aspect of the compartmental organization of the caudate putamen (Graybiel, 1984). In the adult rat, binding is sparse in the globus pallidus and entopeduncular nucleus (Herkenham and McLean, 1986), as it is in the lateral septum. Moderate levels of binding are present in the medial septum and bed

nucleus of the stria terminalis. Certain nuclei of the amygdala contain significant populations of mu receptors.

Mu binding sites are present in moderate to dense concentration in many thalamic nuclei, but not in ventroposterior, parafascicular and reticular nuclei. Binding is especially dense in the medial habenula and subfornical organ, whereas the lateral habenula and zona incerta have no binding sites. In the hypothalamus, mu receptors are most densely localized in suprachiasmatic, dorsomedial, ventromedial and supramammillary nuclei and in anterior and lateral hypothalamic areas.

Mu labeling is moderately dense in the mesencephalon, with a laminar distribution in superior colliculus and dense labeling in the inferior colliculus and periaqueductal gray. The substantia nigra pars compacta and portions of the pars reticulata and the ventral tegmental area contain moderate densities of mu receptors. Dense labeling is also seen in the medial terminal nucleus and several raphe nuclei.

Brainstem nuclei in the pons and medulla also contain mu binding sites, with highest densities in the parabrachial, raphe magnus and spinal trigeminal nuclei, nucleus of the solitary tract and dorsal motor nucleus of the vagus (Xia and Haddad, 1991). Binding sites are also present in locus coeruleus. In contrast, the cerebellum is largely devoid of mu binding sites in the adult.

The superficial layer and the medial portion of lamina V of the spinal cord show moderate densities of mu binding sites (Morris and Herz, 1987; Besse *et al.*, 1990).

(b) Kappa binding sites

Early studies on the distribution of kappa sites in rat brain used conditions which were somewhat selective for kappa$_1$ receptors, which have high affinity and selectivity for the opioid peptide, dynorphin A, and for arylacetamides such as U50,488 (Kornblum *et al.*, 1987a; Mansour *et al.*, 1987; Leslie and

Loughlin, 1992). More recently, it has been shown that only 10% of rat brain kappa sites are of the kappa₁ type (Hurlbut *et al.*, 1988, 1992; Unterwald *et al.*, 1991). The majority of non-mu, non-delta sites in this species have a distinct pharmacological profile, with high affinity for certain benzomorphan drugs (Chang *et al.*, 1984; Zukin *et al.*, 1988; Hurlbut *et al.*, 1988). Although it has been suggested that these represent beta-endorphin-selective epsilon sites (Nock *et al.*, 1990), these have been more commonly designated as kappa₂ (Zukin *et al.*, 1988; Attali and Vogel, 1990). Kappa₁ and kappa₂ sites are found in different brain regions and are differentially distributed in the rat and guinea-pig brain. These differences are thought to account for the complex pharmacological effects of kappa opioids.

In the rat brain, kappa₁ sites are sparse in neocortical structures, though dense binding is observed in the claustrum and endopiriform nucleus. Binding is strikingly low in hippocampus. Kappa₁ sites are heterogeneously distributed in moderate densities in the olfactory tubercle and are present in caudate putamen and nucleus accumbens, with greater density in the medial-dorsal caudate putamen, fundus striata and medial accumbens. Certain nuclei of the amygdala contain moderate densities of kappa₁ sites, including basolateral and medial nuclei. Low densities of binding sites are found in certain thalamic nuclei, especially the midline nuclei, whereas the majority of thalamic nuclei contain no kappa₁ sites. Binding in hypothalamus is moderately dense, especially in the lateral hypothalamic area and ventromedial nucleus. In the mesencephalon, binding sites are of greatest density in superior colliculus, interpeduncular nucleus and central gray.

Kappa₂ sites are present in many brain structures in higher concentrations than kappa₁ sites. In rat brain there are marked similarities between the distribution of these sites and that of the mu receptor (Unterwald *et al.*, 1991; Leslie and Smith, 1992). However, recent pharmacological and developmental data indicate that, though distributed in the same brain regions as mu sites, these kappa₂ sites represent a distinct receptor population (Unterwald *et al.*, 1991; Leslie and Smith, 1992; also see below).

The distribution of kappa₂ sites in adult rat brain has recently been described (Unterwald *et al.*, 1991). Neocortical kappa₂ binding sites are diffusely distributed throughout layers I–VI. In the hippocampus, binding sites are present in moderate density in the pyramidal cell layer and the granular cells of the dentate gyrus. Moderately high densities of kappa₂ sites are observed in the caudate putamen, nucleus accumbens and olfactory tubercle. Dense binding is present in amygdala, especially in basolateral and medial nuclei. In thalamus, kappa₂ sites are moderately dense in lateral, posterior and midline nuclear groups, and particularly dense in the habenula. Diffuse labeling is seen throughout the hypothalamus. In mesencephalon, binding is present in superior colliculus, medial geniculate, central gray, substantia nigra pars reticulata and interpeduncular nucleus. A moderate level of binding is also present in the locus coeruleus, with a low density in the cerebellum.

(c) Delta binding sites

Delta binding sites are also present in high densities in rat brain, particularly in the telencephalon. In the adult cortex, delta binding is relatively diffuse, with moderate density in deep and superficial layers and lower levels of binding in intermediate layers. Delta receptors in hippocampus are present in low concentrations with no marked regional or laminar heterogeneity. In olfactory bulb, the densest labeling is in the external plexiform layer, with moderate levels in granule and internal plexiform layers. Diffuse delta binding of moderate density is present in olfactory tubercle and nucleus accumbens. Binding in the caudate putamen is equivalent in density to that of mu receptors overall but the pattern of labeling differs in that a diffuse distribution of sites increases in density from medial to lateral. The globus pallidus, endopeduncular nucleus and septal nuclei show very low levels of delta binding whereas the bed nucleus of the stria terminalis and the nucleus of the diagonal band show moderate levels. Moderate levels of delta binding sites are also present in those nuclei of the amygdala which express mu sites.

The thalamus and hypothalamus show very low levels of delta binding, except in the ventromedial nucleus. Binding is also sparse throughout the mesencephalon except for moderate concentrations of receptors in the interpeduncular nucleus. Delta sites are diffusely distributed in the brainstem, at levels 2–40 times lower than mu concentrations (Xia and Haddad, 1991). In the spinal cord, some delta binding sites are detectable in the superficial layers of the dorsal horn (Besse *et al.*, 1990).

10.3.3 ONTOGENY OF OPIOID RECEPTORS

Early studies on the developmental appearance of opioid receptors in brain using relatively non-selective antagonists showed that binding sites are detectable by ED14–16 (Coyle and Pert, 1976; Clendennin *et al.*, 1976; Kirby, 1981). They follow a complex course of development with significant regional variation in time of appearance. The ontogeny of opioid binding sites in brain has been reviewed in detail by McDowell and Kitchen (1987) and Leslie and Loughlin (1992) and is summarized briefly here. In rat and mouse, mu and kappa receptors appear in embryonic brain, whereas delta receptors appear later (Leslie *et al.*, 1982; Tavani *et al.*, 1985; Kitchen *et al.*, 1990; Rius *et al.*, 1991a; Leslie and Loughlin, 1992). The time course of appearance in human brain follows a similar profile, with mu and kappa sites appearing earlier than delta (Magnan and Tiberi, 1989).

(a)   Mu binding sites

In the mouse brain, mu binding sites are detectable by ED12.5 (Rius *et al.*, 1991a). In the rat brain, they are present by ED14 (Kent *et al.*, 1982; Kornblum *et al.*, 1987a; Leslie and Loughlin, 1992). Relative to brain protein, mu binding sites decline in concentration during the first postnatal week (Spain *et al.*, 1985; Petrillo *et al.*, 1987; Attali and Vogel, 1990), increase during the second and third weeks, then decline to adult levels. Quantitative autoradiography studies have demonstrated a similar developmental profile (Kent *et al.*, 1982; Unnerstall *et al.*, 1983; Kornblum *et al.*, 1987a). Autoradiograms showing the postnatal development of mu receptor binding in selected regions are shown in Figs. 10.1 and 10.2.

Neocortical labeling of mu sites is first detected at ED20, heterogeneously distributed in both cortical layers and regions. Binding increases throughout early postnatal weeks, especially during the third week. Hippocampal binding is diffuse at birth, becomes laminated by the end of the first postnatal week and achieves near adult levels during the second week. In olfactory bulb and anterior olfactory nucleus, mu binding sites are present in high concentrations at birth. The density of binding decreases markedly in the external plexiform layer between PD2 and PD9, when it begins to increase in other layers. It has been suggested that this developmental profile corresponds to the maturation of specific cell types (Unnerstall *et al.*, 1983).

Prenatally, mu receptor binding is first detected in the striatal anlage at ED14 (Kent *et al.*, 1982; see below). By ED16, diffuse labeling is apparent throughout the developing caudate putamen and nucleus accumbens. Binding remains diffuse until 1–2 days after birth, when it begins to assume the adult patchy distribution (Kent *et al.*, 1982; Murrin and Ferrer, 1984; Kornblum *et al.*, 1987a). Throughout the next two postnatal weeks,

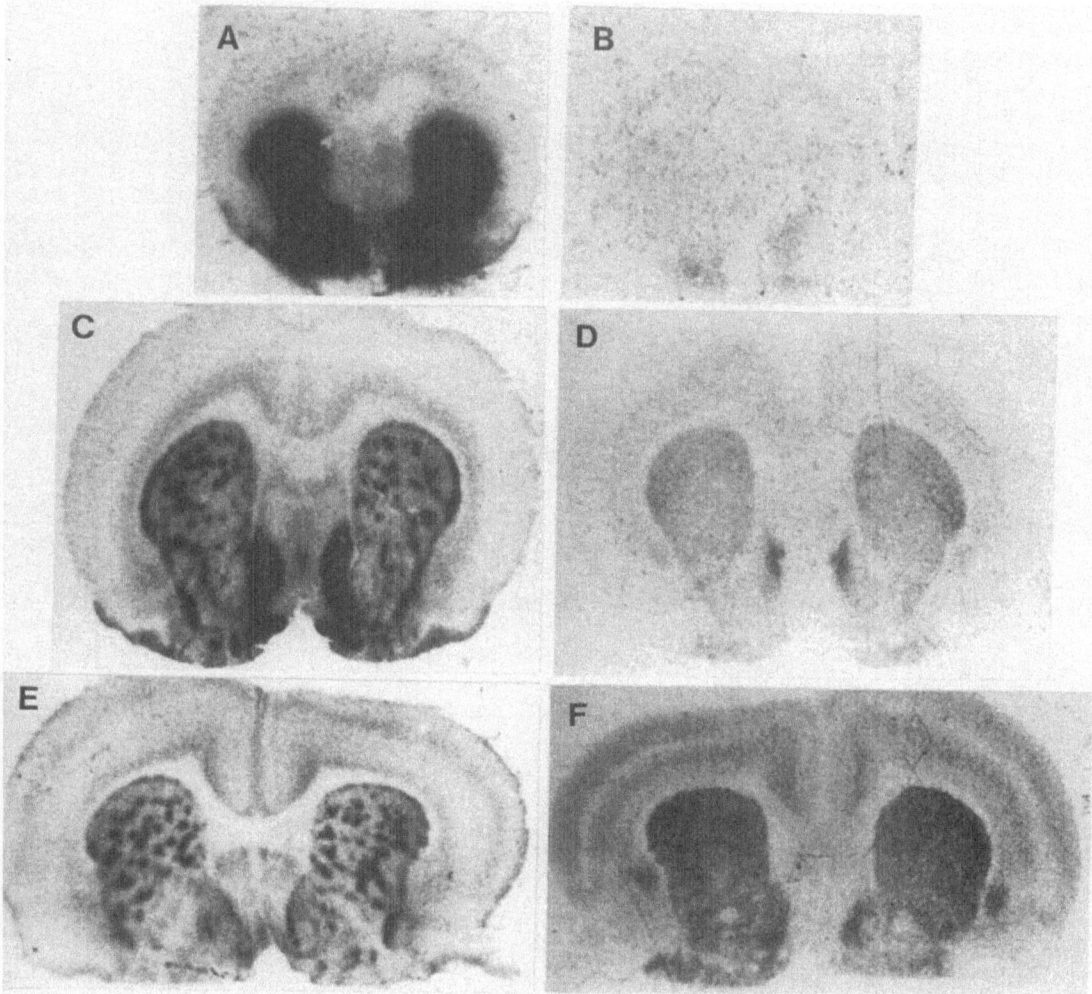

Fig. 10.1. Postnatal development of mu (A, C, E) and delta (B, D, F) binding sites in forebrain. Autoradiograms showing the distribution of mu and delta receptors at PD1 (A, B), PD9 (C, D) and PD17 (E, F) were generated by incubation with [³H]DAGOL and [³H]DADLE under conditions selective for mu and delta binding sites, respectively.

mu binding site density decreases markedly in the surrounding matrix while remaining constant in the patches (Fig. 10.1). In olfactory tubercle and lateral nucleus accumbens, diffuse binding of moderately high density is present at birth which becomes patchy as it decreases to adult levels (Fig. 10.1). In medial nucleus accumbens, binding decreases but remains diffuse (Fig. 10.1). Mu binding sites in the globus pallidus are also dense at birth and decrease to very low levels thereafter. In contrast, binding in the amygdala is present at birth and increases over the next three weeks.

The ontogeny of mu binding sites in the thalamus is largely postnatal, appearing during the second week in adult patterns and increasing to adult levels in the third week

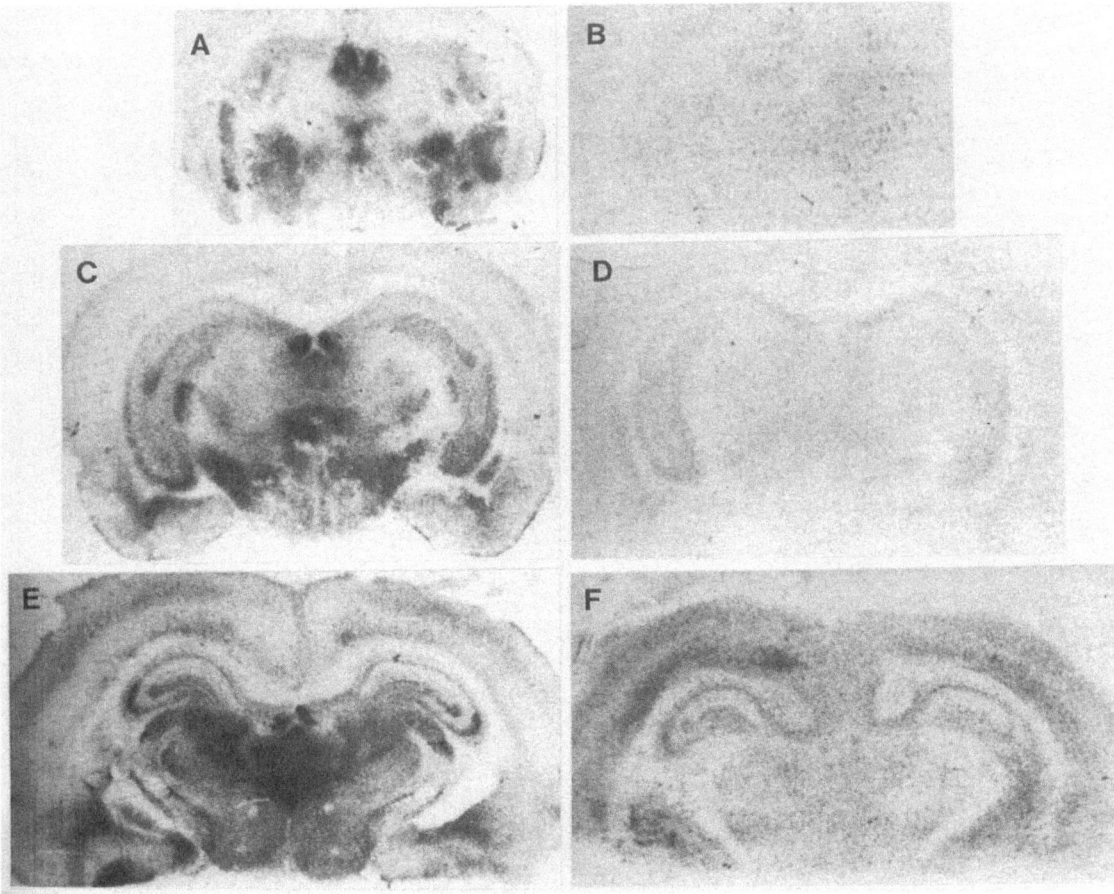

Fig. 10.2.   Postnatal development of mu (A, C, E) and delta (B, D, F) binding sites at the level of the posterior diencephalon. Autoradiograms showing the distribution of mu and delta receptors at PD1 (A, B), PD9 (C, D) and PD17 (E, F) were generated by incubation with [³H]DAGOL and [³H]DADLE under conditions selective for mu and delta binding sites, respectively.

(Fig. 10.2). In the hypothalamus, mu receptors are present in most regions at PD2 and adult levels are reached by the second postnatal week. In the mesencephalon, mu receptors are diffusely distributed throughout the superior colliculus during the first postnatal week, increasing in density and becoming laminated over the next two weeks. Mu sites are detectable in the substantia nigra early in postnatal development and decline two-fold to adult concentrations. Mu sites are also present in brainstem nuclei at birth, especially in

cardiorespiratory nuclei (Xia and Haddad, 1991). Densities increase to adult levels in early postnatal weeks, maturing ahead of forebrain sites. Binding sites in spinal cord achieve adult distribution and densities by the second postnatal week (Besse *et al.*, 1990).

(b)   Kappa binding sites

Recent studies have shown that both kappa₁ and kappa₂ sites are present in rat brain, with the kappa₂ site predominating (Hurlbut *et al.*,

1988; Zukin *et al.*, 1988; Unterwald *et al.*, 1991). In most developmental studies to date, binding conditions have been used which label predominantly kappa$_1$ sites. Minor discrepancies in reported results may reflect labeling of different proportions of the kappa$_2$ receptor population.

In membrane binding studies, kappa$_1$ sites in brain are present in significant densities at birth, and exhibit a small increase in density during the subsequent two postnatal weeks (Spain *et al.*, 1985; Petrillo *et al.*, 1987; Kitchen *et al.*, 1990). Kappa binding sites in spinal cord increase significantly during the first two postnatal weeks, then decline to adult levels (Allerton *et al.*, 1989; Attali and Vogel, 1990).

In autoradiographic studies, a significant density of kappa$_1$ sites is detectable at birth in many brain regions (Kornblum *et al.*, 1987a). In neocortex, kappa receptors do not reach significant levels until PD12, increasing slightly to adult levels in the next few days. The higher levels of kappa binding present in claustrum and endopiriform nucleus appear earlier, suggesting that levels of detection may not be sensitive enough for binding in neocortex to be seen earlier. In hippocampus, kappa receptors are detectable by the end of the second postnatal week.

Binding sites in olfactory tubercle and nucleus accumbens are present at birth and remain constant in density throughout postnatal development. Kappa sites are also present in caudate putamen at birth. In globus pallidus, kappa sites are dense in the first postnatal week, peaking in the second week and decreasing to adult levels by PD17. Kappa sites in amygdala appear differentially across nuclei, with basolateral sites present by PD5 and other nuclei appearing later.

Kappa receptors are not detectable in thalamus until PD6, with sites appearing first in ventral and midline nuclei, and later in others. Within hypothalamus, receptors are present in supraoptic and paraventricular nuclei by PD12 and by PD17 in other nuclei. As for mu sites, kappa binding in superior colliculus is diffuse during the first week after birth, becoming laminated by PD14. Kappa sites are detectable early in the postnatal development of the substantia nigra and remain constant thereafter. Little is known of the ontogeny of kappa sites in brainstem. In spinal cord, kappa binding sites are present in greater density and distribution at PD9–16 than in the adult (Allerton *et al.*, 1989).

Whereas earlier autoradiographic studies have used binding conditions which were somewhat selective for kappa$_1$ sites, binding conditions have recently been used which label both populations of kappa receptors (Leslie and Smith, 1992). In the presence of mu and delta blockers, [^{3}H]diprenorphine labels kappa$_1$ and kappa$_2$ sites with equal affinity (Hurlbut *et al.*, 1988, 1992). By examining [^{3}H]diprenorphine binding in the absence and presence of the kappa$_1$-selective agonist, U50,488, the ontogeny of the two kappa subtypes in rat brain has been compared (Fig. 10.3). Using this approach we have determined that the ontogeny of the kappa$_1$ subtype is largely similar to that described above using more selective labeling conditions. We have also shown that kappa$_2$ sites are present in significant densities at birth in many brain regions, including caudate putamen, nucleus accumbens, olfactory tubercle and habenula (Figs 10.4 and 10.5).

Although there are many similarities in the anatomical distributions of mu and kappa$_2$ sites in adult rat brain (Unterwald *et al.*, 1991; Leslie and Smith, 1992), these receptors exhibit significant differences in their ontogeny. In nucleus accumbens and olfactory tubercle, densities of kappa$_2$ sites are not significantly altered during postnatal development, whereas mu receptor densities decline significantly. In the caudate putamen, kappa$_2$ receptor densities remain constant in the matrix compartment throughout postnatal development, while increasing significantly in the patches. In contrast, mu receptor density remains constant in the patches, while declining significantly in the matrix during the first two postnatal

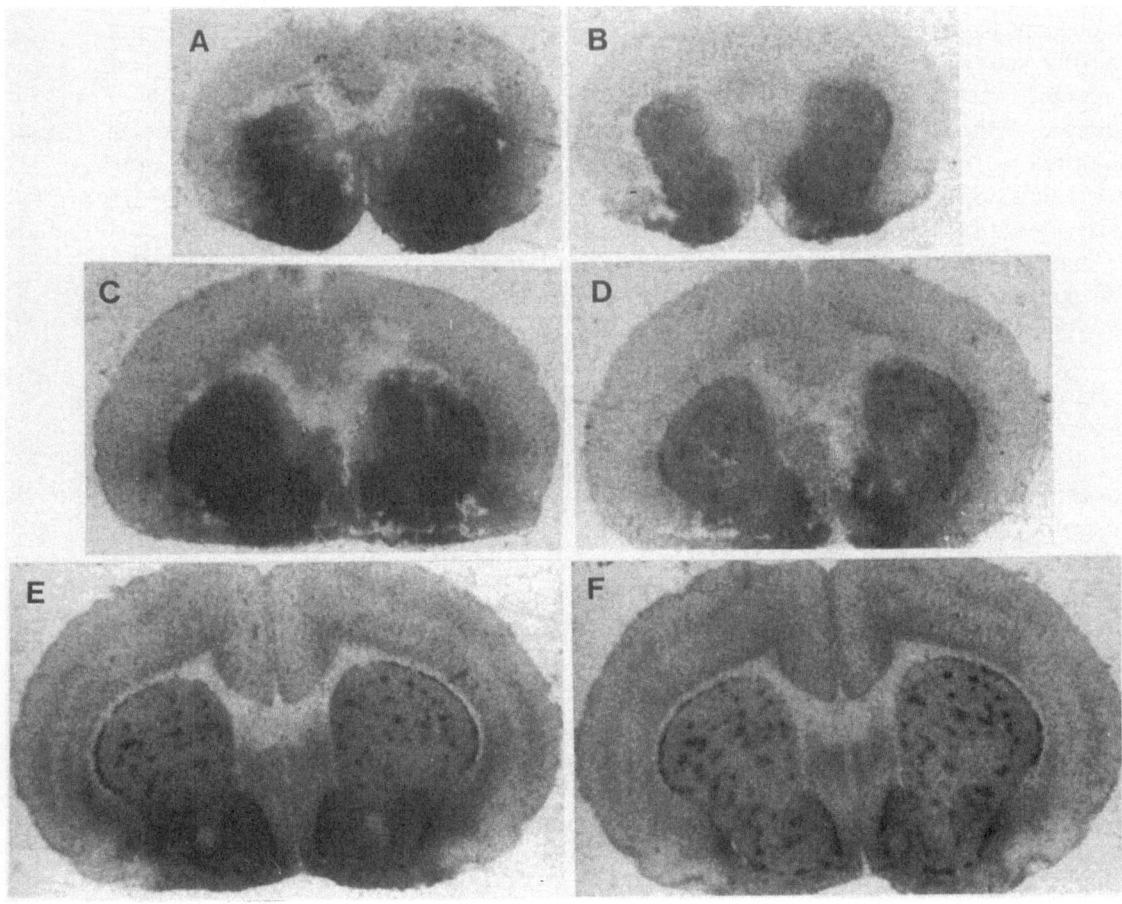

Fig. 10.3. Postnatal development of kappa binding sites in forebrain. The total population of kappa binding sites was localized with [^{3}H]diprenorphine in A, C and E. Sequential sections (B, D, F) were incubated with [^{3}H]diprenorphine in the presence of U50,488 to block binding to kappa$_1$ sites, and thus represent binding to kappa$_2$ sites. Sections A, B are from PD0, C, D are from PD7 and E, F are from PD21 brains.

weeks. Similar differential developmental profiles for mu and kappa$_2$ receptor populations are observed within the habenula. Such data provide strong evidence that these receptors, though distributed in many similar brain regions, are differentially modulated.

(c)   Delta binding sites

The ontogeny of delta binding sites lags significantly behind that of mu and kappa sites

(Figs 10.1 and 10.2). By the end of the first postnatal week, a low density of sites is detectable (Spain *et al.*, 1985; Kornblum *et al.*, 1987a; Milligan *et al.*, 1987; Szucs and Coscia, 1990). These sites are discretely localized to a limited number of regions, particularly nucleus accumbens and olfactory tubercle (Kornblum *et al.*, 1987a). A major increase in density occurs in the second postnatal week, followed by a steady increase which is greater than the rate of increase of protein. It has been

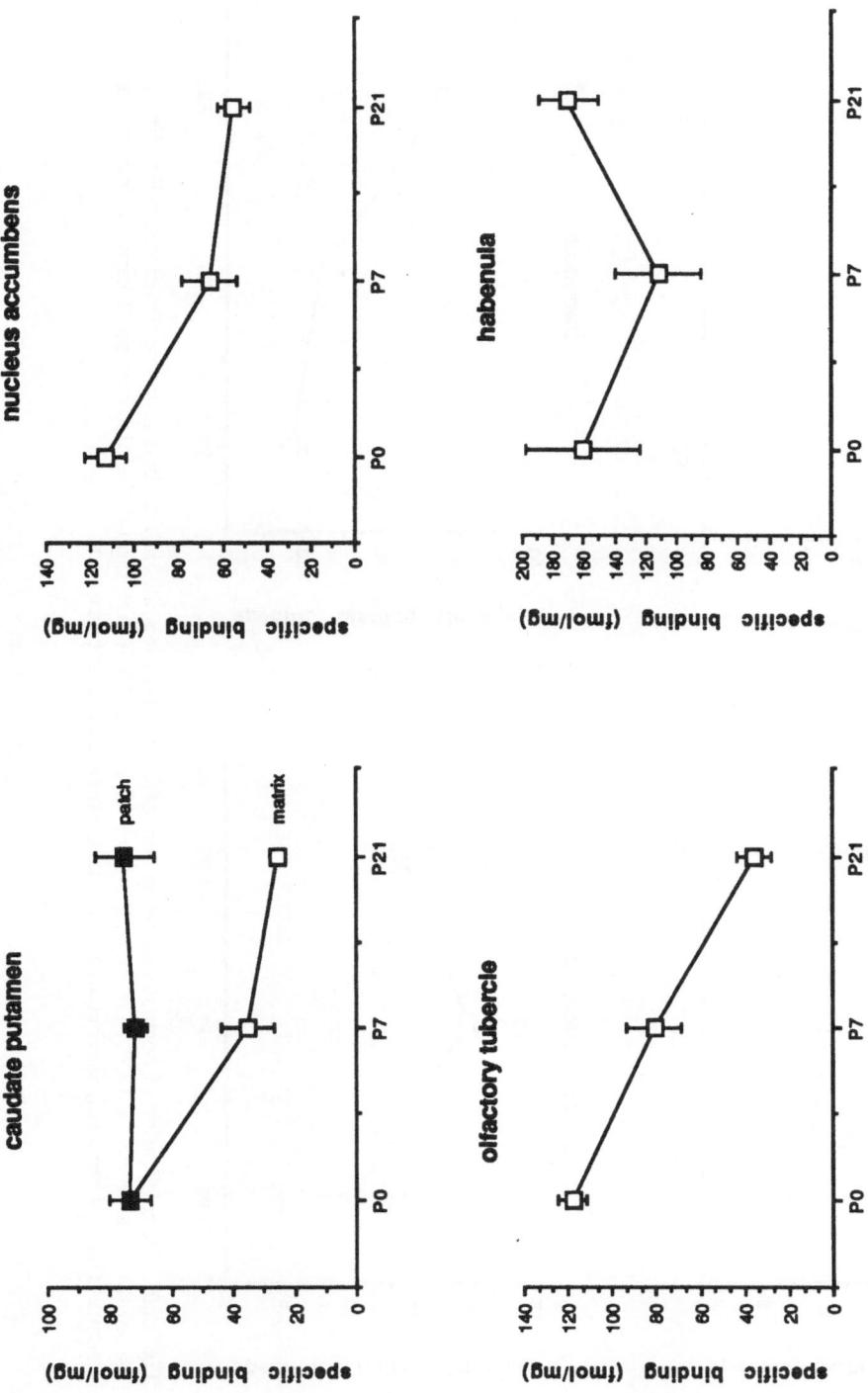

Fig. 10.4. Quantification of postnatal development of mu binding sites in four brain regions. Specific binding (fmol/mg tissue) was determined at PD0, PD7, and PD21 for each region.

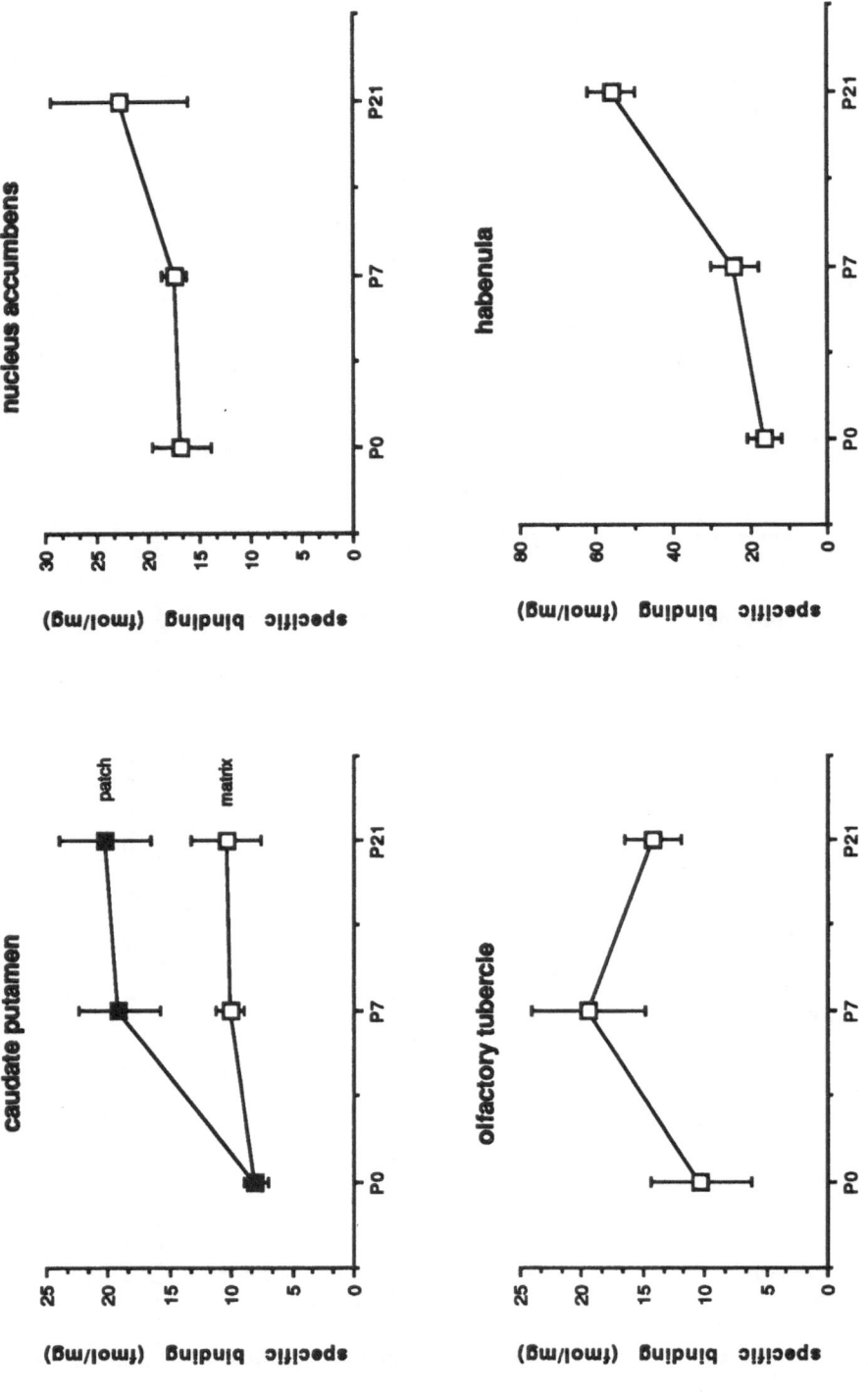

Fig. 10.5. Quantification of postnatal development of kappa$_2$ binding sites in four brain regions. Specific binding (fmol/mg tissue) was determined at PD0, PD7, and PD21 for each region. Note that, though the distribution of kappa$_2$ binding sites resembles that of mu sites, the time course of appearance is different (see text).

suggested that a change in affinity of the delta receptor occurs during ontogeny (Spain *et al.*, 1985), but other investigators have not found such changes (Szucs *et al.*, 1987; Kornblum *et al.*, 1987a; Szucs and Coscia, 1990).

Delta binding sites are detectable in cortex during the first postnatal week, especially in deeper layers and in hippocampus by the end of the second postnatal week. In olfactory bulb, receptors appear between PD9 and PD13, whereas in olfactory tubercle they are present at birth. Both increase throughout development. Some delta labeling is present in the nucleus accumbens at birth, but binding does not reach adult concentrations until PD20. Within the caudate putamen, delta receptors are detectable during the second postnatal week (Fig. 10.1). In globus pallidus, delta sites are present by PD9, at slightly above adult levels, whereas amygdala sites are detectable by the time of birth and increase gradually to adult levels.

Delta receptors are sparse in thalamus, hypothalamus and mesencephalon, except for moderate concentrations in the interpeduncular nucleus (Fig. 10.2). In the brainstem, the ontogeny of delta sites lags significantly behind mu sites, with no detectable binding present at PD1 (Xia and Haddad, 1991). Densities increase differentially across nuclei, showing increases of 2–6-fold in the first postnatal weeks.

10.3.4 OPIOID RECEPTOR EXPRESSION IN GLIA

Whereas glial cells appear to express opioid peptides (see above) and opioid modulation of glial growth has been documented (see below), the expression of opioid receptors by glial cells remains controversial. C_6 glioma cells exhibit binding sites for opioid compounds which may be kappa$_2$ sites (Barg and Simantov, 1991). A number of studies have failed to detect binding sites in primary glial cultures derived from the CNS (Hendrickson and Lin, 1980; Vaysse *et al.*, 1990), and others have

detected opioid binding sites whose affinity is similar to sites on neurons (Maderspach and Solomonia, 1988).

10.3.5 BINDING OF ENDOGENOUS LIGANDS: BETA-ENDORPHIN BINDING

The pharmacology of the endogenous peptide products of the three opioid genes is quite complex (Hollt, 1986). Although some endogenous opioid peptides exhibit higher affinity for a given receptor, none exhibits absolute receptor specificity (Leslie, 1987). Furthermore, different peptide products of the same precursor may have strikingly different pharmacological profiles (Hollt, 1986). In view of this, it is not surprising that no strict relationship has been established between the distributions of any opioid peptide and receptor in adult or developing brain (Loughlin *et al.*, 1985; McDowell and Kitchen, 1987; Mansour *et al.*, 1991).

In order to address functionally relevant opioid systems, it is important to examine the binding and activity profiles of the endogenous peptides, as well as of synthetic opioids. McDowell and Kitchen (1987) have reviewed the ontogeny of binding of a number of endogenous opioid peptides, and Zagon and McLaughlin (1993) have discussed the binding of met-enkephalin, and we have examined the ontogeny of ^{125}I-labeled beta-endorphin binding sites in rat brain (Kornblum *et al.*, 1989; Loughlin *et al.*, 1992b).

In adult rat, ^{125}I-labeled beta-endorphin binding is extensively distributed throughout the brain in a pattern consistent with that of a composite of mu and delta receptor distributions. Figure 10.6 shows that this labeling is differentially displaced by mu-, delta- and kappa-selective compounds. Co-incubation of ^{125}I-labeled beta-endorphin with the mu-selective agonist, D-pro[4]-morphiceptin, results in a pattern of labeling that is similar to that of delta receptors. In particular, binding to the patch compartment of the caudate putamen is significantly blocked, leaving diffuse residual

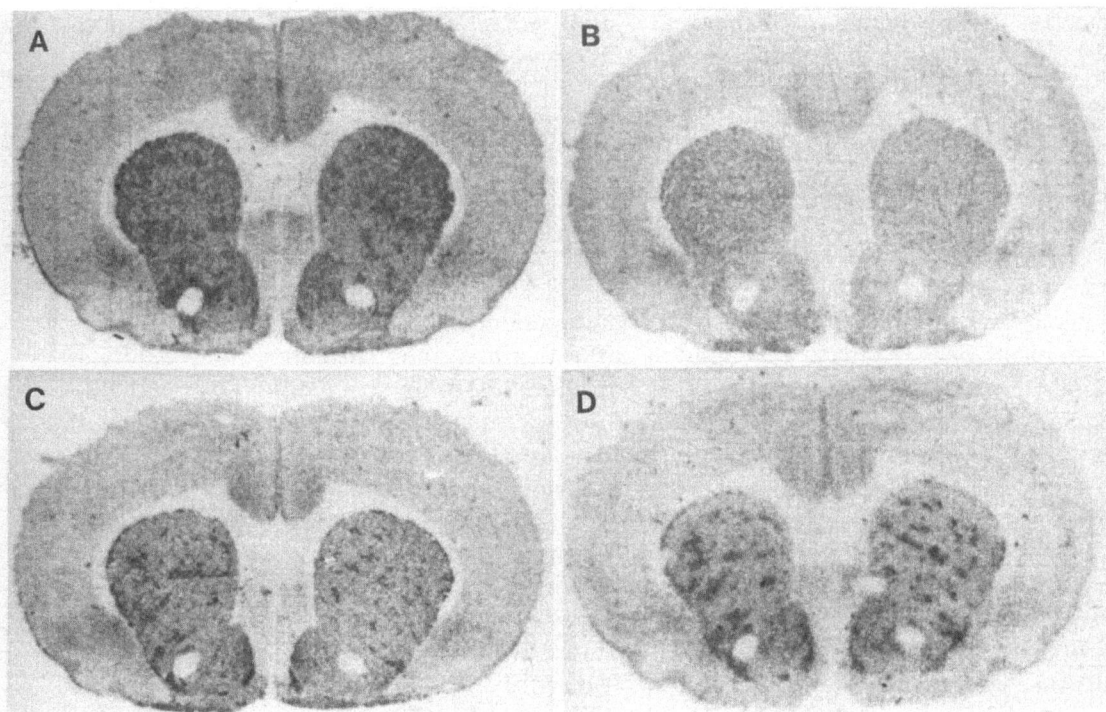

Fig. 10.6. Autoradiograms of ^{125}I-labeled beta-endorphin binding sites in adult rat forebrain. Sections were incubated with ^{125}I-labeled beta-endorphin without blockers (A), and in the presence of D-pro[4]-morphiceptin (B), U50,488 (C) or DPDPE (D) to block binding to mu, kappa₁ or delta binding sites, respectively. Co-incubation with unlabeled beta-endorphin or naloxone blocked all binding (not shown).

labeling. In contrast, co-incubation of radioligand with the delta-selective agonist, DPDPE, produces a pattern of labeling similar to that of mu receptors, with a striking patchy distribution in the caudate putamen. Although the kappa₁-selective agonist, U50,488, does produce some decrease in binding of ^{125}I-labeled beta-endorphin, this inhibition is inconsistent and not regionally selective. Taken together, these data indicate that in adult rat brain the majority of ^{125}I-labeled beta-endorphin binding is to mu and delta receptor populations. This conclusion, consistent with that of Schoffelmeer *et al.* (1990), does not provide evidence for high affinity labeling of a selective epsilon receptor population.

In fetal forebrain, the localization of beta-endorphin binding sites is very similar to that previously shown with mu selective ligands (Figs 10.7 and 10.8). At ED17, mu selective ligands block all but a small population of sites in the ventral forebrain, which is consistent with previous studies suggesting that few delta sites are present before birth (Fig. 10.7). The kappa₁-selective compound U50,488 produces very little, if any, inhibition of binding (Fig. 10.7). These data suggest that, in embryonic brain, ^{125}I-labeled beta-endorphin predominantly labels mu receptor sites.

In both forebrain and hindbrain periventricular regions, ^{125}I-labeled beta-endorphin binding sites are present as early as ED14 (Fig. 10.8). At ED15, dense labeling is present

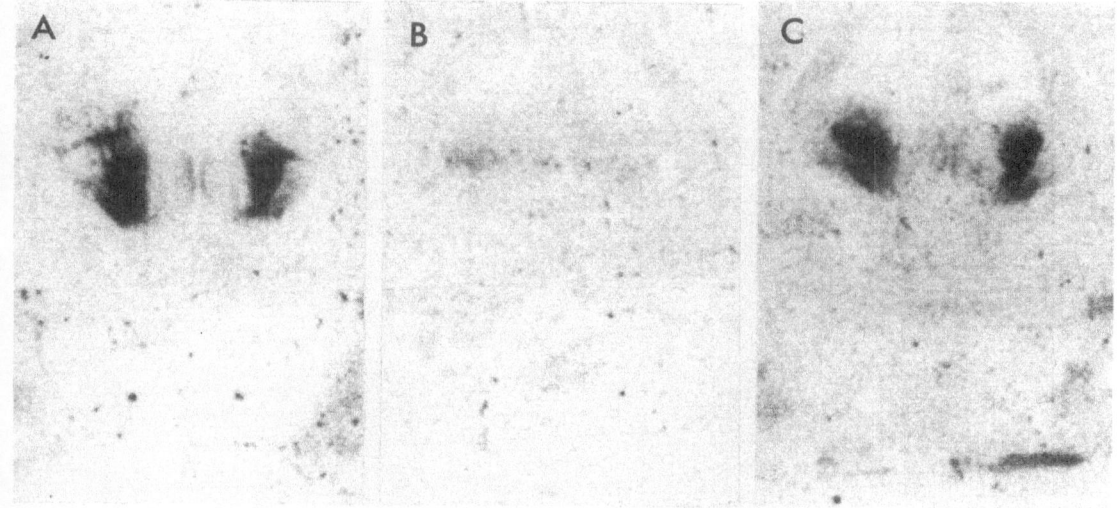

Fig. 10.7. Autoradiograms of ^{125}I-labeled beta-endorphin binding sites in ED17 rat brain. Sections through the developing telencephalon were incubated with ^{125}I-labeled beta-endorphin alone (A), or in the presence of D-pro[4]-morphiceptin (B), or U50,488 (C) to block binding to mu, and kappa[1] binding sites, respectively. Co-incubation with unlabeled beta-endorphin or naloxone blocked all binding (not shown).

in the striatal anlage, with some labeling in the developing cerebral vesicles. Binding sites are detectable in the forebrain germinal zone, especially at its distal poles. Throughout the extent of the midbrain and hindbrain, binding is detectable in the region which surrounds the ventricles. By ED17, this hindbrain labeling is very dense. These data indicate that mu opioid receptors are present at an appropriate time and location to influence the generation and differentiation of many cell groups.

10.4 DEVELOPMENTAL ROLE

It has been suggested that brain opioid binding sites present at early developmental stages differ from those of the adult with regard to selectivity (Barg and Simantov, 1991), molecular weight (McLean *et al.*, 1989), coupling to regulatory proteins (Milligan *et al.*, 1987; Szucs *et al.*, 1987), sensitivity to cation effects (Oetting *et al.*, 1987; Szucs *et al.*, 1987) and association with other opioid receptors (Schoffelmeer *et al.*,

1990). A number of *in vitro* studies, however, suggest that opioid binding sites in developing brain represent functional receptors which bear many similarities to those characterized in adult brain (Chneiweiss *et al.*, 1988; DeVries *et al.*, 1990; Van Vliet *et al.*, 1990; Vaysse *et al.*, 1990; Stiene-Martin and Hauser, 1990, 1991; Hauser and Stiene-Martin, 1991; Stiene-Martin *et al.*, 1991a; Smith *et al.*, 1992a).

A number of reviews (Tempel, 1992; Hammer and Hauser, 1992; Barr, 1992; Kuhn *et al.*, 1992) have recently discussed the effects of opioids on developmental processes. Since more than one in a thousand people in the United States have been exposed to heroin or methadone *in utero* (Tempel, 1992), it is critical to understand the role of opioids in developmental processes. A number of studies support the conclusion that opioids modulate neurogenesis (Kornblum *et al.*, 1987b; Schmahl *et al.*, 1989; Vertes *et al.*, 1982; Zagon and McLaughlin, 1993) and it has been suggested that endogenous opioid systems

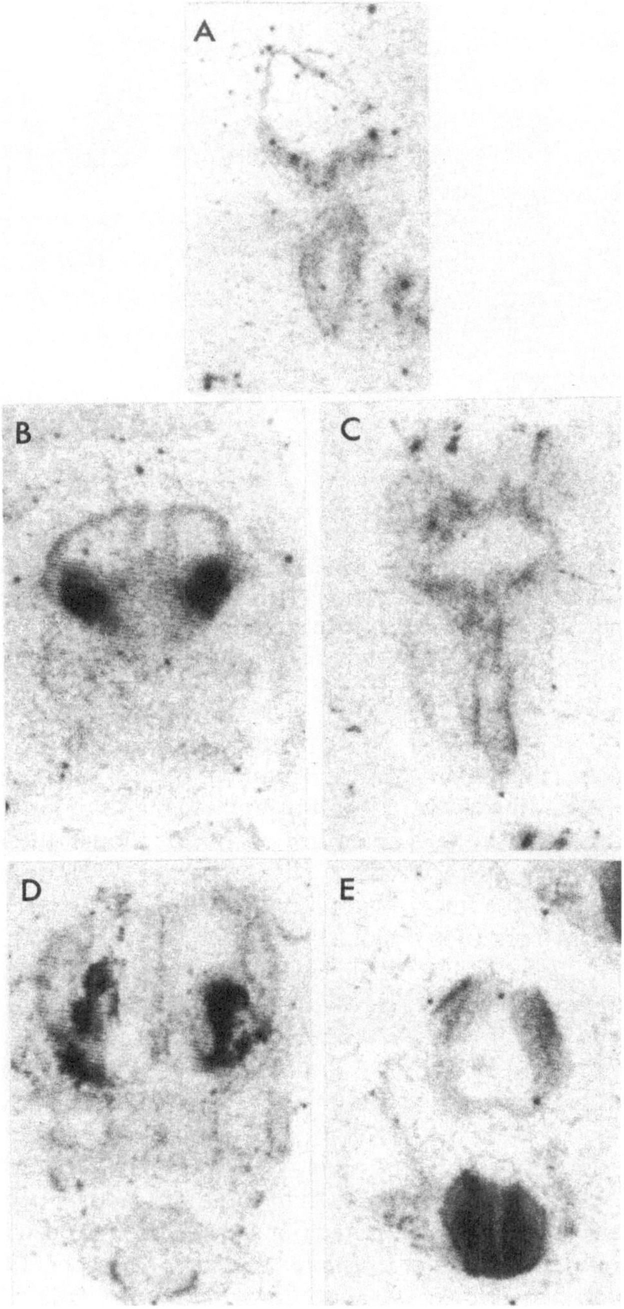

Fig. 10.8. Prenatal appearance of ^{125}I-labeled beta-endorphin binding sites in ED14 (A), ED15 (B,C), and ED17 (D,E) rat brain. Co-incubation with unlabeled beta-endorphin blocked all binding (not shown).

may affect neuronal migration (Kent *et al.*, 1982; Loughlin *et al.*, 1991). Administration of opioids alters dendritic growth and spine formation (Hauser *et al.*, 1989) and a number of studies support the hypothesis that opioids are regulators of glial growth (Stiene-Martin and Hauser, 1991). Opioid peptides also regulate growth-related enzymes (Bartolome *et al.*, 1986, 1987, 1989).

Elucidating the effects of exogenous opioid administration is complicated by the fact that perinatal administration of opioids affects endogenous opioid systems in complex ways (Chapter 12), both at the level of peptide expression (Uhl *et al.*, 1988; Tiong and Olley, 1988) and receptor expression (e.g. Tsang and Ng, 1980; Bardo *et al.*, 1981; Hammer *et al.*, 1991; Tempel, 1991). Opioids also modulate many other neurotransmitter or hormonal systems (Tempel *et al.*, 1990; Mess *et al.*, 1989). Thus, although the evidence to date favors the conclusion that opioids have important developmental functions, the precise effects and the mechanisms of action of opioids in ontogeny remain to be elucidated.

10.4.1 COMPLEXITY OF INTERACTION BETWEEN OPIOIDS AND OTHER NEUROTRANSMITTER SYSTEMS

(a) Role of opioids in nigrostriatal system development and plasticity

Effects of opioids in developmental processes have been studied in whole brain, neocortex and cerebellum (Zagon and McLaughlin, 1993). However, in the nigrostriatal system, opioid peptides and receptors are present very early and in high density. The nigrostriatal system thus offers a good model in which to study the role of opioids in development, especially with regard to interactions with this well-studied monoaminergic system (Fallon and Loughlin, 1985). In the nigrostriatal system, dopaminergic cells in the substantia nigra innervate striatal regions, including the caudate putamen and nucleus accumbens,

very early in embryonic development (Specht *et al.*, 1981; Voorn *et al.*, 1988). Striatal cells reciprocally innervate the substantia nigra. Opioid peptides and receptors are associated with the nigrostriatal system at all levels and are present in early embryonic development, at the time of neurogenesis and throughout the time of process outgrowth. Beta-endorphin binding sites surround the forebrain germinal zone and are present in caudate putamen during striatal neurogenesis (Loughlin *et al.*, 1992b). They are also detectable in the midbrain regions in which substantia nigra cells develop. Mu binding sites are dense throughout the caudate putamen prenatally and are later lost in the matrix compartment (Kornblum *et al.*, 1987a). Kappa binding sites are present in early postnatal development of the nigrostriatal system (Kornblum *et al.*, 1987a; Leslie and Smith, 1992). A transient population of beta-endorphin immunoreactive cells is present in the forebrain germinal zone which sends processes to the lateral edge of the developing caudate putamen (Loughlin *et al.*, 1991). Proenkephalin and prodynorphin mRNAs are detectable in caudate putamen by ED17 (Tecott *et al.*, 1989) and ED15 (Alvarez-Bolado *et al.*, 1990), respectively.

Since mu receptors in caudate putamen develop their characteristic patchy appearance in registration with the developing dopaminergic innervation (Murrin and Ferrer, 1984), it has been suggested that their expression is under the control of the dopaminergic system and that dopaminergic cues may be necessary for development of normal patch-matrix organization (Fishell and van der Kooy, 1987; Gerfen *et al.*, 1987). Indeed, it has been shown that loss of dopaminergic innervation modulates caudate putamen opioid binding sites (Moon-Edley and Herkenham, 1984; Eghbali *et al.*, 1987; Sirinathsinghji and Dunnett, 1989; Smith *et al.*, 1992b; Loughlin *et al.*, 1992a). The time course of these effects is complex. While a significant loss of binding sites occurs shortly after lesions, much greater losses occur in the ensuing months. Interestingly, the secondary loss of opioid receptors which occurs six months after dopaminergic lesions is reversed by transplants of fetal mesencephalon (Sirinathsinghji and Dunnett, 1989; Smith *et al.*, 1991). Sustained loss of striatal dopaminergic innervation decreases dynorphin mRNA (Jiang *et al.*, 1990), and increases enkephalin mRNA in lesion (Mochetti *et al.*, 1985; Young *et al.*, 1986) and mutant (Loughlin *et al.*, 1992c) models, but such changes are not always correlated with receptor down-regulation (Smith *et al.*, 1992b). Pharmacological manipulation of these systems produces many of the same results as lesions (Chapter 12). Whether such changes in opioid systems occur in Parkinson's disease, in which the dopaminergic cells innervating the caudate putamen die, is controversial (Loughlin *et al.*, 1992a, for review). It is tempting to speculate that decreases in opioid binding sites might underlie some of the changes in affect which occur in Parkinson patients. Increases in enkephalin immunoreactivity have been correlated with severity of symptoms in monkeys exposed to MPTP (Dacko and Schneider, 1991).

Striatal opioids, in turn, are important modulators of dopaminergic function. Opioids modulate dopamine effects on adenylate cyclase in adult and developing striatum (Schoffelmeer *et al.*, 1986, 1987; Milligan *et al.*, 1987; Chneiweiss *et al.*, 1988; DeVries *et al.*, 1990; Van Vliet *et al.*, 1990). Dopamine release from striatal slices is itself modulated by kappa opioids (Mulder *et al.*, 1984; Schoffelmeer *et al.*, 1988; Werling *et al.*, 1988) and this effect is first seen in slices obtained from ED17 embryos (DeVries *et al.*, 1990). We have recently shown that kappa$_1$ receptor agonists inhibit dopamine release from dissociated cultures of ED14 substantia nigra cells by eight days *in vitro* (Smith *et al.*, 1992a).

While a number of investigators have examined the effects of opioids on nigrostriatal plasticity and dopaminergic function, the role of endogenous opioid systems in basic developmental processes remains to be elucidated. Schmahl *et al.* (1989) have shown that opioid antagonist treatment increases mitogenesis in the forebrain germinal zone in 4–12-week-old rats. Such effects probably reflect modulation of glial populations. The availability of well-characterized *in vitro* models will allow the precise determination of whether opioids are capable of influencing the development of neurons in this system.

(b)   Role of opioid peptides and receptors in developing CNS

Do endogenous opioids act as neurotrophic factors to mediate basic developmental processes? We have proposed that, in order to be considered as neurotrophic factors, endogenous opioids must meet several criteria, including the presence of endogenous opioids and their receptors at developmentally relevant times and in the appropriate neuronal elements (Fallon and Loughlin, 1992). They might act by regulating cell division or cell survival, or have effects on process outgrowth and/or synthesis of proteins necessary to differentiated functions of neurons (Hefti *et al.*, 1992). Opioid receptors and their endogenous ligands are present early in the development of a number of systems, including the nigrostriatal system, at a time when neurogenesis, neuronal migration, process outgrowth and synaptogenesis are occurring. Opioids and opioid antagonists have been shown to modulate a number of developmental markers (Zagon and McLaughlin, 1993; Chapter 12), including DNA synthesis, enzyme activity, receptor expression, peptide expression, dendritic morphology and cell survival. Whether these are direct or indirect effects remains to be determined. Opioids clearly affect the growth of glial cells (Hauser and Steine-Martin, 1991) and glia may synthesize characterized growth factors, including transforming growth factor alpha (Loughlin *et al.*, 1989, 1992d). Thus, opioids may influence development via interactions with characterized growth factors of glial origin. Interestingly, Schofield *et al.* (1989) have described an opioid binding protein which displays significant homology to a group of cell adhesion and growth factor molecules. Whether opioids might compete for binding to such growth factor receptors *in vivo* remains to be determined.

ACKNOWLEDGEMENTS

The research described here was supported by NIH grants NS 26761 to SEL and NS 19319 to FML. The authors also wish to thank the American Parkinson Disease Association Southern California Chapter for their generous support.

REFERENCES

Allerton, C.A., Smith, J.A.M., Hunter, J.C. *et al.* (1989) Correlation of ontogeny with function of [³H]U69593 labelled kappa opioid binding sites in the rat spinal cord. *Brain Res.*, **502**, 149–57.

Alvarez-Bolado, G., Fairen, A., Douglass, J. and Naranjo, J.R. (1990) Expression of the prodynorphin gene in the developing and adult cerebral cortex of the rat: an in situ hybridization study. *J. Comp. Neurol*, **300**, 287–300.

Attali, B. and Vogel, Z. (1990) Characterization of kappa opiate receptors in rat spinal cord-dorsal root ganglion co-cultures and their regulation by chronic opiate treatment. *Brain Res.*, **517**, 182–8.

Bardo, M.T., Bhatnagar, K.P. and Gebhart, F.F. (1981) Differential effects of chronic morphine and naloxone on opiate receptors, monoamines and morphine-induced behaviors in preweanling rats. *Dev. Brain Res.*, **4**, 139–47.

Barg, J. and Simantov, R. (1991) Transient expression of opioid receptors in defined regions of developing brain: are embryonic receptors selective? *J. Neurochem.*, **57**, 1978–84.

Barr, G.A. (1992) Behavioral effects of opiates during development, in *Development of the Central Nervous System: Effects of Alcohol and Opiates* (ed. M.W. Miller), Wiley-Liss, New York, pp. 221–54.

Bartolome, J.V., Bartolome, M.B., Daltner, L.A. *et al.* (1986) Effects of β-endorphin on ornithine decarboxylase in tissues of developing rats: a potential role for this endogenous neuropeptide in the modulation of tissue growth. *Life Sci.*, **38**, 2355–62.

Bartolome, J.V., Bartolome, M.B., Harris, E.B. and Schanberg, S.M. (1987) N-alpha-acetyl-β-endorphin stimulates ornithine decarboxylase activity in preweanling rat pups: opioid- and non-opioid-mediated mechanisms. *J. Pharmacol. Exp. Ther.*, **240**, 895–9.

Bartolome, J.V., Bartolome, M.B., Harris, E.B. *et al.* (1989) Regulation of insulin and glucose plasma levels by central nervous system beta endorphin in preweanling rats. *Endocrinology*, **124**, 2153–8.

Batter, D.K., Vilijn, M.-H. and Kessler, J. (1991) Cultured astrocytes release proenkephalin. *Brain Res.*, **563**, 28–32.

Bayon, A., Shoemaker, W.J., Bloom, F.E. *et al.* (1979) Perinatal development of the endorphin- and enkephalin-containing systems in the rat brain. *Brain Res.*, **179**, 93–101.

Berry, S. and Haynes, L.W. (1989) The opiomelanocortin peptide family: neuronal expression and modulation of neural cellular development and regeneration in the central nervous system. *Comp. Biochem. Physiol.*, **93A**, 267–72.

Besse, D., Lombard, M.C., Zajac, J.M. *et al.* (1990) Pre- and post-synaptic distribution of mu, delta and kappa opioid receptors in the superficial layers of the cervical dorsal horn of the rat spinal cord. *Brain Res.*, **521**, 15–22.

Bloom, F., Bayon, A., Battenberg, E. *et al.* (1980) Endorphins: developmental, cellular and behavioural aspects. *Adv. Biochem. Psychopharmacol.*, **22**, 619–32.

Chang, K.-J., Blanchard, S.G. and Cuatrecasas, P. (1984) Benzomorphan sites are ligand recognition sites of putative E-receptors. *Mol. Pharmacol.*, **26**, 484–8.

Chneiweiss, H., Glowinski, J. and Premont, J. (1988) Mu and delta opioid receptors coupled negatively to adenylate cyclase on embryonic neurons from the mouse striatum in primary cultures. *J. Neurosci.*, **8**, 3376–82.

Civelli, O., Douglass, J., Goldstein, A. and Herbert, E. (1985) Sequence and expression of the rat prodynorphin gene. *Proc. Natl. Acad. Sci. USA*, **82**, 4291–5.

Clark, J.A., Houghten, R. and Pasternak, G.W. (1988) Opiate binding in calf thalamic membranes: a selective mu₁ binding assay. *Mol. Pharmacol.*, **34**, 308–17.

Clendennin, N.J., Petraitis, M. and Simon, E.J. (1976) Ontological development of opiate receptors in rodent brain. *Brain Res.*, **118**, 157–60.

Coulter, C.L., Browne, C.A. and McMillen, I.C. (1990) The molecular weight profile of enkephalin-containing peptides in the sheep adrenal gland changes during development. *Endocrinology*, **127**, 330–6.

Coyle, J.T. and Pert, C.B. (1976) Ontogenetic development of [^{3}H]naloxone binding in rat brain. *Neuropharmacology*, **15**, 555–60.

Dacko, S. and Schneider, J.S. (1991) Met-enkephalin immunoreactivity in the basal ganglia in symptomatic and asymptomatic MPTP-exposed monkeys: correlation with degree of parkinsonian symptoms. *Neurosci. Lett.*, **127**, 49–52.

Dahl, J.L., Epstein, M.L., Silva, B.L. and Lindberg, I. (1982) Multiple immunoreactive forms of [met] and [leu] enkephalin in fetal and neonatal rat brain and in rat gut. *Life Sci.*, **31**, 1853–6.

De Vries, T.J., Hogenboom, F., Mulder, A.H. and Schoffelmeer, A.N.M. (1990) Ontogeny of mu, delta and kappa opioid receptors mediating inhibition of neurotransmitter release and adenylate cyclase activity in rat brain. *Dev. Brain Res.*, **54**, 63–9.

De Vries, T.J., Van Vliet, B.J., Hogenboom, F. *et al.* (1991) Effect of chronic prenatal morphine treatment on mu-opioid receptor-regulated adenylate cyclase activity and neurotransmitter release in rat brain slices. *Eur. J. Pharmacol.*, **208**, 97–104.

Douglass, J., Civelli, O. and Herbert, E. (1984) Polyprotein gene expression: generation of diversity of neuroendocrine peptides. *Annu. Rev. Biochem.*, **53**, 665–715.

Eghbali, M., Santoro, C., Paredes, W. *et al.* (1987) Visualization of multiple opioid-receptor type in rat striatum after specific mesencephalic lesions. *Proc. Natl. Acad. Sci. USA*, **84**, 6582–6.

Elkabes, S., Loh, Y.P., Nieburgs, A. and Wray, S. (1989) Prenatal ontogenesis of pro-opiomelanocortin in the mouse central nervous system and pituitary gland: an *in situ* hybridization and immunocytochemical study. *Dev. Brain Res.*, **46**, 85–95.

Fallon, J.H. and Ciofi, P. (1990) Dynorphin-containing neurons, in *Handbook of Chemical Neuroanatomy*, Vol.9. *Neuropeptides in the CNS, Part II* (eds A. Bjorklund, T. Hökfelt, and M.J. Kuhar), Elsevier, Amsterdam, pp. 1–130.

Fallon, J.H. and Loughlin, S.E. (1985) The substantia nigra, in *The Rat Nervous System* (ed. G. Paxinos), Academic Press, New York, pp. 353–74.

Fallon, J.H. and Loughlin, S.E. (1992) Functional implications of the anatomical localization of neurotrophic factors, in *Neurotrophic Factors* (eds S.E. Loughlin and J.H. Fallon), Academic Press, New York (in press).

Fishell, G. and van der Kooy, D. (1987) Pattern formation in the striatum: developmental changes in the distribution of striatonigral neurons. *J. Neurosci.*, **7**, 1969–78.

Gall, C. (1984) Ontogeny of dynorphin-like immunoreactivity in the hippocampal formation of the rat. *Brain Res.*, **307**, 327–31.

Gall, C., Brecha, N., Chang, K.-J. and Karten, H.J. (1984) Ontogeny of enkephalin-like immunoreactivity in the rat hippocampus. *Neuroscience*, **11**, 359–80.

Garcin, F. and Coyle, J.T. (1976) Ontogenetic development of [^{3}H]naloxone binding and endogenous morphine-like-factor in rat brain, in *Opiates and Endogenous Opioid Peptides* (ed. H.W. Kosterlitz), Elsevier, North Holland, Amsterdam, pp. 267–73.

Gerfen, C.R., Baimbridge, K.G. and Thibault, J. (1987) The neostriatal mosaic: III. Biochemical and developmental dissociation of patch-matrix mesostriatal systems. *J. Neurosci.*, **7**, 3935–44.

Graybiel, A.M. (1984) Correspondence between the dopamine islands and striosomes of the mammalian striatum. *Neuroscience*, **13**, 1157–87.

Hammer, R.P. and Hauser, K.F. (1992) Consequences of early exposure to opioids on cell proliferation and neuronal morphogenesis, in *Development of the Central Nervous System: Effects of Alcohol and Opiates* (ed. M.W. Miller), Wiley-Liss, New York, pp. 319–39.

Hammer, R.P., Seatriz, J.V. and Ricalde, A.R. (1991) Regional dependence of morphine-induced mu-opiate receptor down-regulation in perinatal rat brain. *Eur. J. Pharmacol.*, **209**, 253–6.

Harlan, R.E., Shivers, B.D., Romano, G.J. *et al.* (1987) Localization of preproenkephalin mRNA in the rat brain and spinal cord by *in situ* hybridization. *J. Comp. Neurol.*, **258**, 159–84.

Hauser, K.F., McLaughlin, P.J. and Zagon, I.S. (1989) Endogenous opioid systems and the regulation of dendritic growth and spine formation. *J. Comp. Neurol.*, **281**, 13–22.

Hauser, K.F., Osborne, J.G., Stiene-Martin, A. and Melner, M.H. (1990) Cellular localization of proenkephalin mRNA and enkephalin peptide products in cultured astrocytes. *Brain Res.*, **522**, 347–53.

Hauser, K.F. and Stiene-Martin, A. (1991) Characterization of opioid-dependent glial development in dissociated and organotypic cultures of mouse central nervous system: critical periods and target specificity. *Dev. Brain Res.*, **62**, 245–55.

Haynes, L.W., Smyth, D.G. and Zakarian, S. (1982) Immunocytochemical localization of beta-endorphin (lipotropin c-fragment) in the developing rat spinal cord and hypothalamus. *Brain Res.*, **232**, 115–28.

Hefti, F., Denton, T.L., Knusel, B. and Lapchak, P.A. (1992) Neurotrophic factors: what are they and what are they doing? in *Neurotrophic Factors* (eds S.E. Loughlin and J.H. Fallon), Academic Press, New York (in press).

Hendrickson, C.M. and Lin, S. (1980) Opiate receptors in highly purified neuronal cell populations isolated in bulk from embryonic chick brain. *Neuropharmacology*, **19**, 731–9.

Herkenham, M. and McLean, S. (1986) Mismatches between receptor and transmitter localizations in the brain, in *Quantitative Receptor Autoradiography* (eds C. Boast, E.W. Snowhill and C.A. Altar), Alan Liss, New York, pp. 131–71.

Hollt, V. (1986) Opioid peptide processing and receptor selectivity. *Annu. Rev. Pharmacol. Toxicol.*, **26**, 59–77.

Hollt, V. (1991) Opioid peptide genes: structure and regulation, in *Neurobiology of Opioids* (eds O.F.X. Almeida and T.S. Shippenberg), Springer-Verlag, Berlin, Heidelberg, pp. 11–51.

Hurlbut, D.E., Broide, R.S. and Leslie, F.M. (1988) Evidence for kappa receptor heterogeneity. *Soc. Neurosci. Abstr.*, **14**, 701.

Hurlbut, D.E., Broide, R.S. and Leslie, F.M. (1992) Pharmacological characteristics of kappa$_2$ opioid binding sites in guinea-pig and rat adult brain. *Brain Res.* (submitted).

Hylka, V.W. and Thommes, R.C. (1991) Avian beta-endorphin: alterations in immunoreactive forms in plasma and pituitary of embryonic and adult chickens. *Comp. Biochem. Physiol.*, **100C**, 643–8.

Imura, I., Kato, Y., Nakai, Y. *et al.* (1985) Endogenous opioids and related peptides: from molecular biology to clinical medicine. *J. Endocrinol.*, **107**, 147–57.

Isayama, T. and Zagon, I.S. (1991) Localization of preproenkephalin A mRNA in the neonatal rat retina. *Brain Res. Bull.*, **27**, 805–8.

Jiang, H.-K., McGinty, J.F. and Hong, J.S. (1990) Differential modulation of striatonigral dynorphin and enkephalin by dopamine receptor subtypes. *Brain Res.*, **507**, 57–64.

Kapcala, L.P. (1983) Discordant changes in immunoreactive ACTH and beta-endorphin in rat brain and pituitary during development. *Clin. Res.*, **31**, 401A.

Kent, J.L., Pert, C.B. and Herkenham, M. (1982) Ontogeny of opiate receptors in rat forebrain: visualization by in vitro autoradiography. *Dev. Brain Res.*, **2**, 487–504.

Keshet, E., Polakiewicz, R.D., Itin, A. *et al.* (1989) Proenkephalin A is expressed in mesodermal lineages during organogenesis. *EMBO J.*, **8**, 2917–23.

Khachaturian, H., Alessi, N.E., Munfakh, N. and Watson, S.J. (1983) Ontogeny of opioid and related peptides in the rat CNS and pituitary: an immunocytochemical study. *Life Sci.*, **33**, 61–4.

Khachaturian, H., Alessi, N.E., Lewis, M.E. *et al.* (1985a) Development of hypothalamic opioid neurons: a combined immunocytochemical and [³H]thymidine autoradiographic study. *Neuropeptides*, **4**, 477–80.

Khachaturian, H., Lewis, M.E., Schafer, M.K.-H. and Watson, S.J. (1985b) Anatomy of the CNS opioid systems. *Trends Neurosci.*, **8**, 111–19.

Kinouchi, K. and Pasternak, G.W. (1991) Evidence for kappa$_1$ opioid receptor multiplicity in the guinea pig cerebellum. *Eur. J. Pharmacol.*, **207**, 135–41.

Kirby, M.L. (1981) Development of opiate receptor binding in rat spinal cord. *Brain Res.*, **205**, 400–4.

Kitchen, I., Kelly, M. and Viveros, P.M. (1990) Ontogenesis of kappa opioid receptors in rat brain using [³H]U-69593 as a binding ligand. *Eur. J. Pharmacol.*, **175**, 93–6.

Kornblum, H.I., Hurlbut, D.E. and Leslie, F.M. (1987a) Postnatal development of multiple opioid receptors in rat brain. *Dev. Brain Res.*, **37**, 21–41.

Kornblum, H.I., Loughlin, S.E. and Leslie, F.M. (1987b) Effects of morphine on DNA synthesis in neonatal rat brain. *Dev. Brain Res.*, **31**, 45–52.

Kornblum, H.I., Loughlin, S.E., Fallon, J.H. and Leslie, F.M. (1989) Developmental appearance of opioid receptors in embryonic and neonatal rat brain using [¹²⁵I] beta-endorphin: an autoradiographic study. *Adv. Biosci.*, **75**, 277–80.

Kuhn, C.M., Windh, R.T. and Little, P.J. (1992) Effects of perinatal opioid addiction on neurochemical development of the brain, in *Development of the Central Nervous System: Effects of Alcohol and Opiates* (ed. M.W. Miller), Wiley-Liss, New York, pp. 341–61.

Leslie, F.M. (1987) Methods used for the study of opioid receptors. *Pharmacol. Rev.*, **39**, 197–249.

Leslie, F.M. (1992) Neurotransmitters as neurotrophic factors, in *Neurotrophic Factors* (eds S.E. Loughlin and J.H. Fallon), Academic Press, New York (in press).

Leslie, F.M. and Loughlin, S.E. (1992) Development of multiple opioid receptors, in *Development of* the Central Nervous System: Effects of Alcohol and Opiates (ed. M.W. Miller), Wiley-Liss, New York, pp. 255–83.

Leslie, F.M. and Smith, T.E. (1992) Autoradiographic localization of kappa₂ opioid binding sites in adult and developing rat brain. *Brain Res.* (submitted).

Leslie, F.M., Tso, S. and Hurlbut, D.E. (1982) Differential appearance of opiate receptor subtypes in neonatal rat brain. *Life Sci.*, **31**, 1393–6.

Loughlin, S.E., Massamiri, T.R., Kornblum, H.I. (1985) Postnatal development of opioid systems in rat brain. *Neuropeptides*, **5**, 469–72.

Loughlin, S.E., Annis, C.M., and Twardzik, D.R. (1989) Growth factors in opioid rich brain regions: distribution and response to intrastriatal transplants. *Adv. Biosci.*, **75**, 403–6.

Loughlin, S.E., Kornblum, H.I., Massamiri, T. and Leslie, F.M. (1991) Transient appearance of beta-endorphin immunoreactive cells within the germinal zone of neonatal rat forebrain. *Int. J. Dev. Neurosci.*, **9**, 493–500.

Loughlin, S.E., An, A. and Leslie, F.M. (1992a) Opioid receptor changes in weaver mouse striatum. *Brain Res.*, **585**, 149–55.

Loughlin, S.E., Kornblum, H.I. and Leslie, F.M. (1992b) Development of [¹²⁵I] beta endorphin binding in pre- and post-natal rat brain (in preparation).

Loughlin, S.E., Reid, S. and Leslie, F.M. (1992c) Opioid peptide expression is changed in weaver mutant mouse striatum: an in situ hybridization study. *Abstr. Soc. Neurosci.*, **282**, 12.

Loughlin, S.E., Annis, C.M., Gentry, L. et al. (1992d) Plasticity of transforming growth factor alpha expression in rat striatum: effects of transplants. *Brain Res. Bull.* (submitted).

Maderspach, K. and Solomania, R. (1988) Glial and neuronal opioid receptors: apparent positive cooperativity observed in intact cultured cells. *Brain Res.*, **441**, 41–7.

Magnan, J. and Tiberi, M. (1989) Evidence for the presence of mu- and kappa- but not of delta-opioid sites in the human fetal brain. *Dev. Brain Res.*, **45**, 275–81.

Mains, R.E., Eipper, E.A. and Ling, N. (1977) Common precursor to corticotropins and endorphins. *Proc. Natl. Acad. Sci. USA*, **74**, 3014–18.

Mansour, A., Khachaturian, H., Lewis, M.E. et al. (1987) Autoradiographic differentation of mu, delta, and kappa opioid receptors in the rat forebrain and midbrain. *J. Neurosci.*, **7**, 2445–64.

Mansour, A., Schafer, M.K.-H., Newman, S.W. and Watson, S.J. (1991) Central distribution of opioid receptors: a cross-species comparison of the multiple opioid systems of the basal ganglia, in *Neurobiology of Opioids* (eds O.F.X. Almeida and T.S. Shippenberg), Springer-Verlag, Berlin, Heidelberg, pp. 169–83.

McDowell, J. and Kitchen, I. (1987) Development of opioid systems: peptides, receptors and pharmacology. *Brain Res. Rev.*, **12**, 397–421.

McLean, S., Rothman, R.B., Chuang, D.-M. et al. (1989) Cross-linking of [¹²⁵I]beta-endorphin to mu-opioid receptors during development. *Dev. Brain Res.*, **45**, 283–9.

McMillen, I.C., Mercer, J.E. and Thorburn, G.D. (1988) Pro-opiomelanocortin mRNA levels fall in the fetal sheep pituitary before birth. *J. Mol. Endocrinol.*, **1**, 141–5.

Merchenthaler, I., Maderdrut, J.L., Altschuler, R.A. and Petrusz, P. (1986) Immunocytochemical localization of proenkephalin-derived peptides in the central nervous system of the rat. *Neuroscience*, **17**, 325–48.

Mess, B., Ruzsas, C. and Hayashi, S. (1989) Impaired thyroid function provoked by neonatal treatment with drugs affecting the maturation of monoaminergic and opioidergic neurons. *Exp. Clin. Endocrinol.*, **94**, 73–81.

Milligan, G., Streaty, R.A., Gierschik, P. et al. (1987) Development of opiate receptors and GTP-binding regulatory proteins in neonatal rat brain. *J. Biol. Chem.*, **262**, 8626–30.

Mochetti, I., Guidotti, A., Schwartz, J.P. and Dosta, E. (1985) Reserpine changes the dynamic state of enkephalin stores in rat striatum and adrenal medulla by different mechanisms. *J. Neurosci.*, **5**, 3379–85.

Moon-Edley, S. and Herkenham, M. (1984) Comparative development of striatal opiate receptors and dopamine revealed by autoradiography and histofluorescence. *Brain Res.*, **305**, 27–42.

Morris, B.J., Haarmann, I., Kempter, B. et al. (1986) Localization of prodynorphin messenger RNA in rat brain by in situ hybridization using a synthetic oligonucleotide probe. *Neurosci. Lett.*, **69**, 104–8.

Morris, B.J. and Herz, A. (1987) Distinct distribution of opioid receptor types in rat lumbosacral spinal cord. *Naunyn-Schmiedebergs Arch. Pharmacol.*, **336**, 240–3.

Mulder, A.H., Burger, D.M., Wardeh, G. et al. (1991) Pharmacological profile of various κ-agonists at κ-, μ- and δ-opioid receptors mediating presynaptic inhibition of neurotransmitter release in rat brain. *Br. J. Pharmacol.*, **102**, 518–22.

Murrin, L.C. and Ferrer, J.R. (1984) Ontogeny of the rat striatum: correspondence of dopamine terminals, opiate receptors and acetylcholinesterase. *Neurosci. Lett.*, **47**, 155–60.

Negri, L., Potenza, R.L., Corsi, R. and Melchiorri, P. (1991) Evidence for two subtypes of delta opioid receptors in rat brain. *Eur. J. Pharmacol.*, **196**, 335–6.

Ng, T.B., Ho, W.K.K. and Tam, P.P.L. (1984) Brain and pituitary beta-endorphin levels at different developmental stages of the rat. *Int. J. Protein Res.*, **24**, 141–6.

Nock, B., Giordano, A.L., Cicero, T.J. and O'Conor, L.H. (1990) Affinity of drugs and peptides for U-69, 593-sensitive and insensitive kappa opiate binding sites: the U-69, 593-insensitive site appears to be the beta endorphin-specific epsilon receptor. *J. Pharmacol. Exp. Ther.*, **254**, 412–19.

Oetting, G.M., Szucs, M. and Coscia, C.J. (1987) Differential ontogeny of divalent cation effects on rat brain delta-, mu-, and kappa-opioid receptor binding. *Dev. Brain Res.*, **31**, 223–7.

Olson, G.A., Olson, R.D. and Kastin, A.J. (1991) Endogenous opiates: 1990. *Peptides*, **12**, 1407–32.

Osborne, J.G., Kindy, M.S. and Hauser, K.F. (1991) Expression of proenkephalin mRNA in developing cerebellar cortex of the rat: expression levels coincide with maturational gradients in Purkinje cells. *Dev. Brain Res.*, **63**, 63–9.

Palmer, M.R., Miller, R.J., Olson, L. and Sieger, A. (1982) Prenatal ontogeny of neurons with enkephalin-like immunoreactivity in the rat central nervous system: an immunohistochemical mapping investigation. *Med. Biol.*, **60**, 2–88.

Pasternak, G.W. and Wood, P.L. (1986) Multiple mu opiate receptors. *Life Sci.*, **38**, 135.

Pasternak, G.W., Zhang, A. and Tecott, L. (1980) Developmental differences between high and low affinity opiate binding sites: their relationship to analgesia and respiratory depression. *Life Sci.*, **27**, 1185–90.

Patey, G., De La Baume, S., Gros, C. and Schwartz, J.-C. (1980) Ontogenesis of enkephalinergic systems in rat brain: post-natal changes in enkephalin levels, receptors and degrading enzyme activities. *Life Sci.*, **27**, 245–52.

Petrillo, P., Tavani, A., Verotta, D. *et al.* (1987) Differential postnatal development of mu-, delta- and kappa-opioid binding sites in rat brain. *Dev. Brain Res.*, **31**, 53–8.

Petrusz, P., Merchenthaler, I. and Maderdrut, J.L. (1985) Distribution of enkephalin-containing neurons in the central nervous system, in *Handbook of Chemical Neuroanatomy*, Vol. 4. *GABA and neuropeptides in the CNS*, Part I (eds A. Bjorklund and T. Hökfelt), Elsevier, Amsterdam, pp. 273–334.

Pittius, C.W., Ellendorff, F., Hollt, V. and Parvizi, N. (1987) Ontogenetic development of proenkephalin A and proenkephalin B messenger RNA in fetal pigs. *Exp. Brain Res.*, **69**, 208–12.

Quinlan, P.E. and Alessi, N.E. (1991) Characterization of beta-endorphin-related peptides in the caudal medulla oblongata and hypothalamus of the prenatal, postnatal and adult rat. *Dev. Brain Res.*, **62**, 1–5.

Rius, R.A., Barg, J., Bem, W.T. *et al.* (1991a) The prenatal developmental profile of expression of opioid peptide and receptors in the mouse brain. *Dev. Brain Res.*, **58**, 237–41.

Rius, R.A., Chikuma, T. and Lo, Y.P. (1991b) Prenatal processing of pro-opiomelanocortin in the brain and pituitary of mouse embryos. *Dev. Brain Res.*, **60**, 179–85.

Rosen, H. and Polakiewicz, R. (1989) Postnatal expression of opioid genes in rat brain. *Dev. Brain Res.*, **46**, 123–9.

Rosen, H., Douglass, J. and Herbert, E. (1984) Isolation and characterization of the rat proenkephalin gene. *J. Biol. Chem.*, **259**, 14309–13.

Rothman, R.B., Bykov, V., De Costa, B.R. *et al.* (1990) Interaction of endogenous opioid peptides and other drugs with four kappa opioid binding sites in guinea pig brain. *Peptides*, **11**, 311–31.

Sato, M., Morita, Y., Saika, T. *et al.* (1991) Localization and ontogeny of cells expressing preprodynorphin mRNA in the cerebral cortex. *Brain Res.*, **541**, 41–9.

Schafer, M.K.-H., Herman, J.P., Day, R. *et al.* (1988) The distribution of prodynorphin mRNA throughout the rat brain: a semi-quantitative mapping study. *Soc. Neurosci. Abst.*, **14**, 545.

Schafer, M.K.-H., Day, R., Watson, S.J. and Akil, H. (1991) Distribution of opioids in brain and peripheral tissues, in *Neurobiology of Opioids* (eds O.F.X. Almeida and T.S. Shippenberg), Springer-Verlag, Berlin, Heidelberg, pp. 53–71.

Schmahl, W., Funk, R., Miaskowski, U. and Plendl, J. (1989) Long-lasting effects of naltrexone, an opioid receptor antagonist, on cell proliferation in developing rat forebrain. *Brain Res.*, **486**, 297–300.

Schoffelmeer, A.N.M., Hansen, H.A., Stoof, J.C. and Mulder, A.H. (1986) Blockade of D-2 dopamine receptors strongly enhances the potency of enkephalins to inhibit dopamine-sensitive adenylate cyclase in rat neostriatum: involvement of delta- and mu-opioid receptors. *J. Neurosci.*, **6**, 2235–9.

Schoffelmeer, A.N.M., Hogenboom, F. and Mulder, A.H. (1987) Inhibition of dopamine-sensitive adenylate cyclase by opioids: possible involvement of physically associated mu- and delta-opioid receptors. *Naunyn-Schmiedebergs Arch. Pharmacol.*, **335**, 278–84.

Schoffelmeer, A.N.M., Rice, K.C., Jacobson, A.E. *et al.* (1988) Mu-, delta- and kappa-opioid receptor-mediated inhibition of neurotransmitter release and adenylate cyclase activity in rat brain slices: studies with fentanyl isothiocyanate. *Eur. J. Pharmacol.*, **154**, 169–78.

Schoffelmeer, A.N.M., Yao, Y.-H., Gioannini, T.L. *et al.* (1990) Cross-linking of human [^{125}I] beta-endorphin to opioid receptors in rat striatal membranes: biochemical evidence for the existence of a mu/delta opioid receptor complex. *J. Pharmacol. Exp. Ther.*, **253**, 419–26.

Schofield, P.R., McFarland, K.C., Hayflick, J.S. *et al.* (1989) Molecular characterization of a new immunoglobulin superfamily protein with potential roles in opioid binding and cell contact. *EMBO J.*, **8**, 489–95.

Schulz, R., Faase, E., Wuster, M. and Herz, A. (1979) Selective receptors for beta-endorphin on the rat vas deferens. *Life Sci.*, **24**, 843–50.

Schulz, R., Wuster, M. and Herz, A. (1981) Pharmacological characterization of the epsilon-opiate receptor. *J. Pharmacol. Exp. Ther.*, **216**, 604–6.

Schwartz, J.P. and Simantov, V. (1988) Developmental expression of proenkephalin mRNA in rat striatum and in striatal cultures. *Dev. Brain Res.*, **40**, 311–14.

Schwartzberg, D.G. and Nakane, P.K. (1982) Ontogenesis of adrenocorticotropin-related peptide determinants in the hypothalamus and pituitary gland of the rat. *Endocrinology*, **110**, 855–64.

Sei, C.A. and Dores, R.M. (1989) Changes in the processing of pro-dynorphin end products in the substantia nigra during neonatal development. *Peptides*, **11**, 89–94.

Seizinger, B.R., Grimm, C. and Herz, A. (1984a) Evidence for a differential postnatal development of proenkephalin B (=prodynorphin)-derived opioid peptides in the rat hypothalamus. *Endocrinology*, **115**, 926–35.

Seizinger, B.R., Hollt, V. and Herz, A. (1984b) Postnatal development of beta-endorphin-related peptides in rat anterior and intermediate pituitary lobes: evidence for contrasting development of proopiomelanocortin processing. *Endocrinology*, **115**, 136–42.

Shaha, C., Margioris, A., Liotta, A.S. *et al.* (1984) Demonstration of immunoreactive beta-endophin- and gamma$_3$-melanocyte-stimulating hormone-related peptides in the ovaries of neonatal, cyclic and pregnant mice. *Endocrinology*, **115**, 378–84.

Sharif, N.A. and Hughes, J. (1989) Discrete mapping of brain mu and delta opioid receptors using selective peptides: quantitative autoradiography species differences and comparison with kappa receptors. *Peptides*, **10**, 499–522.

Sirinathsinghji, D.J.S. and Dunnett, S.B. (1989) Disappearance of the mu opiate receptor patches in the rat neostriatum following lesioning of the ipsilateral nigrostriatal dopamine pathway with 1-methyl-4-phenylpyridinium ion (MPP+): restoration by embryonic nigral dopamine grafts. *Brain Res.*, **504**, 115–20.

Smith, J.A.M., Loughlin, S.E. and Leslie, F.M. (1991) Long-term changes in striatal opioid binding sites after 6-hydroxydopamine (6OHDA) lesion of substantia nigra. *Soc. Neurosci. Abst.*, **183**, 3.

Smith, J.A.M., Loughlin, S.E., and Leslie, F.M. (1992a) Kappa opioid inhibition of [^{3}H] dopamine release from embryonic rat ventral mesencephalic culture. *Mol. Pharmacol.*, **42**, 575–83.

Smith, J.A.M., Leslie, F.M., Broide, R.S. and Loughlin, S.E. (1992b) Alterations in striatal opioid peptide and receptor expression following 6-hydroxydopamine lesions of substantia nigra. *Brain Res.* (submitted).

Spain, J.W., Roth, B.L. and Coscia, C.J. (1985) Differential ontogeny of multiple opioid receptors (mu, delta, and kappa). *J. Neurosci.*, **5**, 584–8.

Specht, L.A., Pickel, V.M., Joh, T.H. *et al.* (1981) Light-microscopic immunocytochemical localization of tyrosine hydroxylase in prenatal rat brain. I. Early ontogeny. *J. Comp. Neurol.*, **199**, 233–53.

Stiene-Martin, A. and Hauser, K.F. (1990) Opioid-dependent growth of glial cultures: suppression of astrocyte DNA synthesis by met-enkephalin. *Life Sci.*, **46**, 91–8.

Stiene-Martin, A. and Hauser, K.F. (1991) Glial growth is regulated by agonists selective for multiple opioid receptor types in vitro. *J. Neurosci. Res.*, **29**, 538–48.

Stiene-Martin, A., Gurwell, J.A. and Hauser, K.F. (1991a) Morphine alters astrocyte growth in primary cultures of mouse glial cells: evidence for a direct effect of opiates on neural maturation. *Dev. Brain Res.*, **60**, 1–7.

Stiene-Martin, A., Osborne, J.G. and Hauser, K.F.

(1991b) Co-localization of proenkephalin mRNA using cRNA probes and a cell-type-specific immunocytochemical marker for intact astrocytes in vitro. *J. Neurosci. Methods,* **36**, 119–26.

Szucs, M. and Coscia, C.J. (1990) Evidence for delta-opioid binding and GTP-regulatory proteins in 5-day-old rat brain membranes. *J. Neurochem.,* **54**, 1419–25.

Szucs, M., Spain, J.W., Oetting, G.M. *et al.* (1987) Guanine nucleotide and cation regulation of μ, δ, and κ opioid receptor binding: evidence for differential postnatal development in rat brain. *J. Neurochem.,* **48**, 1165–70.

Tavani, A., Robson, L.E. and Kosterlitz, H.W. (1985) Differential postnatal development of mu, delta, and kappa opioid binding sites in mouse brain. *Dev. Brain Res.,* **23**, 306–9.

Tecott, L.H., Rubenstein, J.L.R., Paxinos, G. *et al.* (1989) Developmental expression of proenkephalin mRNA and peptides in rat striatum. *Dev. Brain Res.,* **49**, 75–86.

Tempel, A. (1991) Visualization of mu opiate receptor downregulation following morphine treatment in neonatal rat brain. *Dev. Brain Res.,* **64**, 19–26.

Tempel, A. (1992) Regulation of the opioid system by exogenous drug administration, in *Development of the Central Nervous System: Effects of Alcohol and Opiates* (ed. M.W. Miller), Wiley-Liss, New York, pp. 285–318.

Tempel, A., Kessler, J.A. and Zukin, R.S. (1990) Chronic naltrexone treatment increases expression of preproenkephalin and preprotachykinin mRNA in discrete brain regions. *J. Neurosci.,* **10**, 741–7.

Tiong, G.K.L. and Olley, J.E. (1988) Effects of exposure in utero to methadone and buprenorphine on enkephalin levels in the developing rat brain. *Neurosci. Lett.,* **93**, 101–6.

Tsang, D. and Ng, S.C. (1980) Effect of antenatal exposure to opiates on the development of opiate receptors in rat brain. *Brain Res.,* **188**, 199–206.

Uhl, G.R., Ryan, J.P. and Schwartz, J.P. (1988) Morphine alters preproenkephalin gene expression. *Brain Res.,* **459**, 391–7.

Uhler, M. and Herbert, E. (1983) Complete amino acid sequence of mouse proopiomelanocortin derived from the nucleotide sequence of proopiomelanocortin cDNA. *J. Biol. Chem.,* **258**, 257–61.

Unnerstall, J.R., Molliver, M.E., Kuhar, M.J. and Palacios, J.M. (1983) Ontogeny of opiate binding sites in the hippocampus, olfactory bulb and other regions of the rat forebrain by autoradiographic methods. *Dev. Brain Res.,* **7**, 157–69.

Unterwald, E.M., Knapp, C. and Zukin, R.S. (1991) Neuroanatomical localization of kappa₁ and kappa₂ opioid receptors in rat and guinea pig brain. *Brain Res.,* **562**, 57–65.

Van der Kooy, D. and Fishell, G. (1987) Neuronal birthdate underlies the development of striatal compartments. *Brain Res.,* **401**, 155–61.

Van Vliet, B.J., Mulder, A.H. and Schoffelmeer, A.N.M. (1990) μ-Opioid receptors mediate the inhibitory effect of opioids on dopamine-sensitive adenylate cyclase in primary cultures of rat neostriatal neurons. *J. Neurochem.,* **55**, 1274–80.

Vaysse, P.J.-J., Zukin, R.S., Fields, K.L. and Kessler, J.A. (1990) Characterization of opioid receptors in cultured neurons. *J. Neurochem.,* **55**, 624–31.

Vertes, Z., Melegh, G.Y., Vertes, M. and Kovacs, S. (1982) Effect of naloxone and d-met²-pro⁵-enkephalinamide treatment on the DNA synthesis in the development rat brain. *Life Sci.,* **31**, 119–26.

Vilijn, M.-H., Das, B., Kessler, J.A. and Fricker, L.D. (1989) Cultured astrocytes and neurons synthesize and secrete carboxypeptidase E, a neuropeptide–processing enzyme. *J. Neurochem.,* **53**, 1487–93.

Voorn, P., Kalsbeek, B., Jorritsma-Byham, B. and Groenewegen, H.J. (1988) The pre-and postnatal development of the dopaminergic cell groups in the ventral mesencephalon and the dopaminergic innervation of the striatum of the rat. *Neuroscience,* **25**, 857–87.

Watson, S.J., Akil, H., Fischli, W. *et al.* (1982) Dynorphin and vasopressin: common localization in magnocellular neurons. *Science,* **216**, 85–7.

Werling, L.L., Frattali, A., Portoghese, P.S. *et al.* (1988) Kappa receptor regulation of dopamine release from striatum and cortex of rats and guinea pigs. *J. Pharmacol. Exp. Ther.* **246**, 282–6.

Wild, K.D., Vanderah, T., Mosberg, H.I. and Porreca, F. (1991) Opioid delta receptor subtypes are associated with different potassium channels. *Eur. J. Pharmacol.,* **193**, 135–6.

Xia, Y. and Haddad, G.G. (1991) Ontogeny and distribution of opioid receptors in the rat brainstem. *Brain Res.,* **549**, 181–93.

Young, W.S., Bonner, T.I. and Brann, M.R. (1986) Mesencephalic dopamine neurons regulate the expression of neuropeptide mRNAs in the rat forebrain. *Proc. Natl. Acad. Sci. USA,* **83**, 9827–31.

Zagon, I.S., Rhodes, R.E. and McLaughlin, P.J. (1985) Distribution of enkephalin immunoreactivity in germinative cells of developing rat cerebellum. *Science,* **227**, 1049–51.

Zagon, I.S., Goodman, S.R. and McLaughlin, P.J. (1989) Characterization of zeta (ζ): a new opioid receptor involved in growth. *Brain Res.,* **482**, 297–305.

Zagon, I.S., Gibo, D. and McLaughlin, P.J. (1990) Expression of zeta, a growth-related opioid receptor, in metastatic adenocarcinoma of the human cerebellum. *J. Natl. Cancer Inst.,* **82**, 325–7.

Zagon, I.S., Gibo, D.M. and McLaughlin, P.J. (1991) Zeta (ζ), a growth-related opioid receptor in developing rat cerebellum: identification and characterization. *Brain Res.,* **551**, 28–35.

Zagon, I.S. and McLaughlin, P.J. (1983) Increased brain size and cellular content in infant rats treated with an opiate antagonist. *Science,* **221**, 1179–80.

Zagon, I.S. and McLaughlin, P.J. (1990) Ultrastructural localization of enkephalin-like immunoreactivity in developing rat cerebellum. *Neuroscience,* **34**, 479–89.

Zagon, I.S. and McLaughlin, P.J. (1993) Opioid factor receptor in the developing nervous system, in *Receptors in the Developing Nervous System,* Vol. 1, (eds I.S. Zagon and P.J. McLaughlin), Chapman & Hall, London, pp. 39–62.

Zamir, N., Quirion, R. and Segal, M. (1985) Ontogeny and regional distribution of proenkephalin- and prodynorphin-derived peptides and opioid receptors in rat hippocampus. *Neuroscience,* **15**, 1025–34.

Zukin, R.S., Eghbali, M., Olive, D. *et al.* (1988) Characterization and visualization of rat and guinea pig brain κ opioid receptors: Evidence for κ_1 and κ_2 opioid receptors. *Proc. Natl. Acad. Sci. USA,* **85**, 4061–5.

SIGMA RECEPTORS, PCP RECEPTORS AND THE DEVELOPING NERVOUS SYSTEM

Edythe D. London and Stephen R. Zukin

11.1 INTRODUCTION

The existence of σ (sigma) receptors was hypothesized on the basis of work by Martin *et al.* (1976), who studied the physiological and behavioral responses to opioid drugs in the chronic spinal dog. Based on the responses to the prototypical ligands, morphine, ketocyclazocine, and *N*-allylnormetazocine (NANM, SKF 10047), this work led to the hypothesis of multiple opioid receptors (μ, κ and σ). Although this landmark work has greatly advanced the field of opiate research, it was later found that the responses to NANM were non-opioid, inasmuch as they were not reversed by naltrexone (Vaupel, 1983). Furthermore, benzomorphan opioids which show high affinity for σ receptors also interact with receptors for phencyclidine (PCP). Therefore, it is likely that canine delirium, which was originally attributed to interactions with the σ 'opioid' receptor, were due to activity at the PCP receptor.

This chapter gives an overview of current knowledge about the pharmacology and physiological importance of the σ and PCP receptors. The role of these receptors in the developing brain is highlighted.

11.2 IDENTIFICATION AND CHARACTERIZATION OF SIGMA AND PCP RECEPTORS

11.2.1 PHARMACOLOGICAL PROPERTIES

The first report of the biochemical identification of σ binding sites appeared about a decade ago, when racemic [^{3}H]N-allylnormetazocine ([^{3}H]SKF-10047) was used to label 'etorphine-inaccessible sites' in guinea-pig brain (Su, 1981, 1982). The sites showed very low affinity for naloxone and some opioid agonists, including etorphine and morphine, but high affinity for benzomorphan opioids, including pentazocine, NANM and cyclazocine. The receptors showed stereoselectivity, favoring the dextrorotatory over the levorotatory isomers of benzomorphan opioid drugs. Although the D_2 dopamine receptor antagonist, haloperidol, was the most potent inhibitor of σ receptor binding, σ sites were unlike dopamine receptors in that they showed higher affinity for the levorotatory enantiomer of butaclamol than for the dextrorotatory isomer. PCP showed activity at the σ sites, but it had about one-tenth the potency of (+)NANM.

High affinity binding sites with pharma-

Receptors in the Developing Nervous System Vol. 2: Neurotransmitters. Edited by Ian S. Zagon and Patricia J. McLaughlin. Published in 1993 by Chapman & Hall. ISBN 0 412 49400 0. Vols. 1 and 2 (set) ISBN 0 412 54520 9.

cological properties of a PCP receptor were observed in rat brain (Vincent *et al.*, 1979; Zukin and Zukin, 1979), and then characterized (e.g., Quirion *et al.*, 1981; Hampton *et al.*, 1982; Vignon *et al.*, 1982, 1983; Vincent *et al.*, 1983; Zukin *et al.*, 1983; Sircar and Zukin, 1985). The PCP receptor is selective for drugs capable of eliciting PCP-like behavioral effects, and does not show high affinity for other categories of hallucinogens, other drugs, or neurochemicals (reviewed by Zukin and Zukin, 1988). One of the earliest reports on the identification of the PCP receptor noted that NANM inhibited binding to the receptor, with a potency that was about one-third that of PCP (Zukin and Zukin, 1979). Observations that specific [^{3}H]PCP binding can be inhibited by cyclazocine and related benzomorphan opioids (Zukin and Zukin, 1981; Zukin *et al.*, 1983) suggested that the PCP and the σ receptor were the same entity. Despite the fact that benzomorphans do interact with the PCP receptor, subsequent studies have shown that σ receptors are distinct from PCP receptors anatomically, ontogenetically and phylogenetically (Vu *et al.*, 1990).

11.2.2 RADIOLIGAND SPECIFICITY (FIG. 11.1)

Since the initial description of specific σ receptor binding, there have been attempts to develop selective, high affinity radioligands for assay. Early studies of this receptor used racemic mixtures of benzomorphan opioids as radioligands. After the first studies with (±)NANM (Su, 1981, 1982), other assays utilized the benzomorphan (±)ethylketocyclazocine (Tam, 1983). The naloxone-inaccessible binding of this radioligand showed stereoselectivity, favoring the dextrorotatory isomers of benzomorphans, a finding that was consistent with earlier studies of the σ receptor. Based on these observations, subsequent assays used (+)-[^{3}H]NANM (Tam and Cook, 1984; Tam 1985; Su *et al.*, 1988b; McCann and Su, 1990) or (+)-[^{3}H]ethylketocyclazocine (Tam, 1985) as radioligands for σ receptor binding assays. Other studies took advantage of the very high affinity of haloperidol for the σ receptor, and used [^{3}H]haloperidol as the radioligand in the presence of either (+)butaclamol or spiperone to block binding of the radioligand to D$_2$ dopamine receptors (Tam

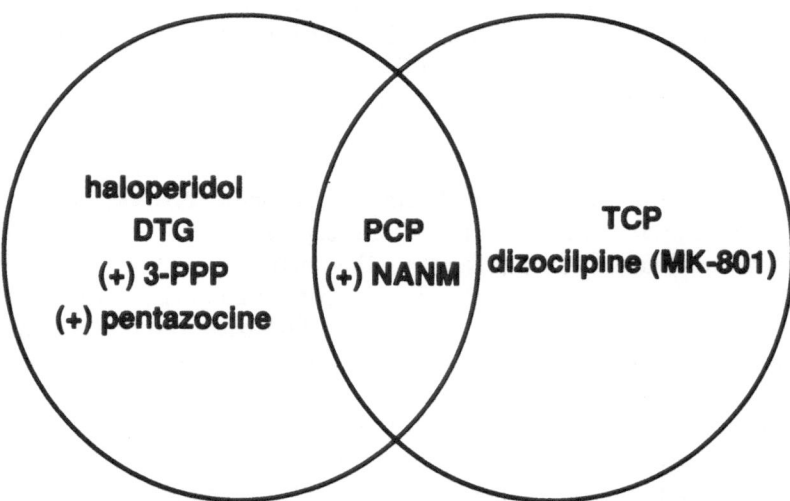

Fig. 11.1. Diagram depicting selectivity of ligands for the σ receptor (left circle) and PCP receptor (right). Area of overlap indicates ligands that interact with both receptors. Abbreviations: DTG, 1,3-di(2-tolyl)guanidine; (+)3-PPP, (+)-3-(3-hydroxyphenyl)-*N*-(1-propyl)piperidine; PCP, phencyclidine; (+)NANM, (+)*N*-allylnormetazocine; TCP, [1-(2-thienyl) cyclohexyl] piperidine.

and Cook, 1984; Su *et al.*, 1988b; McCann and Su, 1990; Vu *et al.*, 1990). Although not as potent at σ sites as haloperidol, the ligand (+)-3-(3-hydroxyphenyl)-*N*-(1-propyl)piperidine ((+)-[³H]3-PPP) has also been used (Largent *et al.*, 1984, 1986a), and it apparently shows higher selectivity for σ sites than many of the other radioligands used previously. Also notable is the finding that [³H]dextromethorphan and σ ligands bind to at least one common high affinity site (Musacchio *et al.*, 1989a,b). The observation that certain symmetrically substituted guanidines can inhibit the binding of (+)-[³H]NANM to membranes from guinea-pig brain led to the development of 1,3-di(2-[5-³H]tolyl)guanidine ([³H]DTG) as a radioligand for the σ receptor (Weber *et al.*, 1986). This ligand had the advantage over the benzomorphans of being highly selective for the σ site, and it yielded a high degree of specific binding in preparations of guinea-pig brain. Most recently, the synthesis of optically pure (+)-[³H]pentazocine, a potent and selective ligand for σ receptors has offered the promise of being a valuable radioligand (de Costa *et al.*, 1989).

[³H]Phencyclidine, the prototypical ligand of the PCP receptor, has proven less than optimally selective for this receptor. Although it displays high affinity for the PCP receptor (K_d = 0.15–0.25 μM) (Zukin and Zukin, 1979; Vincent *et al.*, 1983), it also has significant, albeit lower affinity for σ receptors (IC_{50} values about 1–2 μM against several tritiated ligands for σ receptors (Tam, 1983; Weber *et al.*, 1986)). PCP and related compounds also inhibit dopamine uptake (Vignon *et al.*, 1988), and inhibit the binding of [³H]mazindol to sites on dopamine transporters (Kuhar *et al.*, 1990). *N*-(1-[2-thienyl]cyclohexyl)[³H]piperidine ([³H]TCP), a thienyl derivative of [³H]PCP, displays higher affinity and specificity for the PCP receptor than does PCP itself (Sircar and Zukin, 1985). The most potent and selective ligand currently available for the PCP receptor is dizocilpine (MK-801) (Wong *et al.*, 1986, 1988; Foster and Wong, 1987).

11.2.3 BIOCHEMICAL STUDIES OF SIGMA RECEPTORS

Some of the earliest studies of sigma receptors indicated a protein nature of the sites. When membranes from guinea-pig brain were treated with heat, σ binding, assayed with [³H]NANM or [³H]dextromethorphan, was reduced (Su, 1982; Craviso and Musacchio, 1983a,b). Binding also displayed a marked dependence on pH, with optima achieved above pH 8.0, when either [³H]dextromethorphan or (+)-[³H]-3-PPP were used as radioligands (Craviso and Musacchio, 1983a; Largent *et al.*, 1987). Furthermore, treatment of membranes with proteases reduced the binding of [³H]NANM (Su, 1982) and [³H]dextromethorphan (Craviso and Musacchio, 1983a).

The development of an azido derivative of DTG, [³H]azido-DTG, has allowed photoaffinity labeling and molecular weight determination of a σ receptor protein in guinea-pig brain (Kavanaugh *et al.*, 1988). Irradiation of the ligand–receptor complex with light of 336 nm wavelength produced covalent binding of the azido probe to the receptor; and sodium dodecyl sulfate (SDS)–polyacrylamide gel electrophoresis demonstrated that labeling was to a single polypeptide with a molecular weight of 29 kDa. Pharmacological characterization indicated a ligand specificity consistent with that of the sigma receptor. Similarly, [³H]azido-DTG labels a polypeptide of 29 kDa in NCB-20 cells, which bear σ receptors (Largent *et al.*, 1986a; Adams *et al.*, 1987; Kushner *et al.*, 1988).

In contrast to the findings regarding σ receptors in guinea-pig brain and NCB-20 cells, [³H]azido-DTG labeled two polypeptides, with molecular weights of 18 and 21 kDa in PC12 cells, suggesting heterogeneity of σ receptors (Hellewell and Bowen, 1990). Further evidence for heterogeneity from this study included a reduced affinity for dextrorotatory benzomorphan opioids, as compared to the affinity of the classical σ receptor characterized in the guinea-pig brain. The concept of heterogeneity of σ receptors was consistent with findings of Bowen *et al.* (1989), who noted that irradiation of rat brain membranes with light of 254 nm, a treatment that alters ultraviolet absorbing residues in proteins, had differential effects on the binding of (+)-[³H]NANM as compared to the binding of (+)-[³H]-3-PPP and [³H]DTG. Irradiation increased the binding of [³H] NANM, but decreased binding of the non-benzomorphan radioligands. In treated membranes, DTG and (+)-3-PPP were less potent inhibitors of (+)-[³H]NANM binding than in unirradiated membranes; however, the potencies of the benzomorphans as inhibitors were relatively unaffected. Furthermore, in unirradiated membranes, DTG and haloperidol were competitive inhibitors of (+)-[³H]-3-PPP binding whereas the benzomorphan opioids, (+)NANM and (+)pentazocine, were uncompetitive inhibitors.

Electrophysiological and receptor binding studies in intact NCB-20 cells also indicated the existence of more than one type of σ receptor (Wu *et al.*, 1990). Whole cell voltage clamp studies demonstrated that σ ligands caused an apparent inward current, which was due to blockade of a tonic, outward potassium current. Although the rank order of drugs in producing this effect generally resembled the order of potencies of these drugs in inhibiting binding to σ receptors reported previously, several compounds showed reverse stereoselectivity. Binding assays performed on intact cells under conditions similar to the electrophysiological studies (e.g., Dulbecco's phosphate-buffered saline instead of 50 mM Tris-HCl) revealed the existence of two populations of sites. The high affinity sites showed a pharmacological selectivity similar to that seen previously in membranes from NCB-20 cells (Largent *et al.*, 1986a; Kushner *et al.*, 1988), but the low affinity sites showed a rank order of potencies that correlated well with potencies in the voltage clamp assays, including absence of the usual stereoselectivity for benzomorphan opioids and 3-PPP.

11.2.4 BIOCHEMICAL CHARACTERIZATION OF PCP RECEPTORS

The PCP receptor is proteinaceous in nature, as it is inactivated by heat, and is destroyed by proteases, such as trypsin, pronase, and papain (Zukin and Zukin, 1979; Zukin *et al.*, 1983; Vignon *et al.*, 1982). Pretreatment with *N*-ethylmaleimide or iodoacetamide inhibits activity, suggesting the importance of sulfhydryl groups at or near the binding site, or at a site that influences the tertiary structure of the receptor protein (Zukin and Zukin, 1979).

Photoaffinity labeling of PCP receptors with the photoactive PCP derivative azido-[³H]PCP, followed by SDS–polyacrylamide gel electrophoresis, has been used to determine the polypeptides comprising the binding site. When such a procedure is carried out with membrane homogenates, a number of polypeptides are labeled (Haring *et al.*, 1986, 1987; Sorensen and Blaustein, 1986). However, photoaffinity labeling carried out in lipid vesicles reconstituted with solubilized and partially purified, active PCP receptors results in labeling of only two polypeptides (M_r 98 000 and 59 000), both of which have also been identified in membrane homogenates (Scheideler and Zukin, 1990). Such data suggest that the PCP receptor comprises multiple protein subunits. However, azido-[³H]dizocilpine was reported to label a single polypeptide of M_r 120 000 (Sonders *et al.*, 1990), in agreement with the target size of M_r 118 000 derived from radiation inactivation of the [³H]TCP binding site (Honoré *et al.*, 1989).

11.2.5 ENDOGENOUS LIGANDS

Several research groups have reported preliminary findings on endogenous substances that were extracted from brain, and which appeared to act as ligands for σ and PCP receptors. Acid extraction of porcine brain yielded two factors, termed α- and β-endopsychosin, representing material which selectively inhibited binding to PCP and σ receptors, respectively

(Quirion *et al.*, 1984; Contreras *et al.*, 1987; DiMaggio *et al.*, 1988). α-Endopsychosin was characterized as having a molecular weight of about 3000 Da. It was regionally co-localized in brain with PCP receptors, and it mimicked the effects of PCP upon turning behavior when injected unilaterally into the substantia nigra of the rat. In contrast, β-endopsychosin inhibited the binding of (+)-[³H]NANM and (+)-[³H]-3-PPP without affecting binding to the PCP receptor, opioid receptors or dopamine receptors. Both endogenous factors were sensitive to protease, consistent with a peptide nature. Extracts of human brain have also been found to contain substances that selectively inhibit binding of radioligands to the σ and PCP receptors, as well as factors which inhibit binding to both sites (Zhang *et al.*, 1988). At least two factors, with molecular weights of approximately 700 Da, were partially purified. One of these factors showed activity at σ and PCP receptors whereas another appeared to have preferential activity at the PCP receptor.

Extracts of guinea-pig brain contained two endogenous factors, termed 'sigmaphins', which interacted preferentially with σ receptors (Su *et al.*, 1986). These factors had activity at opioid receptors, but were almost inactive in PCP receptor binding assays. Treatment with trypsin largely reduced inhibitory potencies of the sigmaphins, and, therefore, suggested a protein composition.

Still another factor, isolated from bovine brain, reportedly mimicked the effects of PCP-like drugs upon spontaneous and NMDA-evoked neurotransmitter release (Zukin *et al.*, 1987). The dose–response curves for the endogenous factor were parallel to those of PCP.

Despite these intriguing reports, there has been no report of a definitive characterization of an endogenous ligand for the σ receptor or the PCP receptor. In the case of putative endogenous ligands for the PCP receptor, it is possible that these factors were endogenous substances that antagonize the function of the NMDA receptor at sites distinct from the PCP receptor.

11.3 DISTRIBUTION OF SIGMA AND PCP RECEPTORS IN THE BRAIN

The earliest studies of σ receptors, performed by incubating racemic [³H]NANM or dextro-rotatory isomers of [³H]NANM or [³H]ethylke-tocyclazocine with suspensions from grossly dissected areas of guinea-pig brain, demonstrated the highest level of binding in the brainstem and midbrain, followed by the cerebellum (Su, 1982; Tam, 1985); lower levels of binding were seen in the striatum and cortex. Subsequent autoradiographic studies *in vitro* with (+)-[³H]-3-PPP and (+)-[³H]NANM demonstrated discrete localizations of σ receptors to many limbic, brainstem and cerebellar regions (Largent *et al.*, 1984, 1986b; Gundlach *et al.*, 1985). Particularly high densities of σ sites were observed in the pyramidal cell layer of the hippocampus, hypothalamus, zona incerta, subiculum, superficial layer of the cingulate cortex, Purkinje cell layer of the cerebellum, and pontine and cranial nerve nuclei. A similar distribution was seen when [³H]DTG was used as the radioligand (Weber *et al.*, 1986). Sigma receptors have also been labeled *in vivo* in mouse brain using (+)-[³H] NANM and [³H]haloperidol, with a distribution pattern that was consistent with *in vitro* biochemical and autoradiographic studies (Weissman *et al.*, 1990a).

As was the case for the σ receptor, the earliest reports on identification of the PCP receptor provided information about the distribution of the receptor in brain (Vincent *et al.*, 1979; Zukin and Zukin, 1979). With [³H]PCP as the radioligand in homogenates of regions of rat brain, specific binding was high in telencephalic areas, including hippo-campus, caudate putamen and neocortex; levels of binding were considerably lower in the cerebellum and medulla/pons. Subsequent autoradiographic studies with [³H]PCP and [³H]TCP provided more detailed anatomical information, but generally confirmed the reports of dense labeling of cortical areas and the hippocampal formation, with lower levels of binding in the brainstem and cerebellum (Quirion *et al.*, 1981; Sircar and Zukin, 1985; Largent *et al.*, 1986b). Thus, despite the fact that some benzomorphans and PCP bind to both σ and PCP receptors, the distribution of specific binding using the prototypic ligands for the respective receptors ([³H]NANM for the σ receptor, [³H]PCP for the PCP receptor) showed neuroanatomical divergence. Anatomical studies were also among the key findings that linked the PCP receptor functionally to the *N*-methyl-D-aspartate (NMDA) subset of glutamate receptors in brain. In this regard, autoradiographic studies of [³H]TCP binding showed a high concordance with the binding of [³H]glutamate, assayed under conditions that select for NMDA receptors (Maragos *et al.*, 1986).

11.4 FUNCTIONAL RELEVANCE OF SIGMA AND PCP RECEPTORS

11.4.1 SIGMA RECEPTOR

It has been hypothesized that the σ receptor was the biochemical entity in brain that mediated the psychotomimetic effects of benzomorphan opioids. Evidence for this hypothesis derived, in part, from the fact that the prototypic σ receptor ligand NANM produced psychotomimetic effects in human volunteers (Keats and Telford, 1964) and delirium in the chronic spinal dog (Martin *et al.*, 1976). Racemic NANM was used in both studies. Notably, behavioral effects in the human study were not challenged with an opioid antagonist. Therefore, it is possible that psychotomimesis resulted from interactions with opioid receptors. Some opioids produce naltrexone-reversible psychotomimetic effects (Jasinski *et al.*, 1968; Martin, 1983; Pfeiffer *et al.*, 1986; Musacchio, 1990), which may be attributable to interactions with κ opioid receptors (Pfeiffer *et al.*, 1986; Su, 1991). In this regard, levorotatory benzomorphan opioids are potent ligands for the κ opioid receptor (Su, 1985). Another major

consideration is the fact that NANM interacts with the PCP receptor; therefore, canine delirium in the original report on σ effects may have reflected an interaction of NANM with the PCP receptor.

The aforementioned links between the σ receptor and psychosis seem weak in light of developments over the past 15 years. Nonetheless, the σ hypothesis generated interest in developing σ antagonists, which might prove to be potential agents for the treatment of psychotic disorders. A consistent finding that fueled the search for a σ ligand which would be an antipsychotic drug was the observation that haloperidol was an extremely potent σ receptor ligand. More recent studies demonstrated that several new drugs, with varying affinities for different receptors, and antipsychotic activity in preclinical (and in some cases clinical) trials, have affinity for the σ receptor as a common feature (Largent *et al.*, 1988; Su *et al.*, 1988b). Furthermore, the observation that σ receptors show a lower density in samples of cerebral cortex from patients who died with schizophrenia than in age-matched controls is additional evidence favoring the 'σ hypothesis of psychosis' (Weissman *et al.*, 1990b).

Aside from the potential role of the σ receptor in psychotic disorders and in the development of potential psychotherapeutic medications, the σ receptor has been implicated in numerous functions of the brain and other organ systems (reviewed by Walker *et al.*, 1990; Su, 1991). For example, σ receptors appear to have a role in motor function. Evidence for this view derives from observations that microinjection of σ receptor ligands into the red nucleus produced dystonic effects in the rat, and that circling behavior was induced by intranigral administration of the selective σ receptor ligand di-*o*-tolylguanidine (DTG) (Weber *et al.*, 1986; Walker *et al.*, 1988). It also seems that σ receptors influence the firing pattern of dopaminergic neurons of the substantia nigra (Engberg and Wikström, 1991). The finding that progesterone and other steroids interact with σ receptors in the brain and spleen has led to the hypothesis that this receptor may represent a link between the nervous, endocrine and immune systems (Su *et al.*, 1988a).

11.4.2 PCP RECEPTOR

Numerous lines of evidence indicate that the PCP receptor represents a binding site within the ion channel gated by the NMDA receptor. These include anatomical co-localization of the two sites, as well as correlation of potencies of PCP receptor ligands in PCP receptor binding assays with potencies in use- and voltage-dependent non-competitive antagonism of NMDA receptor-mediated conductances, and sensitivity of specific binding of PCP receptor radioligands to stimulation by NMDA receptor agonists and the co-agonist glycine and to inhibition by NMDA receptor antagonists (reviewed in Javitt and Zukin, 1989b). This functional relationship between PCP and NMDA receptors has several important consequences for performance of PCP receptor binding studies. The NMDA receptor is governed by a multi-state mechanism in which binding of at least one molecule of agonist is required for detection of PCP receptor radioligand binding to the closed conformation of the channel, and binding of two molecules of agonist and of glycine is required for detection of PCP receptor radioligand binding to the open conformation (Javitt *et al.*, 1990). Furthermore, PCP receptor radioligands manifest different binding characteristics depending on whether the channel is in the agonist-associated closed conformation or in the open conformation. Radioligand access to the PCP receptor via the open channel gives rise to a fast exponential of association in kinetic binding assays, or to an apparent high affinity component of binding in saturation studies. By contrast, radioligand access via the agonist-associated, closed channel gives rise to a slow exponen-

tial of association or to an apparent low affinity component of binding. The fast component of binding serves as a biochemical marker of activated NMDA receptor channels (Javitt and Zukin, 1989a). Therefore, it is only in the presence of saturating concentrations of L-glutamate (or NMDA) and glycine that binding properties of PCP receptor ligands can be accurately assessed based upon a simple linear bimolecular model. Under other incubation conditions, total radioligand binding will detect a combination of closed and activated channels, binding may not reach equilibrium for up to 24 h, and only kinetic assay designs, with appropriate non-linear weighted curve-fitting of data, will yield accurate binding parameters. Another important consequence is that the total level of radioligand binding to the PCP receptor, encompassing both fast and slow or high and low affinity components, increases with increasing channel activation (Javitt and Zukin, 1989a,b; Javitt *et al.*, 1990). Thus, an apparent developmental alteration in total receptor density could in principle reflect an altered proportion of activated NMDA channels whereas receptor density actually is unchanged. In fact, the issue of whether developmental alterations in expression and assembly of NMDA receptor subunits may result in alterations in sensitivity to activation has not yet been addressed.

Recently, an NMDA receptor has been cloned and sequenced (Moriyoshi *et al.*, 1991). It consists of 938 amino acids with a calculated molecular mass of 105 500 (M_r value of 105.5 kDa). This single protein appears to contain the PCP receptor, as (+)dizocilpine potently inhibits steady-state currents induced by 100 mM NMDA in *Xenopus* oocytes in which the receptor is expressed.

Another study reported the cloning of a glutamate-binding protein of M_r 57 020 (Kumar *et al.*, 1991). When this protein is combined with several other distinct subunits, not yet cloned, full NMDA receptor-channel function, including PCP receptor binding, is reconstituted. The issue of whether the native PCP receptor corresponds to either of these entities remains to be determined.

The PCP receptor has also been implicated in psychosis. The similarity between the effects of NANM and PCP in the 'canine delirium' model (Vaupel, 1983) suggests that the psychotomimetic effects of σ opioids in the human could be mediated at the PCP receptor. The PCP–NMDA hypothesis of schizophrenia postulates an endogenous deficit in NMDA receptor-mediated neurotransmission in this disorder. It is based on several observations. Single doses of PCP (which would yield serum and brain concentrations at which the PCP receptor would be the predominant molecular target), given to normal human volunteers induces a transient psychotomimetic state incorporating positive and negative symptoms, formal thought disorder, and neuropsychological deficits closely resembling those seen in schizophrenia; the same low dose produces prolonged exacerbation of disease-specific symptoms and signs in schizophrenic subjects. Furthermore, schizophrenia-like abnormalities in event-related potentials may be induced by PCP-like agents, whereas administration of large oral doses of the NMDA receptor co-agonist glycine have been reported to produce clinical improvements in some schizophrenic patients (reviewed by Javitt and Zukin, 1991).

11.5 SIGMA AND PCP RECEPTORS IN THE DEVELOPING BRAIN

Ontogenetic studies of σ and PCP receptors have been performed in rats of several strains. Although methodological differences (binding conditions, radioligands) confound direct comparisons in some cases, the findings generally confirm a separation between these two historically linked but functionally distinct receptors.

11.5.1 ONTOGENY OF THE SIGMA RECEPTOR

Three studies have assessed σ receptor binding in the brains of rats at different ages. In the first of these studies (Majewska *et al.*, 1989), σ receptor binding parameters were assayed in crude, washed membranes from whole brains of Fischer-344 rats. The radioligand was [³H]haloperidol in the presence of spiperone to inhibit radioligand binding to dopamine receptors. Samples were taken from rats at 1, 7, 14, 30, 90 days, 6 months and 1 year postnatally. Because the mass of the rat brain increases through the fourth postnatal month (Rosenberg and Stern, 1966), ages prior to 4 months represent developmental periods. The maximal survival time for the Fischer-344 rats was reported to be approximately 35 months, with a mean survival of 29 months (Coleman *et al.*, 1977). Therefore, rats of this strain can be considered mature, but not senescent at one year of age (London *et al.*, 1981). The results demonstrated no alteration in binding affinity or in the density of σ receptors, expressed per mg protein, in the rat brain from the early postnatal period through the first postnatal year. The lack of postnatal development of σ receptors, as compared to the documented changes in other neurotransmitter receptors, and the fact that σ receptors have high densities in peripheral organs, such as the liver (Samovilova *et al.*, 1988) and immune (Su *et al.*, 1988b; Wolfe *et al.*, 1988) and endocrine tissues (Wolfe *et al.*, 1989) suggested a fundamental role for the σ receptor beyond a function in the brain (Majewska *et al.*, 1989).

Paleos *et al.* (1990) used (+)-[³H]-3-PPP to label σ receptors in the brains of Sprague–Dawley rats taken on prenatal day 2, postnatal days 1, 6 and 28, and adulthood (defined by a body weight of 225 g). The findings indicated a decline in receptor density, expressed as fmol bound per mg protein, from the prenatal period to the early postnatal period, followed by another substantial decline at postnatal day 28, with no subsequent change. Receptor affinity was lower on prenatal day 2 and on postnatal day 1 than at later ages. These findings confirmed the lack of a developmental increase, when results were viewed as a function of protein concentration, although the developmental decline noted here presented an inconsistency. This finding reflected a marked developmental increase in protein content of the brains of the Sprague–Dawley rats assayed. In fact, when σ receptor binding was reported as total pmol bound per brain, there was an increase in receptor binding throughout development.

Another study of σ receptors in the brains of Sprague–Dawley rats used [^{3}H]DTG as the radioligand (Matsumoto *et al.*, 1989). The results indicated an age-related loss of receptor density and affinity, when animals 2–3 months old were compared with rats at 5–6 months of age. The loss of receptors seemed to have functional importance in terms of affecting movement and posture, as unilateral microinjection of DTG into the substantia nigra produced fewer contralateral turns in the older animals. Furthermore, the older animals showed less pronounced postural changes in response to the unilateral injection of DTG into the red nucleus. As no age-related loss of σ receptor binding in brain during the first postnatal year of life was seen in Fischer-344 rats, it is possible that genetic strain may be an important determinant in effects of age on the σ receptor. In this regard, strain differences in age effects on other neurochemical systems in brain have been reported (Waller *et al.*, 1983).

11.5.2 ONTOGENY OF THE PCP RECEPTOR

Seven studies have addressed PCP receptor binding as a function of development in the rat. Of these, two (Majewska *et al.*, 1989; Paleos *et al.*, 1990) have directly compared σ and PCP receptors, whereas two others (McDonald *et al.*, 1990; Tremblay *et al.*, 1990)

have compared development of PCP receptors with development of other components of the NMDA receptor complex.

Sircar and Zukin (1983) characterized prenatal development of PCP receptors labeled with [^{3}H]PCP using a filtration assay of whole brain homogenates derived from Blue Spruce hooded rats from gestational day 13 until parturition, and from adult rats (120 days old). For each age group, Scatchard analysis of binding was carried out with radioligand concentrations ranging from 1 to 500 nM. The key observations were that although K_d values remained unchanged throughout development, B_{max} values (expressed per mg protein) demonstrated marked developmental enhancement. Thus, between gestational days 13 and 15, B_{max} represented 10–15% of adult levels. Beginning at gestational day 17, a rapid and marked increase in B_{max} resulted in the attainment of adult levels by gestational day 21. In order to clarify the functional development of this receptor, stereospecificity of [^{3}H]PCP binding was assayed at each age point by comparing the potencies of the potent PCP-like drug dexoxadrol and its 100-fold less active enantiomer levoxadrol for inhibition of specific binding of radioligand. Stereospecificity was not observed until gestational day 19, suggesting a developmental transition to a functionally adult type of PCP receptor at that point in development.

Majewska *et al.* (1989) undertook a comparative study of the ontogeny of σ receptors labeled with [^{3}H]haloperidol (see above) and PCP receptors labeled with [^{3}H]TCP in Fischer-344 rats. PCP binding assays were accomplished by filtration assays, using radioligand concentrations ranging from 0.5 to 80 nM. From postnatal day 1 to 1 year, K_d values of PCP receptor binding were reported not to change significantly (5.6–15.9 nM). By contrast B_{max} values revealed age-dependent increases, reaching adult levels by 14 days.

Paleos *et al.* (1990) addressed the comparative ontogeny of PCP and σ receptors (see above) in Sprague–Dawley rats. This ontogenetic study was the first to address directly the issue of the dependence of PCP receptor binding on the endogenous concentrations of NMDA receptor agonist (glutamate) and co-agonist (glycine), which were measured and determined to be above the concentrations giving maximal enhancement of [^{3}H]TCP binding throughout the developmental phases examined. Glutamate, 100 μM, was also added to each assay tube. Specific binding of 1 nM [^{3}H]TCP from 2 days prior to parturition to 28 days postnatally increased monotonically, reaching adult levels by postnatal day 21. Scatchard analysis of [^{3}H]TCP binding was conducted at five developmental points from prenatal day 2 to adulthood. Whereas K_d values were stable over this development range (8–10 nM), B_{max} values increased from 350 fmol/mg protein at prenatal day 2 to 1865 fmol/mg protein at adulthood; adult levels were attained by postnatal day 28. Densities of PCP receptors increased more rapidly than developmental increases in brain protein or gross brain mass. Pharmacological selectivity and stereospecificity of the PCP receptor were examined at postnatal day 6, and were found to manifest adult characteristics.

Shinohara *et al.* (1989) examined the ontogeny of [^{3}H]TCP binding in forebrains derived from male Wistar rats from gestational day 18 to postnatal day 50. PCP receptors were also characterized autoradiographically in slide-mounted 10 μM coronal sections, that were exposed to tritium-sensitive film for 4 weeks, or were wiped off and counted by scintillation spectrometry after incubation with [^{3}H]TCP. Assays were carried out at ten age points ranging from gestational day 18 to postnatal day 50. No alteration in K_d of [^{3}H]TCP binding (13.6–14.5 nM) was observed during development. However, B_{max} increased sigmoidally from 183 to 556 fmol/mg protein during this period. This was the first study to investigate whether the sensitivity of PCP receptor binding to inhi-

bition by the direct NMDA receptor antagonist D-(–)2-amino-5-phosphonovaleric acid (D(–)AP5) undergoes developmental alterations. No significant change in the IC$_{50}$ value for D(–)AP5 was observed between postnatal day 7 (376 ± 56 μM) and day 50 (441 ± 11 μM). The autoradiographic portion of the study confirmed the increase in PCP receptor density between postnatal days 7 and 50. Regional localization of binding did not change significantly between these time points, but the distribution of binding in hippocampus at postnatal day 7 was more homogeneous than at postnatal day 50. These investigators also addressed the developmental profile of forebrain concentrations of glutamate and glycine (expressed as nmol/mg protein). Glutamate content declined prenatally, and then increased progressively. Glycine content was stable between gestational day 18 and postnatal day 3, and then decreased. However, these amino acid changes were not sufficient to account for the observed alterations in [^{3}H]TCP binding parameters.

Tremblay *et al.* (1990) studied the postnatal development of [^{3}H]TCP binding sites in hippocampal specimens derived from Wistar rats in both autoradiographic and membrane preparations. They compared the developmental pattern of the PCP receptor with that of the [^{3}H]glycine site of the NMDA receptor complex. Furthermore, they addressed the question of whether there is a developmental alteration in the sensitivity of [^{3}H]TCP binding to added NMDA. Scatchard analysis of [^{3}H]TCP binding was carried out at nine age points ranging from birth to 80 days postnatally. Values of K_d were stable during development, with B_{max} values increasing monotonically 2.4-fold. When incubations were carried out in the presence of 100 μM NMDA, similar increases in B_{max} values of [^{3}H]TCP binding, without significant alterations in K_d values, were observed at all developmental stages. The same was true in the presence of 5 μM glycine, which further potentiated NMDA-stimulated [^{3}H]TCP binding at all developmental stages, and in the presence of 2 mM Mg^{2+}, which inhibited PCP receptor binding to a similar extent at all developmental stages examined. Autoradiograms indicated that density of hippocampal PCP receptors increased progressively from postnatal day 4 to adulthood, with the increase most prominent in strata oriens and radiatum of CA1 and in the molecular layer of the dentate gyrus. By postnatal day 22, the anatomical distribution appeared identical to that in the adult. By contrast, strychnine-insensitive [³H]glycine sites, with an anatomical distribution very similar to that of [³H]TCP sites, reached adult levels of binding at postnatal day 10. It is of interest that developmental patterns of [³H]TCP and [³H]glycine sites in hippocampus are distinct not only from each other, but also from that of the agonist recognition site of the NMDA receptor (Tremblay *et al.*, 1988), which displays a rapid increase after birth to levels exceeding those in the adult, with a decrease after the third postnatal week to adult levels. It is likely that future molecular studies of the NMDA receptor complex will reveal developmental alterations in subunit expression, as have been detected for GABA receptors (Sato and Neale, 1989) and nicotinic acetylcholine receptors (Schuetze, 1986).

The issue of comparative development of multiple receptor domains of the NMDA complex was directly addressed by McDonald *et al.* (1990). Quantitative autoradiography in hippocampal sections derived from Sprague–Dawley rats was carried out at postnatal days 1, 5, 7, 10, 14, 21, 28 and 90 using [³H]glutamate under conditions directing the radioligand to NMDA recognition sites, [³H]TCP to label the PCP receptor, and [³H]glycine to label the glycine recognition site of the NMDA receptor complex. In all of six hippocampal regions examined, ontogenetic profiles of binding to the PCP and glycine sites were very similar, progressively increasing in density and attaining adult levels by postnatal days 14–21. By contrast,

NMDA recognition sites underwent much more rapid increases in density early in postnatal development, and were significantly higher than adult levels in one or more hippocampal regions studied at time points between postnatal days 10 and 28.

Morin *et al.* (1989) undertook a study of the development of [³H]dizocilpine binding sites in Wistar rats at five developmental points from 3 to 20 days postnatally, and in tissue homogenates derived from adult hippocampus, cortex and brainstem. Single-point determinations indicated progressive increases in total [³H]dizocilpine binding (expressed as fmol/mg protein) in all three regions between postnatal days 3 and 10. Levels statistically indistinguishable from adult values were attained by day 10 in hippocampus and cortex. In brainstem, levels significantly greater than those found in adults were apparent from day 3, reaching a maximum of nearly four times the adult level by day 15, and remaining significantly greater than the adult level at day 20. Scatchard analyses were carried out in extensively washed cortical membranes derived from 7-day-old and adult rats, with the goal of eliminating the influences of endogenous glutamate and glycine upon PCP receptor binding. Under these conditions, saturation data could be accurately fit only by a biphasic binding model encompassing distinct low (K_d = 109–185 nM) and high affinity (K_d = 4–6.9 nM) components. A developmental increase in B_{max} was observed only for the high affinity component of binding. The final component of the study was an examination of the abilities of 10 μM glutamate, 100 μM D-(–)AP5, 10 μM glycine, 50 μM PCP and 100 mM Mg^{2+} to modulate specific binding of 3 nM (³H)dizocilpine in homogenates derived from cortex and hippocampus of 7-day-old and adult rat brain. The most significant finding was that in both brain regions the competitive NMDA antagonist D-(–)AP5 failed to reduce [³H]dizocilpine binding in neonatal brain, although it did so robustly in adult brain.

11.6 SUMMARY AND CONCLUSIONS

It is clear from the above discussion that PCP and σ receptors follow sharply distinct courses of development. All studies of PCP receptor ontogeny have revealed patterns of progressive developmental increases in density in forebrain areas, without significant alterations in affinity. There is some disagreement among investigators concerning the developmental stage at which adult levels of PCP receptor binding are attained. The discrepancies may be due to differences in strains of rat, radioligand, or different binding assay methods. By contrast, all studies of σ receptor ontogeny fail to find any developmental increase in receptor density.

The PCP receptor is an integral component of the NMDA receptor complex, and functional coupling between the PCP and NMDA domains appears to exist throughout development. Nonetheless, the transient increases in density of NMDA recognition sites to significantly above adult levels have no parallel in the development of forebrain PCP receptors.

No study of PCP receptor ontogeny has to date addressed the issue of what proportions of the detected radioligand binding are to activated (open) as opposed to resting (closed) conformations of the NMDA receptor-channel complex. It has been shown that under any conditions other than the presence of saturating concentrations of L-glutamate and glycine, radioligand binding to the PCP receptor of the adult rat brain detects a combination of open and closed channels, manifested as apparent high and low affinity binding in saturation studies (Javitt and Zukin, 1989a) or as fast and slow exponentials of association and dissociation in kinetic paradigms (Javitt and Zukin, 1989b). Only kinetic assays are capable of assessing biochemically the crucial question of whether the degree of activatability of NMDA receptors undergoes developmental alteration. The NMDA receptor, encompassing a PCP receptor domain, probably represents only one

form of the native receptor. It is likely that the divergent developmental patterns of PCP as opposed to NMDA recognition sites will ultimately be demonstrated to result from developmental alterations in subunit expression and assembly.

REFERENCES

Adams, J.T., Keana, J.F.W. and Weber, E (1987) Labeling of the haloperidol-sensitive sigma receptor in NCB-20 hybrid neuroblastoma cells with [^{3}H]-(1,3)-di-ortho-tolyl-guanidine and identification of the binding subunit by photoaffinity labeling with [^{3}H]-*m*-azido-di-ortho-tolyl-guanidine. *Soc. Neurosci. Abstr.*, **13**, 1703.

Bowen, W.D., Hellewell, S.B. and McGarry, K.A. (1989) Evidence for a multi-site model of the rat brain σ receptor. *Eur. J. Pharmacol.*, **163**, 309–18.

Coleman, G. L., Barthold, L.S., Osbaldiston, G.W. et al. (1977) Pathological changes during aging in barrier-reared Fischer–344 male rats. *J. Gerontol.*, **32**, 258–78.

Contreras, P.C., DiMaggio, D.A. and O'Donohue, T.L. (1987) An endogenous ligand for the sigma opioid binding site. *Synapse*, **1**, 57–61.

Craviso, G.L. and Musacchio, J.M. (1983a) High-affinity dextromethorphan binding sites in guinea pig brain. I. Initial characterization. *Mol. Pharmacol.*, **23**, 619–28.

Craviso, G.L. and Musacchio, J.M (1983b) High-affinity dextromethorphan binding sites in guinea pig brain. II. Competition experiments. *Mol. Pharmacol.*, **23**, 629–40.

de Costa, B.R., Bowen, W.D., Hellewell, S.B. et al. (1989) Synthesis and evaluation of optically pure [^{3}H]-(+)-pentazocine, a highly potent and selective radioligand for σ receptors. *FEBS Lett.*, **251**, 53–8.

DiMaggio, D.A., Contreras, P.C and O'Donohue, T.L. (1988) Biological and chemical characterization of the endopsychosins: distinct ligands for PCP and sigma sites, in *Sigma and Phencyclidine-like Compounds as Molecular Probes in Biology* (eds E.F. Domino and J.-M. Kamenka), NPP Books, Ann Arbor, pp. 157–71.

Engberg, G. and Wikström, H. (1991) σ-receptors: implication for the control of neuronal activity of nigral dopamine-containing neurons. *Eur. J. Pharmacol.*, **201**, 199–202.

Foster, A.C. and Wong, E.H.F. (1987) The novel anticonvulsant MK801 binds to the activated state of the N-methyl-D-aspartate receptor in rat brain. *Br.J. Pharmacol.*, **91**, 403–9.

Gundlach, A.L., Largent, B.L. and Snyder, S.H. (1985) Phencyclidine and σ opiate receptors in brain: biochemical and autoradiographical differentiation. *Eur.J. Pharmacol.*, **113**, 465–6.

Hampton, R.Y., Medzihradsky, F., Woods, J.H. *et al.* (1982) Stereospecific binding of [³H]-phencyclidine in brain membranes. *Life Sci.*,**30**, 2147–54.

Haring, R.H., Kloog, Y. and Sokolovsky, M. (1986) Identification of polypeptides of the phencyclidine receptor of rat hippocampus by photoaffinity labeling with [³H]azidophencyclidine. *Biochemistry*, **25**, 612–20.

Haring, R.H., Kloog, Y., Kalir, A. *et al.* (1987) Binding studies and photoaffinity labeling identify two classes of phencyclidine (PCP) receptors in rat brain. *Biochemistry*, **26**, 5854–61.

Hellewell, S.B. and Bowen, W.D. (1990) A sigmalike binding site in rat pheochromocytoma (PC12) cells: decreased affinity for (+)-benzomorphans and lower molecular weight suggest a different sigma receptor form from that of guinea pig brain. *Brain Res.*, **527**, 244–53.

Honoré, T., Drejer, J., Nielsen, E.Ø. *et al.* (1989) Molecular target size analyses of the NMDA-receptor complex in rat cortex. *Eur. J. Pharmacol.*, **172**, 239–47.

Jasinski, D.R., Martin, W.R. and Sapira, J.D. (1968) Antagonsim of the subjective, behavioral, pupillary, and respiratory depressant effects of cyclazocine by naloxone. *Clin. Pharmacol. Ther.*, **9**, 215–22.

Javitt, D.C. and Zukin, S.R. (1989a) Biexponential kinetics of [³H]MK-801 binding: evidence for access to closed and open N-methyl-D-aspartate receptor channels. *Mol. Pharmacol.*, **35**, 387–93.

Javitt, D.C. and Zukin, S.R. (1989b) Interaction of [³H]MK-801 with multiple states of the N-methyl-D-aspartate receptor complex of rat brain. *Proc. Natl. Acad. Sci. USA*, **86**, 740–4.

Javitt, D.C. and Zukin, S.R. (1991) Recent advances in the phencyclidine model of schizophrenia. *Am. J. Psychiatry*, **148**, 1301–8.

Javitt, D.C., Frusciante, M.J. and Zukin, S.R. (1990) Rat brain N-methyl-D-aspartate receptors require multiple molecules of agonist for activation. *Mol. Pharmacol.*, **37**, 603–7.

Kavanaugh, M.P., Tester, B.C., Scherz, M.W. *et al.* (1988) Identification of the binding subunit of the σ-type opiate receptor by photoaffinity labeling with 1-(4-azido-2-methyl[6-³H]phenyl)-3-(2-methyl[4,6-³H]phenyl) guanidine. *Proc. Natl. Acad. Sci. USA.*, **85**, 2844–8.

Keats, A. and Telford, J. (1964) Narcotic antagonists as analgesics. Clinical aspects, in *Molecular Modification in Drug Design, Advances in Chemistry*, 45th edn (ed R.F. Gould), American Chemical Society, Washington, DC., pp 170–6.

Kuhar, M.J., Boja, J.W. and Cone, E.J. (1990) Phencyclidine binding to striatal cocaine receptors. *Neuropharmacology*, **29**, 295–7.

Kumar, K.N., Tilakaratne, N., Johnson, P.S. *et al.* (1991) Cloning of cDNA for the glutamate-binding subunit of an NMDA receptor complex. *Nature*, **354**, 70.

Kushner, L., Zukin, S.R. and Zukin, R.S. (1988) Characterization of opioid, σ, and phencyclidine receptors in the neuroblastoma-brain hybrid cell line NCB-20. *Mol. Pharmacol.*, **34**, 689–94.

Largent, B.L., Gundlach, A.L. and Snyder, S.H. (1984) Psychotomimetic opiate receptors labelled and visualized with (+)-[³H]-3-(3hydroxyphenyl)-N-(1-propyl)piperidine. *Proc. Natl. Acad. Sci. USA.*, **81**, 4983–7.

Largent, B.L., Gundlach, A.L. and Snyder, S.H. (1986a) σ receptors on NCB-20 hybrid neurotumor cells labeled with (+)[³H]SKF 10,047 and (+)[³H]3-PPP. *Eur. J. Pharmacol.*, **124**, 183–7.

Largent, B.L., Gundlach, A.L. and Snyder, S.H. (1986b) Pharmacological and autoradiographic discrimination of sigma and phencyclidine receptor binding sites in brain with (+)-[³H]SKF 10047, (+)-[³H]-3-[3-hydroxyphenyl]-N-(1-propyl) piperidine and [³H]-1-[1-(thienyl)-cyclohexyl] piperidine. *J. Pharmacol. Exp. Ther.*, **238**, 739–48.

Largent, B.L., Wikström, H., Gundlach, A.L. *et al.* (1987) Structural determinants of σ receptor affinity. *Mol. Pharmacol.*, **32**, 772–84.

Largent, B.L., Wikström, H., Snowman, A.M. *et al.* (1988) Novel antipsychotic drugs share high affinity for σ receptors. *Eur. J. Pharmacol.*, **155**, 345–7.

London, E.D., Nespor, S.M., Ohata, M. *et al.* (1981) Local cerebral glucose utilization during development and aging of the Fischer-344 rat. *J. Neurochem.*, **37**, 217–21.

Majewska, M.D., Parameswaran, S. and London, E.D. (1989) Divergent ontogeny of sigma and phencyclidine binding sites in the rat brain. *Dev. Brain Res.*, **47**, 13–18.

Maragos, W.F., Chu, D.C.M., Greenamyre, J.T. *et al.* (1986) High correlation between the localization of [³H]TCP binding and NMDA receptors. *Eur.J. Pharmacol.*, **123**, 173–4.

Martin, W.R., Eades, C.G., Thompson, J.A. *et al.* (1976) The effects of morphine- and nalorphine-like drugs in the non-dependent and morphine-dependent chronic spinal dog. *J. Pharmacol. Exp. Ther.*, **197**, 517–32.

Martin, W.R. (1983) Pharmacology of opioids. *Pharmacol. Rev.*, **35**, 283–323.

Matsumoto, R.R., Bowen, W.D. and Walker, J.M. (1989) Age-related differences in the sensitivity of rats to a selective sigma ligand. *Brain Res.*, **504**, 145–8.

McCann, D.J. and Su, T.-P. (1990) Haloperidol-sensitive (+)-[³H]SKF-10,047 binding sites (σ sites) exhibit a unique distribution in rat brain subcellular fractions. *Eur. J. Pharmacol.*, **188**, 211–18.

McDonald, J.W., Johnston, M.V. and Young, A.B. (1990) Differential ontogenic development of three receptors comprising NMDA receptor/channel complex in the rat hippocampus. *Exp. Neurol.*, **110**, 237–47.

Morin, A.M., Hattori, H., Wasterlain, C.G. *et al.* (1989) [³H]MK-801 binding sites in neonate rat brain. *Brain Res.*, **487**, 376–9.

Moriyoshi, K., Masu, M., Ishii, T. *et al.* (1991) Molecular cloning and characterization of the rat NMDA receptor. *Nature*, **354**, 31–7.

Musacchio, J.M. (1990) The psychotomimetic effects of opiates and the σ receptor. *Neuropsychopharmacology*, **3**, 191–200.

Musacchio, J.M., Klein, M. and Canoll, P.D. (1989a) Dextromethorphan and sigma ligands: common sites but diverse effects. *Life Sci.*, **45**, 1721–32.

Musacchio, J.M., Klein, M. and Paturzo, J.J. (1989b) Effects of dextromethorphan site ligands and allosteric modifiers on the binding of (+)-[³H]3-(-3-hydroxyphenyl)-*N*-(1-propyl)piperidine. *Mol. Pharmacol.*, **35**, 1–5.

Paleos, G.A., Yang, Z.W. and Byrd, J.C. (1990) Ontogeny of PCP and sigma receptors in rat brain. *Dev. Brain Res.*, **51**, 147–52.

Pfeiffer, A., Brantl, V., Herz, A. *et al.* (1986) Psychotomimesis mediated by κ opiate receptors. *Science*, **233**, 774–6.

Quirion, R., Hammer, R.P., Herkenham, M. *et al.* (1981) Phencyclidine (angel dust)/σ 'opiate' receptor: visualization by tritium-sensitive film. *Proc. Natl. Acad. Sci. USA*, **78**, 5881–5.

Quirion, R., DiMaggio, D.A., French, E.D. *et al.* (1984) Evidence for an endogenous peptide ligand of the phencyclidine receptor. *Peptides*, **5**, 967–73.

Rosenberg, A. and Stern, N. (1966) Changes in sphingosine and fatty acid components of the gangliosides in developing rat and human brain. *J. Lipid Res.*, **7**, 122–31.

Samovilova, N.N., Nagornaya, L.V. and Vinogradov, V.A. (1988) (+)-[³H]SK&F 10,047 binding sites in rat liver. *Eur. J. Pharmacol.*, **147**, 259–64.

Sato, T.N. and Neale, J.H. (1989) Type I and type II γ-aminobutyric acid/benzodiazepine receptors: purification and analysis of novel receptor complex from neonatal cortex. *J. Neurochem.*, **52**, 1114–22.

Scheideler, M.A. and Zukin, R.S. (1990) Reconstitution of solubilized delta-opiate receptor binding sites in lipid vesicles. *J. Biol. Chem.*, **265**, 15176–82.

Schuetze, S. (1986) Embryonic and adult acetylcholine receptors: molecular basis of developmental changes in ion channel properties. *Trends Neurosci.*, **9**, 346–8.

Shinohara, K., Nishikawa, T., Ishii, T. *et al.* (1989) Embryonic and postnatal development of *N*-(1-[2-thienyl]cyclohexyl)[³H]piperidine binding sites in rat forebrain homogenates and slices. *Neurosci. Lett.*, **107**, 307–12.

Sircar, R. and Zukin, S.R. (1983) Ontogeny of sigma opiate/phencyclidine-binding sites in rat brain. *Life Sci.*, **33**, 255–8.

Sircar, R. and Zukin, S.R. (1985) Quantitative localization of [³H]TCP binding in rat brain by light microscopy autoradiography. *Brain Res.*, **344**, 142–5.

Sonders, M.S., Barmettler, P., Lee, J.A. *et al.* (1990) A novel photoaffinity ligand for the phencyclidine site of the *N*-methyl-D-aspartate receptor labels a M_r 120,000 polypeptide. *J. Biol. Chem.*, **265**, 6776–81.

Sorensen, R.G. and Blaustein, M.P. (1986) m-Azido-phencyclidine covalently labels the rat brain PCP receptor, a putative K channel. *J. Neurosci.*, **6**, 3676–81.

Su, T.-P. (1981) Psychotomimetic opioid binding: specific binding of [³H]SKF-10047 to etorphine-inaccessible sites in guinea-pig brain. *Eur. J. Pharmacol.*, **75**, 81–2.

Su, T.-P. (1982) Evidence for sigma opioid receptor: binding of [³H]SKF-10047 to etorphine-inacessible sites in guinea-pig brain. *J. Pharmacol. Exp. Ther.*, **223**, 284–90.

Su, T.-P. (1985) Further demonstration of kappa opioid binding sites in the brain: evidence for heterogeneity. *J. Pharmacol. Exp. Ther.*, **232**, 144–8.

Su, T.-P. (1991) σ receptors: putative links between nervous, endocrine and immune systems. *Eur. J. Biochem.*, **200**, 633–42.

Su, T.-P., Weissman, A.D. and Yeh, S.-Y (1986) Endogenous ligands for sigma opioid receptors in the brain ('sigmaphin'): evidence from binding assays. *Life Sci.*, **38**, 2199–210.

Su, T.-P., London E.D. and Jaffe, J.H. (1988a) Steroid binding at σ receptors suggests a link between endocrine, nervous and immune systems. *Science*, **240**, 219–21.

Su, T.-P., Schell, S.E., Ford-Rice, F.Y. *et al.* (1988b) Correlation of inhibitory potencies of putative antagonists for σ receptors in brain and spleen. *Eur. J. Pharmacol.*, **148**, 467–70.

Tam, S.W. (1983) Naloxone-inacessible σ receptor in rat central nervous system. *Proc. Natl. Acad. Sci. USA*, **80**, 6703–7.

Tam, S.W. (1985) (+)-[³H]SKF 10,047, (+)-[³H]ethylketocyclazocine, μ, κ, δ and phencyclidine binding sites in guinea pig brain membranes. *Eur. J. Pharmacol.*, **109**, 33–41.

Tam, S.W. and Cook, L. (1984) σ opiates and certain antipsychotic drugs mutually inhibit (+)-[³H]SKF 10047 and [³H]haloperidol binding in guinea pig brain membranes. *Proc. Natl. Acad. Sci. USA*, **81**, 5618–21.

Tremblay, E., Roisin, M.P., Represa, A. *et al.* (1988) Transient increased density of NMDA binding sites in the developmental rat hippocampus. *Brain Res.*, **461**, 393–6.

Tremblay, E., Roisin-Lallemand, M.-P. and Ben-Ari, Y. (1990) Developmental study of [³H]TCP and [³H]glycine binding sites in the rat hippocampus. *Dev. Brain Res.*, **57**, 21–8.

Vaupel, D.B. (1983) Naloxone fails to antagonize the σ effects of PCP and SKF 10,047 in the dog. *Eur. J. Pharmacol.*, **92**, 269–74.

Vignon, J., Vincent, J.P., Bidard, J.N. *et al.* (1982) Biochemical properties of the brain phencyclidine receptor. *Eur. J. Pharmacol.*, **81**, 531–43.

Vignon, J., Chicheportiche, R., Chicheportiche, M. *et al.* (1983) [³H]TCP: a new tool with high affinity for the PCP receptor in rat brain. *Brain Res.*, **280**, 194–7.

Vignon, J., Cerruti, C., Chaudieu, I. *et al.* (1988) Interaction of molecules in the phencyclidine series with the dopamine uptake system: correlation with their binding properties to the phencyclidine receptor. Binding properties of [³H]BTCP, a new PCP analog, to the dopamine uptake complex, in *Sigma and Phencyclidine-like Compounds as Molecular Probes in Biology* (eds E.F. Domino and J.-M. Kamenka), NPP Books, Ann Arbor, pp. 199–208.

Vincent, J.-P., Kartalovski, B., Geneste, P. *et al.* (1979) Interaction of phencyclidine ('angel dust') with a specific receptor in brain membranes. *Proc. Natl. Acad. Sci. USA*, **76**, 4678–82.

Vincent, J.-P., Bidard, J. -N., Lazdunski, M. *et al.* (1983) Identification and properties of phencyclidine-binding sites in nervous tissues. *Fed. Proc.*, **42**, 2570–3.

Vu, T.H., Weissman, A.D. and London, E.D. (1990) Pharmacological characteristics and distributions of sigma and phencyclidine binding sites in the animal kingdom. *J. Neurochem.*, **54**, 598–604.

Walker, J.M., Matsumoto, R.R., Bowen, W.D. *et al.* (1988) Evidence for a role of haloperidol-sensitive σ-'opiate' receptors in the motor effects of antipsychotic drugs. *Neurology*, **38**, 961–5.

Walker, J.M., Bowen, W.D., Walker, F.O. *et al.* (1990) Sigma receptors: biology and function. *Pharmacol. Rev.*, **42**, 355–402.

Waller, S.B., Ingram, D.K., Reynolds, M.A. *et al.* (1983) Age and strain comparisons of neurotransmitter synthetic enzyme activities in the mouse. *J. Neurochem.*, **41**, 1421–8.

Weber, E., Sonders, M., Quarum, M. *et al.* (1986) 1,3-Di(2-[5-³H]tolyl) guanidine: a selective ligand that labels σ-type receptors for psychotomimic opiates and antipsychotic drugs. *Proc. Natl. Acad. Sci. USA*, **83**, 8784–8.

Weissman, A.D., Broussolle, E.P. and London , E.D. (1990a) In vivo binding of [³H]*d*-N-allylnormetazocine and [³H]haloperidol to sigma receptors in the mouse brain. *J. Chem. Neuroanat.*, **3**, 347–54.

Weissman, A.D., Casanova, M.F., Kleinman, J.E. *et al.* (1990b) Selective loss of cerebral cortical sigma, but not PCP binding sites in schizophrenia. *Biol. Psychiatry*, **29**, 41–54.

Wolfe, S.A., Kulsakdinun, C., Battaglia, G. *et al.* (1988) Initial identification and characterization of sigma receptors on human peripheral blood leukocytes. *J. Pharmacol. Exp. Ther.*, **247**, 1114–19.

Wolfe, S.A., Culp, S.G. and De Souza, E.B. (1989) σ-Receptors in endocrine organs: identification, characterization, and autoradiographic localization in rat pituitary, adrenal, testis, and ovary. *Endocrinology*, **124**, 1160–72.

Wong, E.H.F., Kemp, J.A., Priestly, T. *et al.* (1986) The anticonvulsant MK-801 is a potent N-methyl-D-aspartate antagonist. *Proc. Natl. Acad. Sci. USA*, **83**, 7104–8.

Wong, E.H.F., Knight, A.R. and Woodruff, G.N. (1988) [³H]MK-801 labels a site on the N-methyl-D-aspartate receptor complex in rat brain membranes. *J. Neurochem.*, **50**, 274–81.

Wu, X.-Z., Bell, J.A., Spivak, C.E. *et al.* (1990) Electrophysiological and binding studies on intact NCB-20 cells suggest presence of a low affinity sigma receptor. *J. Pharmacol. Exp. Ther.*, **257**, 351–9.

Zhang, A.-Z., Mitchell, K.N., Cook, L. and Tam, S.W. (1988) Human endogenous brain ligands for sigma and phencyclidine receptors, in *Sigma and Phencyclidine-like Compounds as Molecular Probes in Biology* (eds E.F. Domino and J.-M. Kamenka), NPP Books, Ann Arbor, pp. 335–43.

Zukin, R.S. and Zukin, S.R. (1981) Demonstration of [³H]-cyclazocine binding to multiple opiate receptor sites. *Mol. Pharmacol.*, **20**, 246–54.

Zukin, R.S. and Zukin, S.R. (1988) The σ receptor, in *The Opiate Receptors* (ed. G.W. Pasternak), The Humana Press, Clifton, NJ, pp. 143–63.

Zukin, S.R. and Zukin, R.S. (1979) Specific [³H]phencyclidine binding in rat central nervous system. *Proc. Natl. Acad. Sci. USA*, **76**, 5372–6.

Zukin, S.R., Fitz-Syage, M.L., Nichtenhauser, R. *et al.* (1983) Specific binding of [³H]phencyclidine in rat central nervous tissue: further characterization and technical considerations. *Brain Res.*, **258**, 277–84.

Zukin, S.R., Zukin R.S., Vale, W. *et al.* (1987) An endogenous ligand of the brain σ/PCP receptor antagonizes NMDA-induced neurotransmitter release. *Brain Res.*, **416**, 84–9.

OPIOID RECEPTORS AND PEPTIDE SYSTEM REGULATION IN THE DEVELOPING NERVOUS SYSTEM

Ann Tempel

12.1 INTRODUCTION

The biochemical demonstration of opioid receptors was first established in 1973 (Pert and Snyder, 1973). Earlier attempts were difficult due to low levels of specific binding and high background binding. Although these early studies identified only a single type of receptor, pharmacological studies had already suggested the existence of multiple opiate receptor types (Martin *et al.*, 1967). Evidence from behavioral, pharmacological and biochemical studies demonstrates the existence of several opiate receptor classes: mu (μ), delta (δ) and kappa (κ) (Lord *et al.*, 1977; Chang and Cuatrecasas, 1979). The sigma (σ) receptor is no longer considered an opioid receptor and it has been suggested that it is the PCP site of the NMDA receptor (for reviews see Snyder, 1984; Zukin and Zukin, 1988). These receptor classes mediate diverse behavioral effects, exhibit different ligand selectivity patterns and have different distributions throughout the central and peripheral nervous systems. The μ receptor is operationally defined as the high affinity site at which morphine-like opiates produce analgesia and a variety of other classical opiate effects. This receptor has been further subdivided into μ_1 and μ_2 (Pasternak *et al.*, 1983). The δ receptor exhibits a higher affinity for the naturally occurring enkephalins (a class of shorter opioid peptides) than for morphine and was originally found in peripheral tissue such as the mouse vas deferens (Lord *et al.*, 1977). The κ receptor is that site at which ketocyclazocine-like opiates produce analgesia, as well as their unique ataxic and sedative effects (Martin *et al.*, 1976). It is also defined as a receptor highly selective for dynorphin (a 17-amino acid opioid peptide). Some researchers have demonstrated the existence of several κ receptor subtypes as well (Attali *et al.*, 1982; Gouarderes *et al.*, 1983; Zukin *et al.*, 1988). Actions at all three of these sites are reversible by the opioid antagonist naloxone or naltrexone with increasing doses required going from μ to δ to κ receptors. The complex neuropharmacological actions of a given opioid would appear to reflect its interaction at a combination of these and other opioid receptor sites with varying potencies.

One of the aims in anatomically differentiating the opioid receptor types is to gain insight into the functional significance of these receptors. Some of the possible functions are briefly outlined by Mansour *et al.* (1987) and Herkenham and McLean (1986). High densities of μ and κ opioid receptors have been localized to areas involved in processing of all types of sensory information, i.e. visual (superior colliculus, lateral geniculate, optic tract), auditory (inferior colliculus and medial genic-

Receptors in the Developing Nervous System Vol. 2: Neurotransmitters. Edited by Ian S. Zagon and Patricia J. McLaughlin. Published in 1993 by Chapman & Hall. ISBN 0 412 49400 0. Vols. 1 and 2 (set) ISBN 0 412 54520 9.

ulate), olfactory (olfactory bulb and amygdala) and nociceptive (thalamus, central gray, spinal cord). In contrast, δ receptors appear to be localized primarily to more recently developed areas of the forebrain, which may indicate a need for more specialized receptors or neuromodulators. High levels of κ binding in the hypothalamus and neural lobe of the pituitary suggest that they play a role in neuroendocrine regulation. κ opioid receptors have also been implicated in the regulation of feeding and drinking behaviors suggesting a role in motivational behaviors.

It is now well established that there are three major classes of opioid peptides: β-endorphin, the enkephalins, and dynorphin-related peptides, which serve as the endogenous ligands for the μ, δ and κ receptors. These arise from three different precursor molecules in three independent biosynthetic pathways (for a review see Weber *et al.*, 1983). The first, β-endorphin occurs together with adrenocorticotrophic hormone (ACTH), a peptide which stimulates the adrenal cortex, in the 31 kDa precursor molecule proopiomelanocortin (POMC) (Mains *et al.*, 1977). POMC, β-endorphin, and ACTH are found in highest concentrations in the pars intermedia and pars distalis of the pituitary. Under conditions of severe stress, β-endorphin and ACTH are co-released.

The second class includes methionine- and leucine-enkephalin (Hughes *et al.*, 1975), pentapeptides that arise from pro-enkephalin, an approximately 50 kDa protein that has been identified in the adrenal medulla and in the striatum (Gubler *et al.*, 1982). The pro-enkephalin molecule contains six copies of met-enkephalin and one of leu-enkephalin. The distribution of the enkephalins differs from that of β-endorphin in that they appear to be much more widely distributed throughout the brain.

Dynorphin A, a 17-amino acid peptide that contains the sequence of leu-enkephalin at its N-terminal end, is the most recently discovered opioid peptide (Goldstein *et al.*, 1981). It arises together with dynorphin B, its 1–13 counterpart, from a 4000 dalton peptide which in turn is found in the much larger prodynorphin molecule (James *et al.*, 1982). Dynorphin A and its (1–8) fragment occur in approximately equal concentrations in the brain. Whereas β-endorphin displays equipotency at μ and δ receptors, the enkephalins show a much greater affinity for the δ receptor. It has been speculated that the more stable β-endorphin molecule functions as a neurohormone in both pathways, whereas the enkephalins play a more specific role, acting as neurotransmitters or neuromodulators over shorter distances (Lord *et al.*, 1977). The longer dynorphin forms (1–13 and 1–17) appear to be more κ-selective and are more stable, whereas the shorter fragments (e.g., 1–8) which are potent at both κ and δ receptors and are less stable, are more likely candidates for transmitter-like function (Corbett *et al.*, 1983).

12.2 HISTORY OF SUBSTANCE ABUSE

Descriptions of human neonates undergoing withdrawal following maternal opioid abuse during pregnancy date back to the latter part of the nineteenth century (Goodfriend *et al.*, 1956). Methadone was first synthesized by the Germans during World War II. Favorable preliminary findings by Dole and Nyswander (1965) were followed by the wide-scale use in heroin addiction drug treatment programs (for review, see Hutchings, 1985a). By 1975, there were some 70 000–80 000 heroin addicts in methadone maintenance programs throughout the USA and a significant proportion of these women were of childbearing age. At the time, it was estimated that in the New York City metropolitan area alone, 10 000–12 000 such women were enrolled in methadone programs yet little was known of possible risk to the fetus and the newborn. With the advent of methadone maintenance as an experimental treatment for heroin addiction in the early 1970s, attention turned to the questions of reproductive hazards and developmental toxicity. There are confounding variables in clinical experiments such as women coming from low socioeconomic levels, long histories of drug abuse, poor diets, heavy smokers, use of marijuana, cocaine, barbiturates, tranquilizers or alcohol (for review see Householder *et al.*, 1982; Hutchings and Fifer, 1986).

The type of opioid abused has changed over the years. Until the 1950s, morphine appeared to be the drug of choice but a change to heroin usage was reported by Goodfriend *et al.* (1956). Following Dole and Nyswander's (1965) advocation of the methadone maintenance treatment program as an alternative to heroin dependency, numerous reports have documented methadone-dependent offspring.

In the early 1980s, cocaine became the drug of choice. Cocaine abuse rose sharply during the 1980s to the point that the drug subculture represented a large portion of the population. An estimated 50% or more of pregnancies at some of the inner-city hospitals in New York are cocaine exposed (Dow-Edwards, 1991). Interest in the biological effects of maternal cocaine and/or opioid abuse on the development of the unborn child has been increasing. As a result, several laboratories attempted to develop animal models of prenatal opioid exposure. Laboratory environments would provide controlled experiments alleviating the confounding variables in the clinical studies. However, there are a host of methodological and interpretive problems in animal studies as well (for a comprehensive bibliography see Zagon *et al.*, 1982). Although most studies have used the rat, procedures have differed with respect to strain, dose level, dosing regimen, route of administration, gestational age at treatment and fostering techniques. These factors make it difficult to compare results from one laboratory to another. Because of the importance of these variables in interpretation of data and conclusions drawn, they are briefly discussed below.

12.2.1 SURROGATE FOSTERING

The evidence is sufficiently convincing that prenatal manipulations of the pregnant dam can alter maternal behavior and the mother–infant dyad to produce effects in her offspring, even when treatment is terminated before the last week or so of gestation. Daily drug treatment can disrupt circadian rhythms, cause acute nutritional deficits and either inhibit or interfere with hormones that mediate maternal behavior. Recent studies demonstrate the mother's integral role as a regulator of a host of developmental parameters in her offspring that include activity level, sleep–wake states, milk ingestion, heart rate, oxygen consumption, and growth hormone (see Hofer, 1984 for review). If not included, interpretation of the data will be compromised in that one cannot rule out with any degree of confidence that offspring effects have not been maternally mediated during the postnatal period.

12.2.2 DOSE–RESPONSE RELATIONSHIP

Prenatal drug studies involve two mutually interacting biological systems – the mother and fetoplacental unit. Dose–response relationships are very complex and involve interactive, pharmacological, and toxic effects in the mother and offspring (Hutchings, 1985a). The general principle is that as dose is increased from subpharmacological levels through the pharmacological range of the compound, there is a corresponding increase in toxicity culminating in death. For example, thalidomide at subtoxic, pharmacological levels in the mother, is highly embryotoxic during morphogenesis. Another compound with the same sort of profile and embryotoxic response in humans is the vitamin A derivative, isotretinoin (Hutchings, 1985b). In animal studies, what is frequently seen is that within the pharmacological range of the compound, and at levels that are not toxic to the dam, there is no embryotoxic response. Embryotoxicity is seen only at levels that produce maternal toxicity (this is reviewed and detailed in Hutchings, 1985a). One example in the rat is phencyclidine (PCP), which has been shown to be teratogenic, but only at doses that are highly toxic to the dam. In the case of PCP, Hutchings (1985a) administered two doses that were pharmacologically potent as measured by behavioral effects in the dam (5 or 10 mg/kg) but of relatively low maternal toxicity based on maternal weight gain. Yet they observed normal birthweights and no postnatal behavioral effects on two independent behavioral measures (Hutchings *et al.*, 1984).

Dose–response relationships, though complex, are critical for a meaningful description and understanding of effects. It is helpful that every study includes some measure of maternal dose–response. This is of particular importance when embryotoxicity is found only at doses that produce maternal toxicity. Under these circumstances, it is important to determine whether the effects produced in the offspring are primary effects of the compound or secondary to maternal toxicity.

A variety of drug-induced deficits have been observed in developing rats neonatally addicted to opioids. These deficits have been shown to persist into adulthood (Davis and Lin, 1972; Zagon and McLaughlin, 1977; Freeman, 1980). In general, these deficits have included impaired growth, behavioral, and neuroendocrine responses. This spectrum of drug-related developmental deficits may be a general pattern found in the young animal exposed to neurotropic drugs early in life. Neurotropic drugs such as morphine alter both the levels and the biosynthesis of the catecholamines (Slotkin and Anderson, 1975; McGinty and Ford, 1980; Rech *et al.*, 1980; Slotkin *et al.*, 1982) and such effects may be the neurochemical basis of the alterations and deficits in development (Weiner, 1974).

Little is known about mechanisms regulating the initial expression of opioid receptors. However, receptor development may be influenced quantitatively by a variety of factors in the neuronal environment. It is well established that perinatal exposure to drugs may influence the ontogenesis of brain opioid systems. However, results from a variety of studies are contradictory and complicated apparently due to differences in exposure time, dose and route of administration and the age at which animals are tested. This chapter will attempt to review the studies on prenatal opioid (morphine and cocaine) treatment and the biochemical and cellular alterations in normal development of the central nervous system (CNS) as a result of addiction. Studies in the adult CNS are included for comparison.

12.3 EFFECTS OF CHRONIC ADMINISTRATION IN THE ADULT CNS

12.3.1 OPIOID AGONISTS

A major unresolved issue is whether opioid agonist-induced tolerance and dependence *in vivo* and desensitization *in vitro* are associated with altered receptor numbers. It has been reported that chronic administration of morphine produces either no significant change in the adult CNS in opioid receptor number (Pert *et al.*, 1973; Klee and Streaty, 1976; Hitzemann *et al.*, 1974; Simon and Hiller, 1978; Bardo *et al.*, 1982; Holaday *et al.*, 1982; Perry *et al.*, 1982) or a modest receptor up-regulation (Brady *et al.*, 1989; Rothman *et al.*, 1986). More recent studies report that chronic morphine treatment of rats produces an up-regulation of the low affinity [^{3}H]D-Ala^2, D-Leu^5-enkephalin (DADLE) site in brain (Rothman *et al.*, 1986; Danks *et al.*, 1988; Brady *et al.*, 1989). In contrast, chronic etorphine produces down-regulation of μ and δ receptors *in vivo* (Tao *et al.*, 1988) and chronic enkephalin produces down-regulation of δ receptors *in vivo* (Tao *et al.*, 1987; Steece *et al.*, 1986) and in neurotumor cells (Hazum *et al.*, 1981; Chang *et al.*, 1982; Blanchard *et al.*, 1983). Intrathecal administration of high concentrations of the μ-selective peptide [*N*-Me-Phe^3, D-

Pro⁴] morphiceptin leads to a reduction in the number of spinal cord, but not brain, μ receptors (Nishino *et al.*, 1990). Only in the neonatal rat, however, is there a correlation between morphine-induced receptor down-regulation and the development of tolerance (Tempel *et al.*, 1988; Tempel, 1991). Although numerous studies have examined the effects of chronic μ and δ opioids on brain receptors, few have addressed the effects of chronic κ agonist administration. Rats treated chronically with κ opioids develop tolerance, as determined in behavioral paradigms measuring analgesia (Bhargava *et al.*, 1989a). Chronic administration of bremazocine leads to a reduction in [³H]bremazocine binding under conditions in which μ and δ binding has been suppressed (Morris and Herz, 1989). Moreover, chronic treatment of rats with U-50,488H (κ opioid) leads to a decrease in [³H]ethylketocyclazocine binding to brain membranes (Bhargava *et al.*, 1989b). Because ethylketocyclazocine is a nonselective μ opioid and because [³H]bremazocine labels non-opioid sites in addition to μ, δ and κ receptors (Zukin *et al.*, 1988), it is not clear whether the observed decreases are in κ opioid receptors.

An important question is why chronic treatment with etorphine and opioid peptides, but not with morphine, produce opioid receptor down-regulation in the adult animal. Several explanations are plausible. For example, it may be that morphine is a partial agonist. Alternatively, interaction of etorphine and enkephalin-like peptides with δ receptors may be critical to long-term changes at the cellular level. Yet another possibility is that opioid peptides bind differently than do opioid narcotic agonists to the same receptor, thereby producing different effects.

12.3.2 CHRONIC OPIOID ANTAGONIST ADMINISTRATION AND THE ADULT CNS

Antagonist-induced opioid supersensitivity and opioid receptor up-regulation are well correlated in the adult CNS. Long-term *in vivo* administration of the opioid antagonist naloxone or naltrexone results in enhanced morphine-induced analgesia (Lahti and Collins, 1978; Tang and Collins, 1978; Tempel *et al.*, 1985; Yoburn *et al.*, 1985) and enhanced effects of morphine on neurons of the locus coeruleus (Bardo *et al.*, 1983b) and the myenteric plexus (Schulz *et al.*, 1979). This functional supersensitivity reflects both an increased number of μ and δ opioid receptors (Zukin *et al.*, 1982; Brunello *et al.*, 1984; Tempel *et al.*, 1985; Danks *et al.*, 1988; Millan *et al.*, 1988; Yoburn *et al.*, 1989) and an increased coupling of receptor to the inhibitory guanyl nucleotide binding protein G_i (Zukin *et al.*, 1982; Tempel *et al.*, 1985). The changes in opioid receptor density vary heterogeneously throughout the brain; highest increases in μ receptor density occur in the nucleus accumbens, amygdala, striatum, layers I and III of the neocortex, and the periaqueductal gray region (Tempel *et al.*, 1984).

Up-regulation of μ opioid receptors has also been observed *in vitro* using explant cultures of fetal mouse spinal cord with attached dorsal root ganglia (DRG) (Tempel *et al.*, 1986). Long-term exposure of the explant cultures to naloxone produces an increase in μ opioid receptor density relative to control cultures, even in the presence of the protein synthesis inhibitor cycloheximide at a concentration that blocks >95% protein synthesis. This finding suggests that antagonist-induced opioid receptor up-regulation does not require the synthesis of new receptor molecules.

12.3.3 REGULATION OF OPIOID PEPTIDE GENE EXPRESSION IN THE ADULT CNS

A number of studies in the adult CNS have shown that chronic opioid drugs regulate opioid peptide levels and that this regulation occurs at the level of gene expression. Chronic naltrexone treatment increases met-enkephalin-like immunoreactivity in the striatum and nucleus accumbens (Tempel *et al.*, 1984). Chronic naltrexone also increases substance P immunoreactivity in the striatum (Tempel *et al.*, 1990) and decreases β-endorphin immunoreactivity in the hypothalamus, thalamus and amygdala (Ragavan *et al.*, 1983). In an effort to determine the mechanism by which opioid antagonists stimulate enkephalin and substance P production, the effects of naltrexone on the mRNA levels of preproenkephalin (PPE) and preprotachykinin (PPT) were studied (Tempel *et al.*, 1990). This study indicated that long-term blockade of opioid receptors by naltrexone leads to large increases in both PPE and PPT mRNA in the striatum. Chronic morphine treatment, on the other hand, decreases striatal PPE mRNA levels (Uhl *et al.*, 1988). These findings suggest that activation or blockade of opioid receptors influence PPE and PPT gene expression.

12.3.4 THE ROLE OF G PROTEINS AND SECOND MESSENGER SYSTEMS IN THE DEVELOPMENT OF TOLERANCE

It has been suggested that opioid-receptor density is controled by either the inhibition or excitation of the cAMP system via G-binding proteins (Costa *et al.*, 1988; Sibley and Lefkowitz, 1985; Mayorga *et al.*, 1989). Guanine nucleotide binding regulator proteins (G proteins) transduce a variety of extracellular signals across the cell membranes into changes in the levels of intracellular second messengers (Gilman, 1987). Members of this protein family are all heterotrimers consisting of α, β and γ subunits. In most biological systems, the GTP-binding α subunits serve as the transducer of signals between receptors and effectors. In contrast, the role of βγ is less well understood. The βγ complex has been demonstrated to facilitate the interaction of the subunits to receptors (Fung, 1983; Florio and Sternwies, 1985) and to mediate the binding of α subunit to membranes (Sternwies, 1986). The α subunit of G protein exhibits variation in structure classifying it into $G\alpha_s$, $G\alpha_i$, $G\alpha_o$, $G\alpha_t$ and $G\alpha_z$ (Stryer, 1986; Gilman, 1987; Provost *et al.*, 1988). $G\alpha_s$ can stimulate the activity of adenylate cyclase whereas $G\alpha_i$ type inhibits adenyl

cyclase activity. The βγ complex does not show any variation in structure and is interchangeable between different Gα proteins.

Acute treatment with opioids inhibits adenylate cyclase, and thereby decreases cAMP levels. Such changes have been described in cultured neuroblastoma × glioma hybrid cells (NG108 cells) (Sharma *et al.*, 1975; Traber *et al.*, 1975) and in brain regions like the locus coeruleus (Duman *et al.*, 1988; Beitner *et al.*, 1989), neostriatum and cerebral cortex (Collier and Roy, 1974; Tsang *et al.*, 1978; Law *et al.*, 1981; Schoffelmeer *et al.*, 1986). In addition, chronic opioid treatment increases levels of $G_{i\alpha}$ and $G_{o\alpha}$, adenylate cyclase activity, cAMP-dependent protein kinase activity in the locus coeruleus (LC) but not in other brain regions (Duman *et al.*, 1988; Nestler and Tallman, 1988; Guitart and Nestler, 1989, 1990a; Nestler *et al.*, 1989; Guitart *et al.*, 1990). It has been proposed that such an upregulated G-protein/cAMP system in the locus coeruleus contributes to opioid tolerance, and dependence in the adult CNS (Nestler *et al.*, 1990a,b; Rasmussen *et al.*, 1990). In addition, exposure of rat spinal cord dorsal rat ganglion co-cultured neurons to κ agonists causes a 60–70% reduction in the $G_{\alpha i}$ subunit (Attali and Vogel, 1989). These results suggest that treatment with opioids alters the G-protein/cAMP levels in different brain regions, however, a causal relationship between opioid receptor density change and G-protein/cAMP levels under such conditions has not yet been demonstrated. Sakellaridis and Vernadakis (1987) have shown in the chick embryo that morphine inhibited forskolin-stimulated adenylate cyclase at embryonic days (ED) 6 and 8, but that morphine had no effect on ED10. Naloxone produced the same effect with the same time schedule. A combination of both morphine and naloxone also decreased the adenylate cyclase activity. These investigators interpreted their findings to indicate that the conventional opioid receptor was not responsible for the observed effects since the inhibitory effects were not naloxone reversible and that naloxone itself produced this effect.

Electrophysiological studies in dorsal rat ganglia spinal cord explant cultures (Crain *et al.*, 1988; Shen and Crain, 1989, 1990) suggest that during sustained morphine exposure, a gradual enhancement of excitatory opioid actions coincides with a reduction in inhibitory effects. These authors hypothesize that the effect of morphine is the result of a balance between inhibitory and excitatory actions. In addition, biochemical studies in the guinea-pig longitudinal muscle with adherent myenteric plexus (Gintzler and Xu, 1991) supports this hypothesis. Their data indicate that different classes of G proteins appear to mediate the opioid enhancement or inhibition of stimulated enkephalin release. In addition, they suggest that a pertussis-toxin (PTX)-sensitive G protein (G_i or G_o) and a cholera-toxin (CTX)-sensitive G protein (G_s) are integral components of the mechanism that mediates opioid inhibition and opioid enhancement, respectively, of evoked enkephalin release. Recently, reciprocal effects of chronic morphine administration on stimulatory (G_s) and inhibitory (G_i) G-protein α subunits were demonstrated in primary cultures of rat striatal neurons (Van Vliet *et al.*, 1991). These data, taken together, suggest that opioid addiction and tolerance involve a crucial balance between inhibitory and stimulatory G proteins. Studies investigating the relationship of G proteins to opiate tolerance are just beginning to emerge. Results from these studies should elucidate cellular mechanisms underlying opiate addiction.

12.4 REGULATION OF OPIOID RECEPTORS IN NEONATAL RAT BRAIN

The first appearance of opioid binding sites in the rat occurs prior to ED14, however, their neuroanatomical distribution is different from that of adult brain (Clendeninn *et al.*, 1976; Coyle and Pert, 1976; Young and Kuhar, 1979; Herkenham and Pert, 1981; for a review see Chapter 10). The different opioid receptor types are expressed at different times during brain development, although in each case the largest increases are observed between 3 and 15 days after birth (Petrillo *et al.*, 1987). Mu opioid receptors are present in the mouse and rat brain at the time of birth and receptor density increases rapidly postnatally (Tavani *et al.*, 1985; Petrillo *et al.*, 1987). By contrast, δ receptors are not detectable until 7–10 days after birth (Tavani *et al.*, 1985; Petrillo *et al.*, 1987; Kent *et al.*, 1982; Leslie *et al.*, 1982; Spain *et al.*, 1985; Tempel *et al.*, 1988), and by 15 days after birth they have reached 50% of adult levels (Tavani *et al.*, 1985). Kappa receptors are present at birth and their density reaches adult levels one week after birth (Leslie *et al.*, 1982; Spain *et al.*, 1985; Barr *et al.*, 1986; Loughlin *et al.*, 1985). These diverse patterns of development suggest that different mechanisms may regulate the expression of the various opioid receptor types.

12.4.1 CHRONIC OPIOID ANTAGONIST ADMINISTRATION

Little is known about the mechanisms regulating the initial expression of opioid receptors. However, receptor development may be influenced quantitatively by a variety of factors in the neuronal environment. Antenatal or perinatal exposure to opioid agonists or antagonists alters the apparent number of opioid receptors in the brain with concomitant changes in antinociceptive responses (Handelmann and Quirion, 1983; Tempel *et al.*, 1988). In addition, stress, pain, and a variety of drugs and toxins alter receptor number (Torda, 1978; Kirby *et al.*, 1982; Watanabe *et al.*, 1983; Moon, 1984), however, the intracellular factors that regulate opioid receptor synthesis, insertion and degradation are unknown.

When naloxone is administered chronically during the prenatal period, the caudal brainstem is more likely than the forebrain to exhibit an increase in opioid receptors (Tsang and Ng, 1980). In contrast, when naloxone is administered chronically during the postnatal period, the rostral brain regions are more likely than the caudal brain regions to exhibit an increase in opioid receptors (Bardo *et al.*, 1982, 1983a). These researchers suggest that opioid receptor systems which are undergoing rapid ontogenetic proliferation may be particularly susceptible to receptor supersensitivity following chronic opioid receptor blockade. Rats injected with naloxone during the first three postnatal weeks showed significant increases in [³H]naloxone binding in the cortex, striatum, hypothalamus and spinal cord. Body and brain weights for naloxone and saline-treated rats were similar indicating that naloxone did not produce a general developmental or nutritional deficit. There were no significant changes in opioid binding one week after cessation of naloxone treatment. These results are similar to what was seen in the adult CNS (Tempel *et al.*, 1984, 1985). Naloxone-treated infants displayed an enhanced response to the anti-nociceptive efficacy of morphine which was also seen in adult rats (Tempel *et al.*, 1985).

12.4.2 CHRONIC AGONIST ADMINISTRATION

The ontogenesis of the opioid system is sensitive to pre- and postnatal exposure to drugs. Results from these studies are contradictory and complicated apparently due to differences in exposure time, dose and route of administration and the age at which animals are tested. In some studies, morphine administration to pups increases μ-ligand binding in the striatum and nucleus accumbens (Handelmann and Quirion, 1983), whereas other studies have failed to find any effect on binding following chronic prenatal morphine treatment (Coyle and Pert, 1976; Bardo *et al.*, 1982). Morphine administration to the dam during pregnancy has also been reported to decrease binding during the first week of life (Kirby and Aronstam, 1983) and increase binding later in life (Iyengar and Rabii, 1982; Tsang and Ng, 1980). Chronic administration of opioid agonists during pre-and/or postnatal development may alter opioid receptor ontogeny, and concomitantly, sensitivity to opioid drugs. Down-regulation of opioid receptors has been demonstrated in whole brain homogenates after chronic morphine treatment (Tempel *et al.*, 1988). One week of prenatal morphine treatment via the dam produced a statistically significant 35% decrease in brain μ-opioid receptors of offspring, on the day of birth. There was no significant change in the affinity or in GABA receptors. Interestingly, this down-regulation was no longer evident by postnatal day (PD) 14. Table 12.1 illustrates the time course of the down-regulation of brain μ-opioid receptors after chronic prenatal treatment with morphine.

Table 12.1. Changes in brain μ-opioid receptor densities of infant rats after chronic prenatal treatment with morphine

Postnatal day	[³H] DAGO binding				% change
	Control (fmol/mg protein)		Morphine-treated (fmol/mg protein)		
0	(4)	77 ± 2.3	(4)	50 ± 3.5	−35[*]
5	(4)	81 ± 4.3	(4)	59 ± 5.3	−27
14	(3)	81 ± 4.1	(3)	83 ± 4.7	−
28	(4)	132 ± 12.5	(4)	129 ± 11.7	−

[*]Statistically significant difference (two-tailed *t*-test, $P < 0.05$). Pregnant female rats were exposed to morphine. Rat pups were killed by decapitation on the day of parturition (day zero), and 5, 14 and 28 days postnatal. Values were generated by computer-assisted linear regression analysis. Receptor density values are reported as means ± S.E.M. from a minimum of three independent experiments. The number of animals per group for each time point are indicated in parentheses. (Reprinted from Tempel *et al.* (1988), with permission from the publishers.)

(a) Postnatal treatment

Four days of daily postnatal (PD1–4) morphine treatment produced a significant 30% decrease in brain μ-opioid receptors. No change in δ or κ receptors was observed. Further treatment with morphine did not result in any significant changes in μ-opioid receptors relative to saline-treated (control) animals. In fact, the degree of μ-opioid receptor down-regulation diminished over time (Table 12.2) with the age of the animal. At PD14, there was no statistically significant difference in receptor density between brains of control and morphine treated pups, as in adult animals. When rat pups were chronically treated (for 7 or 14 days) with morphine, beginning on PD14 and assayed on either PD22 or PD29, no significant differences in opioid receptors were detected relative to saline treated (control) animals (Tempel *et al.*, 1988). Thus, morphine-induced receptor down-regulation appears to occur only during the first week of life.

In order to visualize the neuroanatomical pattern of opiate receptor changes in specific brain regions, light microscopy receptor autoradiography was carried out on rat brain sections from control and chronic morphine treated pups (Tempel, 1991). In the case of brains exposed to morphine from PD1 to PD4, μ-receptor density on PD5 in the striatal patches was virtually non-existent. There were also significant decreases in the surrounding matrix area (Fig. 12.1A, B). In contrast to what is seen after 4 days of morphine treatment, (PD1–4) 8 days of postnatal (PD 1–8) morphine treatment does not produce a significant loss in striatal μ-opioid receptors (Fig. 12.1C, D). Thus, during a brief period of development, there is a unique plasticity of the immature opioid receptor system that is lost with maturation, however, the significance of these findings is as yet unclear.

Several hypotheses can be proposed to explain our developmental results relative to adult rat CNS data. For this, we reference work on other neurotransmitter systems. In the adrenergic system, it has been shown that agonists produce down-regulation of β-adrenergic receptors. This down-regulation was shown to be associated with an internalization mechanism (Chang and Costa, 1979). Conversely, chronic treatment with β-adrenergic antagonists leads to up-regulation or an increase in β-adrenergic receptor number in adult animals (Galant *et al.*, 1978) and in

Table 12.2. Changes in brain μ-opioid receptor densities of infant rats following chronic postnatal treatment with morphine

Duration of treatment post-parturition (days)	[^{3}H]DAGO binding: Control (fmol/mg protein)		[^{3}H]DAGO binding: Morphine-injected (fmol/mg protein)		% change
4	(6)	81 ± 3.8	(6)	57 ± 3.3	−30*
8	(6)	74 ± 8.5	(6)	61 ± 11.5	−18
21	(4)	156 ± 5.7	(4)	135 ± 10.4	−13
28	(3)	132 ± 16.5	(3)	128 ± 8.4	0

* Statistically significant difference (two-tailed *t*-test, $P < 0.05$). Neonatal rats were each given one daily s.c. injection of morphine or saline beginning on PD1. Rat pups were killed by decapitation at the times indicated. Values were generated by computer-assisted linear regression analysis. Receptor density values are reported as means ± S.E.M. from a minimum of three experiments. The numbers of animals per group for each time point are indicated in parentheses. (Reprinted from Tempel *et al.* (1988), with permission from the publishers.)

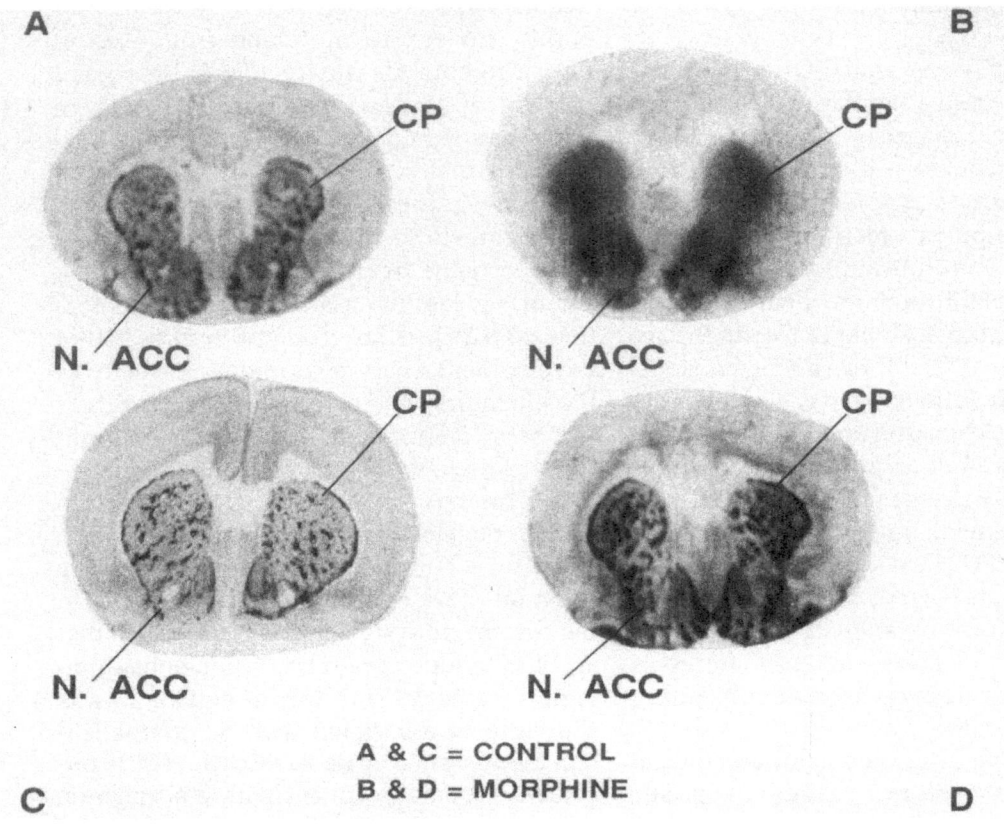

Fig. 12.1. Autoradiographic visualization of alterations in μ-opiate receptor binding at the level of the caudate putamen. Note the lack of μ-receptor 'patches' in the striatum after four days of postnatal morphine treatment (B) relative to control brain (A). Eight days of postnatal morphine treatment (D) induced no significant changes in μ receptor distribution relative to controls (C). Abbreviations; CP, caudate putamen; N. ACC, nucleus accumbens. (Reprinted from Tempel (1991), with permission of the publishers.)

humans (Glaubiger and Lefkowitz, 1977). More recently, several laboratories working with the opioid system have shown that long-term exposure of neurotumor cell lines (Hazum *et al.*, 1981; Chang *et al.*, 1982; Simantov *et al.*, 1982a; Blanchard *et al.*, 1983) and adult rats (Steece *et al.*, 1986) to enkephalin results in a decrease in δ-receptor density. Studies of [^{3}H]DADLE uptake by N4TG1 cells (Hazum *et al.*, 1981; Chang *et al.*, 1982; Simantov *et al.*, 1982b; Blanchard *et al.*, 1983) suggest that enkephalin is internalized via receptor-mediated endocytosis. One hypothesis to explain our findings is that long-

term exposure of the system to opioid agonists or peptides leads to coupling of the receptors to cyclase, followed by internalization. As active receptors disappear from the membrane surface, they would be replaced by inactive or spare receptors. Two possibilities then exist. (1) If the rate of internalization exceeded reactivation, an apparent down-regulation in receptor density would be observed. (2) If the rate of reactivation paralleled internalization, no apparent change in receptor density would be observed. If down-regulation involves internalization, degradation of receptors, and recycling or reactivation, it is possible that the internalization process is developed by birth but that the recycling process is not fully developed until 2–3 weeks postnatal. When the recycling process is developed, down-regulation is no longer seen because recycling takes place as quickly as internalization; therefore no difference in receptor number is observed.

Another possibility is that opioids affect second messenger systems. Recent data (section 12.3.4) suggest that the overall effect of morphine is the result of a balance/reciprocal relationship between inhibitory and excitatory G proteins. In the developing CNS, it is not yet known which of the G proteins are present, what proportions they are present in and whether or not their interaction with the receptor and/or cyclase system is operational. All of these factors may be important in elucidating the mechanisms underlying addiction in the developing CNS in comparison to adult CNS tolerance. Future studies in this area will elucidate the exact nature of this interaction.

12.4.3 CHANGES IN OPIOID PEPTIDE SYNTHESIS IN NEONATAL RAT BRAIN

The effect of prolonged opioid use by pregnant women on opioid peptide levels in their newborns has received considerable attention. Opioid drugs cross the placental barrier (Blinick *et al.*, 1975) and their effects on the health of the pregnant woman and developing fetus are great. Pregnancies of opioid-dependent women are often complicated by problems such as toxemia, maternal syphilis and hepatitis and placental problems (Finnegan, 1975; Perlmutter, 1974; Rementeruia and Lotongkhum, 1977; Hans, 1989). Opioid-exposed infants are especially likely to show signs of fetal distress and to be born at low birthweight even when receiving adequate prenatal care (Connaughton *et al.*, 1973; Strauss *et al.*, 1974; Kandall *et al.*, 1975; Hans, 1989). The effects of prenatal opioid exposure on the behavior of newborn infants are also

dramatic. Babies born to morphine- and methadone-dependent mothers have elevated plasma β-endorphin levels for at least 40 days after birth and exhibit behavioral abnormalities (Lesser-Katz, 1982; Bauman and Levine, 1986; Deren, 1986), but fail to exhibit signs of withdrawal (Panerai *et al.*, 1983; Genazzani *et al.*, 1986). More recent studies have shown that during the first week after birth, most opioid-exposed infants show a well-documented neonatal abstinence syndrome that includes a variety of behaviors associated with central and autonomic nervous system hyperarousal such as tremors, hypertonus, hyperactive reflexes, high-pitched crying, poor sleeping and feeding, fever and rapid respiration (Jeremy *et al.*, 1985; Hans, 1989).

Numerous studies have investigated the effects of chronic opioid treatment on brain enkephalin levels, but they have produced conflicting results. Implantation of morphine pellets in mature rats has been reported to reduce brain enkephalin levels after 3 days (Shani *et al.*, 1979) or 11 days (Bergstrom and Terenius, 1983), but to have no effect after 5 days (Childers *et al.*, 1977; Fratta *et al.*, 1977) or 21 days (Shani *et al.*, 1979). Some studies report marked decreases in β-endorphin as well as enkephalin in several discrete areas of the rat brain following treatment in adults with morphine pellets for 30 days (Hollt *et al.*, 1978; Przewlocki *et al.*, 1979). Reduced met-enkephalin levels have also been reported in the hippocampus of the monkey after 10 days of morphine treatment (Elsworth *et al.*, 1986).

In studies of the adult rat CNS, researchers have failed to detect any significant change in opioid peptide levels (Childers *et al.*, 1977; Fratta *et al.*, 1977; Wesche *et al.*, 1977; Shani *et al.*, 1979) in rat brain following chronic morphine treatment. Little is known about the molecular alterations in the functioning of endogenous brain opioid peptide systems as a result of chronic exogenous opioid administration. Neither brain opioid receptors, second messenger systems, nor opioid peptide levels have been shown to vary consistently as a result of drug treatment (Wuster *et al.*, 1983; Akil *et al.*, 1984; Redmond and Krystal, 1984; Collier, 1985; Rothman *et al.*, 1986). In a recent study by Uhl *et al.* (1988), the effect of chronic morphine treatment on opioid peptide gene expression in adult CNS was examined. It was shown that rats made tolerant to morphine via subcutaneous implantation of pellets displayed a significant decrease in striatal preproenkephalin mRNA that persisted during the withdrawal period. In contrast, levels of met-enkephalin were normal at the end of opioid treatment but reduced after withdrawal.

In an effort to gain more insight into these mechanisms clonal cell lines have been studied (Schwartz, 1988). In the NG108-15 neuroblastoma–glioma hybrid cell line, it has been demonstrated that enkephalin peptide content increases after exposure to cAMP (Braas *et al.*, 1983; Yoshikawa and Sabol, 1986). The opioid receptor is known to be linked to adenylate cyclase in an inhibitory fashion (Sharma *et al.*, 1975b). Tolerance and dependence that are produced in animals by chronic exposure to morphine can be reproduced in cells (Sharma *et al.*, 1975a; Sharma *et al.*, 1977). Schwartz (1988) has shown that opioid agonists stimulate proenkephalin synthesis with increases seen in mRNA, precursor forms and enkephalin peptides. However, these effects appear to occur independent of changes in adenylate cyclase activity or cAMP content. It is suggested that proenkephalin synthesis in NG108 cells can be regulated by two different mechanisms, one involving cAMP, and the other regulated by the opioid receptor, but this is yet to be determined.

In the developing CNS, there are very few data and these are complicated to interpret. In a preliminary study, Tempel *et al.* (unpublished) have found significant and contrasting effects of morphine treatment on preproenkephalin mRNA levels. Four days of postnatal (PD1–4) morphine treatment produced a 24% increase in striatal proenkephalin mRNA

levels. However, longer treatments with morphine (PD1–14) decrease striatal preproenkephalin mRNA levels by 39%. This decrease is similar to what was observed in the adult striatum after chronic morphine treatment (Uhl *et al.*, 1988). It is noteworthy that chronic naltrexone treatment produced the exact opposite effect. Four days of postnatal (PD1–4) naltrexone treatment induced a 33% decrease in proenkephalin mRNA levels in striatal tissue whereas 8 days of treatment (PD1–8) produced an increase (+23%) in opioid peptide gene expression. The increase in gene expression is in the same direction, but smaller than that observed in the adult striatum (Tempel *et al.*, 1990). These data suggest that the mechanisms underlying opioid addiction and withdrawal in the developing CNS differ from those in the adult brain. These differences may be due to the interaction of the opioid receptor system with the G-protein/cAMP system.

Future research along these lines will provide valuable information into the mechanisms of addiction and tolerance which will in turn provide insight into the diagnosis and treatment of drug-related effects observed in children born to addicted mothers. Moreover, such studies offer a unique view into the function of the opiate system in normal animals.

12.5 COCAINE AND THE OPIOID SYSTEM

Cocaine use has increased dramatically during the past few years (Clayton, 1985). Recent clinical studies indicate that maternal cocaine use can cause a variety of adverse effects. Cocaine has complex pharmacological effects on the mother and fetus. Cocaine is known to be a potent anorexic agent in both humans and animals and poor nutrition during pregnancy is known to place the infant at risk of developing many physical and functional abnormalities. Many clinical studies report that cocaine produces a premature separation of the placenta from the wall of the uterus with or without premature delivery (Acker *et al.*,

1983; Chasnoff *et al.*, 1985, 1987; Bingol *et al.*, 1987; Ryan *et al.*, 1987; Cherukuri *et al.*, 1988; Critchley *et al.*, 1988; Townsend *et al.*, 1988; Fulroth *et al.*, 1989; Kaye *et al.*, 1989; Valencia *et al.*, 1989).

Several clinical studies report an increase in stillbirth rate in pregnancies complicated by cocaine (Chasnoff *et al.*, 1985; Ryan *et al.*, 1987; Cherukuri *et al.*, 1988; Critchley *et al.*, 1988; Bingol *et al.*, 1989). Neonatal effects include cerebral infarction (Chasnoff *et al.*, 1986), depressed interactive behavior, intrauterine growth retardation, cardiac anomalies as well as fetal deaths (for review see Church *et al.*, 1990). Although these data suggest that cocaine abuse causes problems for the unborn child, there are also contradictory reports. There are reports of cocaine-abusing mothers delivering infants of normal weight and gestational age (Chasnoff *et al.*, 1986, 1987; Madden *et al.*, 1986; Ryan *et al.*, 1987). In addition, the human data are confounded by teenage pregnancy, undernutrition, multiple drug abuse, variability in the concentration of cocaine used and the presence of other chemicals that are sometimes mixed with cocaine (such as heroin, strychnine, talc, amphetamine, etc.). While placental abruption has been identified in one animal study, cocaine generally has no effect on gestational length in the rat (Church *et al.*, 1988; Hutchings *et al.*, 1989; Spear *et al.*, 1989). Animal studies have shown that cocaine reduces litter size but only at the most toxic doses (80–100 mg/kg s.c.) (Church *et al.*, 1988). These doses of cocaine result in a large proportion of maternal lethality and therefore may not be relevant to the human situation.

Although there is a need to establish good animal models for assessment of neurobehavioral and neurochemical consequences of early cocaine exposure, most laboratory studies have focused on maternal toxicity and fetal teratogenicity following gestational cocaine exposure (Mahalik *et al.*, 1980, 1984; Fantel and MacPhail, 1982; Church *et al.*, 1988). Record work has focused on neurobehavioral

and chemical consequences of early cocaine exposure in rodents (Dow-Edwards *et al.*, 1986, 1988; Foss and Riley, 1988; McGivern, 1988; Dow-Edwards, 1989, 1991; Smith *et al.*, 1989; Spear *et al.*, 1989).

The presence of psychoactive drugs can affect any of several neurotransmitter systems which can result in long-term alterations in the structure and function of these systems. Cocaine has been shown to alter activity at the dopaminergic, noradrenergic and serotonergic level. D_1 receptor binding increases by as much as 72% in the caudal portions of caudate nucleus of adult females exposed to cocaine (50 mg/kg) between PD11 and PD20 (Dow-Edwards, 1989). In addition, brain glucose metabolism, an index of functional activity, is differentially affected in males and females postnatally exposed to cocaine. Female treated rats show significantly increased rates of brain functional activity with specific highly dopaminergic regions showing the greatest percentage changes. For example, the cingulate cortex and ventral tegmental area show the largest increases in glucose activity. In contrast to female rats, neonatal cocaine treatment has little effect on male brain glucose metabolism, indicating that females may be more susceptible than males to the toxic effects of cocaine.

A variety of evidence suggests a 'dopamine hypothesis' for the reinforcing properties of cocaine, in particular, the dopamine (DA) mesolimbocortical neurons. D_1 receptors (3H SCH23390) are increased in the caudate of adult females exposed to cocaine between PD 11 and PD20 (for review see Dow-Edwards, 1991). An essential step for understanding the mechanism of action of any drug is to identify its receptor site. Receptor binding studies suggest that the cocaine receptor is the dopamine transporter. Recently, two independent laboratories have cloned the DA transporter (Shimada *et al.*, 1991; Kilty *et al.*, 1991) which should aid the understanding of the pharmacological actions of cocaine. Cocaine first binds to the transporter and blocks DA uptake, which results in a potentiation of dopaminergic neurotransmission in the limbic pathways. However, the existence of another important site or neurotransmitter that is involved in cocaine's actions cannot be ruled out. Serotonin and noradrenergic systems have also been implicated (for a review see Kuhar *et al.*, 1991).

Functional studies of adult rats exposed to cocaine during specific periods of development have shown that several important neuronal circuits are permanently altered by exposure to moderate (not toxic) doses (Dow-Edwards, 1989; Dow-Edwards *et al.*, 1990). Prenatal exposure produces persistent decrements in function in the nigrostriatal pathway which, due to its role in regulation of fine motor coordination, may be involved in the production of tremors seen in newborn humans exposed to cocaine *in utero*. Early postnatal exposure (which approximates the third trimester in human brain development) in the female rat increases function in the mesolimbic dopaminergic system. Since this pathway is involved in the reinforcing effects of cocaine, one might predict that developmental exposure to cocaine could alter drug-seeking behavior. However, the appropriate studies to test this hypothesis have not been carried out.

One of the other systems that may be involved in the reinforcing effects of cocaine is the opioid system. Very few data exist in this area, which is summarized in section 12.5.1.

In the adult CNS, chronic cocaine exposure increases [³H]naloxone binding in regions such as the nucleus accumbens, ventral pallidum and lateral hypothalamus (Hammer, 1989). In addition, cocaine treatment has been shown to increase dynorphin levels (Sivam, 1989; Smiley *et al.*, 1990) with no change in met-enkephalin or substance P levels (Sivam, 1989). These data suggest that cocaine may interact with the κ-opioid receptor, at least, in the adult CNS. Clow *et al.* (1991) examined opiate receptor binding in offspring of cocaine-exposed dams from ED8-20. Gesta-

tional cocaine exposure caused a generalized, dose-dependent increase of [³H]naloxone binding with highest increases in dopaminergic terminal regions as well as limbic and cortical regions. This effect is present 23–24 days after cessation of treatment. The effect is more extensive, and less specific on the opioid system than is seen in the adult after cocaine exposure. However, these studies just touch on the interaction of cocaine and the opioid system. Much more work is needed in order to characterize fully the mechanism of action of cocaine and its interaction with the opioid/dopamine systems.

12.5.1 SECOND MESSENGER SYSTEMS AND COCAINE TREATMENT

To date, there has been very little work on the effects of cocaine treatment on G-protein/second messenger systems. Two research groups have recently begun to explore this interaction in the adult CNS. Nestler *et al.* (1990b) have reported a decrease in $G_{o\alpha}$ and $G_{i\alpha}$ in the A10 DA region and the nucleus accumbens following 14 days of daily cocaine treatment (15 mg/kg). In order to assess further the role of G proteins in cocaine abuse, Steketee *et al.* (1991) examined the effects of PTX administration and ADP-ribosylation of G_i and G_o in these brain regions. The capacity of acute cocaine treatment to increase dopamine release in the nucleus accumbens was significantly increased in rats pretreated 14 days earlier with PTX. These data suggest that injection of PTX into the A10 cell group produces alterations in mesolimbic dopamine function as well as G proteins in cocaine abuse. Although these data may shed light on the mechanism of action of cocaine in the adult CNS, no work has been done in the developing CNS. It is possible that the actions of cocaine in the neonatal CNS may be quite different from those seen in the adult CNS. Future research along these lines is needed to fully comprehend the long-term effects in offspring exposed to cocaine prenatally so that they may best overcome the toxic effects of this drug.

ACKNOWLEDGEMENT

I would like to thank Annette Snoddy for her careful and patient typing of this chapter.

REFERENCES

Acker, D., Sachs, B.P., Tracey, K.J. and Wise, W.E. (1983) Abruptio placentae associated with cocaine use. *Am. J. Obstet. Gynecol.*, **146**, 220–1.

Akil, H., Watson, S.J., Young, E. *et al.* (1984) Endogenous opioids: biology and function. *Annu. Rev. Neurosci.*, **7**, 223–55.

Attali, B., Gouarderes, C., Mazaguil, H. *et al.* (1982) Evidence of multiple kappa binding sites by use of opioid peptides in the guinea pig lumbrosacral spinal cord. *Neuropeptides*, **3**, 53.

Attali, B. and Vogel, Z. (1989) Long-term opiate exposure leads to reduction of the α_i-1 subunit of GTP-binding proteins. *J. Neurochem.*, **53**, 1636–9.

Bardo, M.T., Bhatnagar, R.K. and Gebhart, G.F. (1982) Differential effects of chronic morphine and naloxone on opiate receptors, monoamines and morphine-induced behaviors in preweanling rats. *Dev. Brain Res.*, **4**, 139–47.

Bardo, M.T., Bhatnagar, R.K. and Gebhart, G.F. (1983a) Age-related differences in the effect of chronic administration of naloxone on opiate binding in rat brain. *Neuropharmacology*, **22**, 453–61.

Bardo, M.T., Bhatnagar, R.K. and Gebhart, G.F. (1983b) Chronic naltrexone increases opiate binding in brain and produces supersensitivity to morphine in the locus coeruleus of the rat. *Brain Res.*, **289**, 223–34.

Barr, G.A., Paredes, W., Erickson, K.L. and Zukin, R.S. (1986) κ-opioid receptor-mediated analgesia in the developing rat. *Dev. Brain Res.*, **28**, 145–52.

Bauman, P.S. and Levine, S.A. (1986) The development of children of drug addicts. *Int. J. Addict.*, **21**, 849–63.

Beitner, D.B., Duman, R.S. and Nestler, E.J. (1989) A novel action of morphine in the rat locus coeruleus: persistent decrease in adenylate cyclase. *Mol. Pharmacol.*, **35**, 559–64.

Bergstrom, L. and Terenius, L. (1983) Enkephalin levels decrease in rat striatum during morphine abstinence. *Eur. J. Pharmacol.*, **60**, 349–52.

Bhargava, H.N., Ramarao, P. and Gulati, A. (1989a) Effects of morphine in rats treated chronically with U-50,488H, a kappa opioid receptor agonist. *Eur. J. Pharmacol.*, **162**, 257–64.

Bhargava, H.N., Gulati, A. and Ramarao, P. (1989b) Effect of chronic administration of U-50,488H on tolerance to its pharmacological actions and on multiple opioid receptors in rat brain regions and spinal cord. *J. Pharmacol. Exp. Ther.*, **251**, 21–6.

Bingol, N., Fuchs, M., Diaz, V. *et al.* (1987) Teratogenicity of cocaine in humans. *J. Pediatr.*, **110**, 93–6.

Blanchard, S.G., Chang, K.J. and Cuatrecasas, P. (1983) Characterization of the association of tritiated enkephalin with neuroblastoma cells under conditions optimal for receptor down-regulation. *J. Biol. Chem.*, **258**, 1092–7.

Blinick, G., Inturrisi, C.E., Jerez, E. and Wallach, R.C.(1975) Methadone assays in pregnant women and progeny. *Am. J. Obstet. Gynecol.*, **121**, 617–21.

Braas, K.A., Childers, S.R. and U'Prichard, D.C. (1983) Induction of differentiation increases Met^5-enkephalin and Leu^5-enkephalin content in NG108-15 hybrid cells: an immunocytochemical and biochemical analysis. *J. Neurosci.*, **3**, 1713–27.

Brady, L.S., Herkenham, M., Long, J.B. and Rothman, R.B. (1989) Chronic morphine increases mu-opiate receptor binding in rat brain: a quantitative autoradiographic study. *Brain Res.*, **477**, 382–6.

Brunello, N., Volterra, A., DiGiulio, A.M. *et al.* (1984) Modulation of opioid system in C57 mice after repeated treatment with morphine and naloxone: biochemical and behavioral correlates. *Life Sci.*, **34**, 1669–78.

Chang, D.M. and Costa, E. (1979) Evidence for internalization of the recognition site of beta-adrenergic receptors during receptor subsensitivity induced by (-)isoproterenol. *Proc. Natl. Acad. Sci. USA*, **76**, 3024–8.

Chang, K.J. and Cuatrecasas, P. (1979) Multiple opiate receptors: enkephalins and morphine bind to receptors of different specificity. *J. Biol. Chem.*, **254**, 2610–18.

Chang, K.J., Eckel, R.W. and Blanchard, S.G. (1982) Opioid peptides induce reduction of enkephalin receptors in cultured neuroblastoma cells. *Nature*, **296**, 446–8.

Chasnoff, I.J., Burns, W.J., Schnoll, S.H. and Burns, K.A. (1985) Cocaine use in pregnancy. *N. Engl. J. Med.*, **313**, 666–9.

Chasnoff, I.J., Burns, K.A. and Burns, W.J. (1987) Cocaine use in pregnancy: perinatal morbidity and mortality. *Neurotoxicol. Teratol.*, **9**, 291–3.

Chasnoff, I.J., Bussey, M.E., Sarich, R. and Stack, C.M. (1986) Perinatal cerebral infarction and maternal cocaine use. *J. Pediatr.*, **108**, 456–9.

Cherukuri, R., Minkoff, H., Feldman, J. *et al.* (1988) A cohort study of alkaloidal cocaine ('crack') in pregnancy. *Obstet. Gynecol.*, **72**, 147–51.

Childers, S.R., Simantov, R. and Snyder, S.H. (1977) Enkephalin: radioimmunoassay and radioreceptor assay in morphine dependent rats. *Eur. J. Pharmacol.*, **46**, 289–93.

Church, M.W., Dintcheff, B.A. and Gessner, P.K. (1988) Dose-dependent consequences of cocaine on pregnancy outcome in the Long-Evans Rat. *Neurotoxicol. Teratol.*, **10**, 51–8.

Church, M.W., Overbeck, G.W. and Andrzejczak, A.L. (1990) Prenatal cocaine exposure in the Long-Evans Rat: I. Dose-dependent effects on gestation, mortality, and postnatal maturation. *Neurotoxicol. Teratol.*, **12**, 327–34.

Clayton, R.R. (1985) Cocaine use in the U.S.: in a blizzard or just being snowed, in *Cocaine Use in America: Epidemiologic and Clinical Perspectives, NIDA Monograph* (eds J. Kozel and E.H. Adams), US Dept. Health Human Services, Rockville, MD, Vol. 61, pp. 8–34.

Clendeninn, N.J., Petraitis, M. and Simon, E.J. (1976) Ontological development of opiate receptors in rodent brain. *Brain Res.*, **118**, 157–60.

Clow, D.W., Hammer, R.P., Kirstein, C.L. and Spear, L.P. (1991) Gestational cocaine exposure increases opiate receptor binding in weanling offspring. *Dev. Brain Res.*, **59**, 179–85.

Collier, H.O.J. (1985) A general theory of the genesis of drug dependence by induction of receptors. *Nature*, **205**, 181–3.

Collier, H.O.J., and Roy, A.C. (1974) Morphine-like drugs inhibit the stimulation of E prostaglandins of cyclic AMP formation by rat brain homogenate. *Nature*, **248**, 24–7.

Connaughton, J.F., Finnegan, L.P., Schur, J. and Emich, J.P. (1973) Current concepts in the management of the pregnant opiate addict. *Addictive Dis.*, **2**, 21–36.

Corbett, A.D., Paterson, S.J., McKnight, A.T. *et al.* (1983) Dynorphin 1-8 and dynorphin 1-9 are ligands for the κ-subtype of opiate receptor. *Nature*, **299**, 79–81.

Costa, T., Klinz, F.J., Vachon, L. and Herz, A. (1988) Opioid receptors are coupled tightly to G proteins but loosely to adenylate cyclase in NG108-15 cell membranes. *Mol. Pharmacol.*, **34**, 744–54.

Coyle, J.T. and Pert, C.B. (1976) Ontogenetic development of [^{3}H] naloxone binding in rat brain. *Neuropharmacology*, **15**, 555–60.

Crain, S.M., Shen, K.F. and Chalazonitis, A. (1988) Opioids excite rather than inhibit sensory neurons after chronic opioid exposure of spinal cord-ganglion cultures. *Brain Res.*, **455**, 99–109.

Critchley, H.O.D., Woods, S.M., Barson, A.J. *et al.* (1988) Fetal death in utero and cocaine abuse. Case report. *Br. J. Obstet. Gynaecol.*, **95**, 195–6.

Danks, J.A., Tortella, F.C., Bykov, V. *et al.* (1988) Chronic administration of morphine and naltrexone up-regulate [^{3}H][D-Ala^2, D-leu^5]enkephalin binding sites by different mechanisms. *Neuropharmacology*, **27**, 965–74.

Davis, M.W. and Lin, C.H. (1972) Prenatal morphine effects on survival and behavior of rat offspring. *Res. Commun. Chem. Pathol. Pharmacol.*, **3**, 205–14.

Deren, S. (1986) Children of substance abusers: a review of the literature. *J. Subst. Abuse Treat.*, **3**, 77–94.

Dole, V.P. and Nyswander, M.A. (1965) Medical treatment for diacetyl-morphine (heroin) addiction. *J. Am. Med. Assoc.*, **193**, 646–50.

Dow-Edwards, D.L. (1989) Long-term neurochemical and neurobehavioral consequences of cocaine use during pregnancy. *Ann. NY Acad. Sci.*, **562**, 280–9.

Dow-Edwards, D.L. (1991) Cocaine effects on fetal development: a comparison of clinical and animal research findings. *Neurotoxicol. Teratol.*, **13**, 347–52.

Dow-Edwards, D.L., Freed, L.A. and Milorat, T.H. (1986) The effects of cocaine on development. *Soc. Neurosci. Abstr.*, **12**, 59.12.

Dow-Edwards, D.L., Fico, T.A. and Hutchings, D.E. (1988) Functional effects of cocaine given during critical periods of development. *Teratology*, **37**, 518.

Dow-Edwards, D.L., Freed, L.A. and Fico, T.A. (1990) Structural and functional effects of prenatal cocaine exposure in adult rat brain. *Dev. Brain Res.*, **57**, 263–8.

Duman, R.S., Tallman, J.F. and Nestler, E.J. (1988) Acute and chronic opiate-regulation of adenylate cyclase in brain: specific effects in locus coeruleus. *J. Pharmacol. Exp. Ther.*, **246**, 1033–9.

Elsworth, J.D., Redmond, D.E. and Roth, R.H. (1986) Effect of morphine treatment and withdrawal on endogenous methionine- and leucine-enkephalin levels in primate brain. *Biochem. Pharmacol.*, **35**, 3415–17.

Fantel, A.G. and MacPhail, B.J. (1982) The teratogenicity of cocaine. *Teratology*, **26**, 17–19.

Finnegan, L.P. (1975) Narcotics dependence in pregnancy. *J. Psychedelic Drugs*, **7**, 299–311.

Florio, V.A. and Sternwies, P.C. (1985) Reconstitution of resolved muscarinic cholinergic receptors with purified GTP-binding proteins. *J. Biol. Chem.*, **260**, 3477–83.

Foss, J.A. and Riley, E.A. (1988) Behavioral evaluation of animals exposed prenatally to cocaine. *Teratology*, **37**, 517.

Fratta, W., Yang, H.Y.T., Hong, J. and Costa, E. (1977) Stability of met-enkephalin content in brain structures of morphine-dependent or footshock stressed rats. *Nature*, **268**, 452–3.

Freeman, P.R. (1980) Methadone exposure in utero: effects on open-field activity in weanling rats. *Int. J. Neurosci.*, **11**, 295–300.

Fulroth, R., Phillips, B. and Durand, D.J. (1989) Perinatal outcome of infants exposed to cocaine and/or heroin in utero. *Am. J. Dis. Child.*, **143**, 905–10.

Fung, B.K.-K. (1983) Characterization of transducin from bovine retinal rod outer segments. I. Separation and reconstitution of the subunits. *Biol. Chem.*, **258**, 10495–502.

Galant, S.P., Durisetti, L., Underwood, S. and Insel, P.A. (1978) Decreased beta-adrenergic receptors on polymorphonuclear leukocytes after adrenergic therapy. *N. Engl. J. Med.*, **299**, 933–6.

Genazzani, A.R., Petraglia, F., Guidetti, R., *et al.* (1986) Neonatal β-endorphin secretion in babies passively addicted to opiates. *Int. Congr. Ser. Excerpta Med.*, **369**, 379–82.

Gilman, A.G. (1987) G proteins: transducers of receptor-generated signals. *Annu. Rev. Biochem.*, **56**, 615–49.

Gintzler, A.R. and Xu, H. (1991) Different G proteins mediate the opioid inhibition or enhancement of evoked [5-methionine] enkephalin release. *Proc. Natl. Acad. Sci. USA*, **88**, 4741–5.

Glaubiger, G. and Lefkowitz, R.J. (1977) Elevated beta-adrenergic receptor number after chronic propranolol treatment. *Biochem. Biophys. Res. Commun.*, **78**, 720–5.

Goldstein, A., Fischli, W., Lowney, L.I. *et al.* (1981) Porcine pituitary dynorphin: complete amino acid sequence of the biologically active heptadecapeptide. *Proc. Natl. Acad. Sci. USA*, **78**, 7219–23.

Goodfriend, M.J., Shey, I.A. and Klein, M.D. (1956) The effects of maternal narcotic addiction on the newborn. *Am. J. Obstet. Gynecol.*, **71**, 29–36.

Gouarderes, C., Attali, B., Audigier, Y and Cros, J. (1983) Interaction of selective mu and delta ligands with the kappa$_2$ subtype of opiate binding sites. *Life Sci.*, **33**, 175–8.

Gubler, U., Seeberg, P., Hoffman, B.J. *et al.* (1982) Molecular cloning establishes proenkephalin as precursor of enkephalin-containing peptides. *Nature*, **295**, 206–8.

Guitart, X. and Nestler, E.J. (1989) Identification of morphine and cyclic AMP-regulated phosphoproteins (MARPPs) in the locus coeruleus and other regions of rat brain. Regulation by acute and chronic morphine. *J. Neurosci.*, **9**, 4371–87.

Guitart, X. and Nestler, E.J. (1990) Identification of MARPP (14–20), morphine- and cyclic AMP-regulated phosphoproteins of 14–20 kDa, as myelin basic proteins: evidence for their acute and chronic regulation by morphine in rat brain. *Brain Res.*, **516**, 57–65.

Guitart, X., Hayward, M.D., Nissenbaum, L.K. *et al.* (1990) Identification of MARPP-58, a morphine- and cAMP-regulated phosphoprotein of 58 kDa, as tyrosine hydroxylase: evidence for regulation of its expression by chronic morphine in the rat locus coeruleus. *J. Neurosci.*, **10**, 2649–59.

Hammer, R.P. (1989) Cocaine alters opiate receptor binding in critical brain reward regions. *Synapse*, **3**, 55–60.

Handelmann, G.E., and Quirion, R. (1983) Neonatal exposure to morphine increases mu opiate binding in the adult forebrain. *Eur. J. Pharmacol.*, **94**, 357–8.

Hans, S.L. (1989) Development consequences of prenatal exposure to methadone. *Ann. NY Acad. Sci.*, **562**, 195–207.

Hazum, E., Chang, K.J. and Cuatrecasas, P. (1981) Receptor redistribution induced by hormones and neurotransmitters: possible relationship to biological functions. *Neuropeptides*, **1**, 217–30.

Herkenham, M. and Pert, C.B. (1981) Mosaic distribution of opiate receptors, parafascicular projections and acetylcholinesterase in rat striatum. *Nature*, **291**, 415–18.

Herkenham, M. and McLean, S. (1986) Mismatches between receptor and transmitter localizations in the brain, in *Quantitative Receptor Autoradiography* (eds C. Boast, E.W. Altar and C.A. Snowhill), Alan Liss, New York, p. 131–71.

Hitzemann, R., Hitzemann, B. and Loh, H. (1974) Binding of [^{3}H]naloxone in the mouse brain: effect of ions and tolerance development. *Life Sci.*, **14**, 2393–404.

Hofer, M.A. (1984) Relationships as regulators: a psychobiologic perspective on bereavement. *Psychosom. Med.*, **46**, 183–97.

Holaday, J.W., Hitzemann, R.J., Curell, J. *et al.* (1982) Repeated electroconvulsive shock or chronic morphine treatment increases the number of [^{3}H]D-Ala^2 D-Leu^5-enkephalin binding sites in rat brain membranes. *Life Sci.*, **31**, 2359–62.

Hollt, V., Przewlocki, R. and Herz, A. (1978) β-Endorphin-like immunoreactivity in plasma, pituitaries and hypothalamus of rats following treatment with opiates. *Life Sci.*, **23**, 1057–66.

Householder, J., Hatcher, R., Burns, W. and Chasnoff, I. (1982) Infants born to narcotic-addicted mothers. *Psychol. Bull.*, **92**, 453–68.

Hughes, J., Smith, T.W., Kosterlitz, H.W. *et al.* (1975) Identification of two related pentapeptides from the brain with potent opiate agonist activity. *Nature*, **258**, 577–80.

Hutchings, D.E. (1985a) Issues of methodology and interpretation in clinical and animal behavioral teratology studies. *Neurobehav. Toxicol. Teratol.*, **7**, 639–42.

Hutchings, D.E. (1985b) Prenatal opioid exposure and the problem of casual inference, in *Current Research on the Consequences of Maternal Drug Use. National Institute on Drug Abuse Research Series* (ed. T.M. Pinkert) DHHS Pub. No. (ADM) 85-1400. Supt of Docs, US Govt Printing Office, Washington, DC, pp. 6–9.

Hutchings, D.E. and Fifer, W.P. (1986) Neurobehavioral effects in human and animal offspring following prenatal exposure to methadone, in *Handbook of Behavioral Teratology* (eds E.P. Riley and C.V. Vorhees), Plenum Press, New York, pp. 141–60.

Hutchings, D.E., Bodnarenko, S.R. and Diaz-DeLeon, R. (1984) Phencyclidine during pregnancy in the rat: effects on locomotor activity in the offspring. *Pharmacol. Biochem. Behav.*, **20**, 251–4.

Hutchings, D.E., Fico, T.A. and Dow-Edwards, D.L. (1989) Prenatal cocaine: maternal toxicity, fetal effects, and locomotor activity in rat offspring. *Neurotoxicol. Teratol.*, **11**, 65–9.

Iyengar, S. and Rabii, J. (1982) Effect of prenatal exposure to morphine on the postnatal development of opiate receptors. *Fed. Proc.*, **41**, 354–7.

James, I.F., Chavkin, C. and Goldstein, C. (1982) Selectivity of dynorphin for kappa opioid receptors. *Life Sci.*, **31**, 1331–4.

Jeremy, R.J. and Hans, S.L. (1985) Behavior of neonates exposed in utero to methadone as assessed on the Brazelton scale. *Infant Behav. Dev.*, **8**, 323–36.

Kandall, S.R., Album, S., Dreyer, E. *et al.* (1975) Differential effects of heroin and methadone on birth weights. *Addictive Dis.*, **2**, 347–55.

Kaye, K., Elkind, L., Goldberg, D. and Tytun, A. (1989) Birth outcomes for infants of drug abusing mothers. *NY State J. Med.*, **89**, 256–61.

Kent, J.L., Pert, C.B. and Herkenham, M. (1982) Ontogeny of opiate receptors in rat forebrain: visualization by *in vitro* autoradiography. *Dev. Brain Res.*, **2**, 487–504.

Kilty, J.E., Lorang, D. and Amara, S.G. (1991) Cloning and expression of a cocaine-sensitive rat dopamine transporter. *Science*, **254**, 578–9.

Kirby, M.L. and Aronstam, R.S. (1983) Levor-phanol-sensitive [³H] naloxone binding in developing brainstem following prenatal morphine exposure. *Neurosci. Lett.*, **35**, 191–5.

Kirby, M.L., Gale, T.F. and Mattio, T.C. (1982) Effects of prenatal capsaicin treatment on fetal spontaneous activity, opiate receptor binding, and acid phosphatase in the spinal cord. *Exp. Neurol.*, **76**, 298–308.

Klee, W.A. and Streaty, R.A. (1974) Narcotic receptor sites in morphine-dependent rats. *Nature*, **248**, 61–3.

Kuhar, M.J., Ritz, M.C. and Boja, J.W. (1991) The dopamine hypothesis of the reinforcing properties of cocaine. *Trends Neurol. Sci.*, **14**, 229–302.

Lahti, R.A. and Collins, R.J. (1978) Chronic naloxone results in prolonged increases in opiate binding sites in brain. *Eur. J. Pharmacol.*, **51**, 185–6.

Law, P.Y., Wu, J., Koehler, J. and Loh, H.H.J. (1981) Demonstration and characterization of opiate inhibition of the striatal adenylate cyclase. *J. Neurochem.*, **36**, 1834–6.

Leslie, F.M., Tso, S. and Hurlbut, D.E. (1982) Differential appearance of opiate receptor subtypes in neonatal rat brain. *Life Sci.*, **31**, 1393–6.

Lesser-Katz, M. (1982) Some effects of maternal drug addiction on the neonate. *Int. J. Addict.*, **17**, 887–96.

Lord, J.A.H., Waterfield, A.A., Hughes, J. and Kosterlitz, H.W. (1977) Endogenous opioid peptides: multiple agonists and receptors. *Nature*, **267**, 495–9.

Loughlin, S.E., Massamiri, T., Kornblum, H.I. and Leslie, F.M. (1985) Postnatal development of opioid systems in rat brain. *Neuropeptides*, **5**, 469–72.

Madden, J.D., Payne, T.F. and Miller, S. (1986) Maternal cocaine abuse and effect on the newborn. *Pediatrics*, **77**, 209–11.

Mahalik, M.P., Gautieri, R.F. and Mann, D.E. (1980) Teratogenic potential of cocaine hydrochloride in CS-1 mice. *J. Pharm. Sci.*, **69**, 703–6.

Mahalik, M.P., Gautieri, R.F. and Mann, D.E. (1984) Mechanisms of cocaine-induced teratogenesis. *Res. Commun. Subst. Abuse*, **5**, 279–302.

Mains, R.E., Eipper, B.A. and Ling, N. (1977) Common precursor to corticotropins and endorphins. *Proc. Natl. Acad. Sci. USA*, **74**, 3014–8.

Mansour, A., Khachaturian, H., Lewis, M.E. *et al.* (1987) Autoradiographic differentiation of mu, delta and kappa opioid receptors in the rat forebrain and midbrain. *J. Neurosci.*, **7**, 2445–64.

Martin, W.R., Eades, C.G., Thompson, J.A. *et al.* (1976) The effects of morphine- and nalorphine-like drugs in the nondependent and morphine-dependent chronic spinal dog. *J. Pharmacol. Exp. Ther.*, **197**, 517–32.

Mayorga, L.S., Diaz, R. and Stahl, P.D. (1989) Regulatory role for GTP-binding proteins in endocytosis. *Science*, **244**, 1475–7.

McGinty, J.F. and Ford, D.H. (1980) Effects of prenatal methadone on rat brain catecholamines. *Dev. Neurosci.*, **3**, 224–34.

McGivern, R.F., Raum, W.J., Sokol, R.Z. and Peterson, P. (1988) Long-term effects of prenatal cocaine exposure on scent marking and the HPG axis in male rats. *Teratology*, **37**, 518.

Millan, M.J., Morris, B.J. and Herz, A. (1988) Antagonist-induced opioid receptor upregulation. I. Characterization of supersensitivity to selective mu and kappa agonists. *J. Pharmacol. Exp. Ther.*, **247**, 721–8.

Moon, S.L. (1984) Prenatal haloperidol alters striatal dopamine and opiate receptors. *Brain Res.*, **323**, 109–13.

Morris, B.J. and Herz, A. (1989) Control of opiate receptor number in vivo: simultaneous kappa-receptor down-regulation and mu-receptor up-regulation following chronic agonists/antagonist treatment. *Neuroscience*, **29**, 433–42.

Nestler, E.J. and Tallman, J.F. (1988) Chronic morphine treatment increases cyclic AMP-dependent protein kinase activity in the rat locus coeruleus. *Mol. Pharmacol.*, **33**, 127–32.

Nestler, E.J., Erdos, J.J., Terwilliger, R. *et al.* (1989) Regulation of G-proteins by chronic morphine in the rat locus coeruleus. *Brain Res.*, **476**, 230–9.

Nestler, E.J., Beitner, D.B., Hayward, M. *et al.* (1990a) A general role for adaptations in G-proteins and the cyclic AMP system in mediating the chronic actions of morphine and cocaine on brain function. *Soc. Neurosci. Abstr.*, **16**, 928.

Nestler, E.J., Terwilliger, R.Z., Walker, J.R. *et al.* (1990b) Chronic cocaine treatment decreases levels of the G protein subunits $G_{i\alpha}$ and $G_{o\alpha}$ in discrete regions of rat brain. *J. Neurochem.*, **55**, 1079–82.

Nishino, K., Su, Y.F., Wong, C.S. *et al.* (1990) Dissociation of mu opioid tolerance from receptor down-regulation in rat spinal cord. *J. Pharmacol. Exp. Ther.*, **253**, 67–72.

Panerai, A.E., Martini, A., DiGiulio, A.M. *et al.* (1983) Plasma β-endorphin, β-lipotropin, and met-enkephalin concentrations during preg-

nancy in normal and drug-addicted women and their newborn. *J. Clin. Endocrinol. Metab.*, **57**, 537–43.

Pasternak, G.W., Gintzler, A.R., Houghton, R.A. *et al.* (1983) Biochemical and pharmacological evidence for opioid receptor multiplicity in the central nervous system. *Life Sci.*, **33** (Suppl 1), 167.

Perlmutter, J.F. (1974) Heroin addiction and pregnancy. *Obstet. Gynecol. Surv.*, **29**, 439–46.

Perry, D.C., Rosenbaum, J.S. and Sadee, W. (1982) In vitro binding of [^{3}H]etorphine in morphine-dependent rats. *Life Sci.*, **31**, 1405–8.

Pert, C.B., Pasternak, G. and Snyder, S.H. (1973) Opiate agonists and antagonists discriminated by receptor binding in brain. *Science*, **182**, 1359–61.

Pert, C.B. and Snyder, S.H. (1973) Opiate receptor demonstration in nervous tissue. *Science*, **179**, 1011–14.

Petrillo, P., Tavani, A., Verotta, D. *et al.* (1987) Differential postnatal development of mu-, delta- and kappa-opioid binding sites in rat brain. *Brain Res.*, **428**, 53–8.

Provost, N.M., Somers, D.E. and Hurley, J.B. (1988) A *Drosophila melanogaster* G protein alpha subunit gene is expressed primarily in embryos and pupae. *J. Biol. Chem.*, **263**, 12070–76.

Przewlocki, R., Hollt, V., Duka, T. *et al.* (1979) Long-term morphine treatment decreases endorphin levels in rat brain and pituitary. *Brain Res.*, **174**, 357–61.

Ragavan, V.V., Wardlaw, S.L., Kreek, M.J. and Frantz, A.G. (1983) Effect of chronic naltrexone and methadone administration on brain immunoreactive beta-endorphin in the rat. *Neuroendocrinology*, **37**, 266–8.

Rasmussen, K., Beitner-Johnson, D.B., Krystal, J.H. *et al.* (1990) Opiate withdrawal and the rat locus coeruleus: behavioral, electrophysiological and biochemical correlates. *J. Neurosci.*, **10**, 2308–17.

Rech, R.H., Lomuscio, G. and Algeri, S. (1980) Methadone exposure in utero: effects on brain biogenic amines and behavior. *Neurobehav. Toxicol.*, **2**, 75–8.

Redmond, D.E. and Krystal, J.H. (1984) Multiple mechanisms of withdrawal from opioid drugs. *Annu. Rev. Neurosci.*, **7**, 443–78.

Rementeria, J.L. and Lotongkhum, K. (1977) The fetus of the drug-addicted woman: conception, fetal wastage, and complications, in *Drug Abuse in Pregnancy and Neonatal Effects* (ed. J.L. Rementeria), Mosby, St Louis, MO, pp. 1–18.

Rothman, R.B., Danks, J.A., Jacobson, A.E. *et al.* (1986) Morphine tolerance increases μ-noncompetitive binding sites. *Eur. J. Pharmacol.*, **124**, 113–19.

Ryan, L., Ehrlich, S. and Finnegan, L. (1987) Cocaine abuse in pregnancy: effects on the fetus and newborn. *Neurotoxicol. Teratol.*, **9**, 295–9.

Sakellaridis, N. and Vernadakis, A. (1987) The chick embryo vs. neural tissue culture as models for the study of opiate neurotoxicity in development, in *Model Systems in Neurotoxicology: Alternative Approaches to Animal Testing* (eds A. Shahar and A.M. Goldberg), Alan R. Liss, New York, pp. 85–100.

Schoffelmeer, A.N.M., Hansen, H.A., Stoff, J.C. and Mulder, A.H. (1986) Blockade of D-2 dopamine receptors strongly enhances the potency of enkephalins to inhibit dopamine-sensitive adenylate cyclase in rat neostriatum: involvement of delta- and mu-opioid receptors. *J. Neurosci.*, **6**, 2235–9.

Schulz, R., Wuster, M. and Herz, A. (1979) Supersensitivity to opioids following the chronic blockade of endorphin action by naloxone. *Naunyn Schmiedebergs Arch. Pharmacol.*, **306**, 93–6.

Schwartz, J.P. (1988) Chronic exposure to opiate agonists increases proenkephalin biosynthesis in NG108 cells. *Mol. Brain Res.*, **3**, 141–6.

Shani, J., Azov, R. and Weissman, B.A. (1979) Enkephalin levels in rat brain after various regimens of morphine administration. *Neurosci. Lett.*, **12**, 319–22.

Sharma, S.K., Klee, W.A. and Nirenberg, M. (1975a) Dual regulation of adenylate cyclase accounts for narcotic dependence and tolerance. *Proc. Natl. Acad. Sci. USA*, **72**, 590–4.

Sharma, S.K., Nirenberg, M. and Klee, W.A. (1975b) Morphine receptors as regulators of adenylate cyclase activity. *Proc. Natl. Acad. Sci. USA*, **72**, 3092–6.

Sharma, S.K., Klee, W.A. and Nirenberg, M. (1977) Opiate-dependent modulation of adenylate cyclase. *Proc. Natl. Acad. Sci. USA*, **74**, 3365–3369.

Shen, K.F. and Crain, S.M. (1989) Dual opioid modulation of the action potential duration of mouse dorsal root ganglion neurons in culture. *Brain Res.*, **491**, 227–42.

Shen, K.F. and Crain, S.M. (1990) Cholera toxin-A subunit blocks opioid excitatory effects on sensory neuron action potentials indicating mediation by G_s-linked opioid receptors. *Brain Res.*, **525**, 225–31.

Shimada, S., Kitayama, S., Lin, C.L. *et al.* (1991) Cloning and expression of a cocaine-sensitive dopamine transporter complementary DNA. *Science*, **254**, 576–8.

Sibley, D.R., and Lefkowitz, R.J. (1985) Molecular mechanisms of receptor desensitization using the beta-adrenergic receptor-coupled adenylate cyclase system as a model. *Nature*, **317**, 124–9.

Simantov, R., Levy, R. and Baram, D. (1982a) Down-regulation of enkephalin (delta) receptors – demonstration in membrane-bound and solubilized receptors. *Biochim. Biophys. Acta*, **721**, 478–84.

Simantov, D., Baram, D., Levy, R. and Hadler, H. (1982b) Enkephalins and α-adrenergic receptors: evidence for both common and differentiable regulatory pathways and down-regulation of the enkephalin receptor. *Life Sci.*, **31**, 1323–6.

Simon, E.J. and Hiller, J.M. (1978) In vitro studies on opiate receptors and their ligands. *Fed. Proc.*, **37**, 141–6.

Sivam, S.P. (1989) Cocaine selectively increases striatonigral dynorphin levels by a dopaminergic mechanism. *J. Pharmacol. Exp. Ther.*, **250**, 818–24.

Slotkin, T.A. and Anderson, T.R. (1975) Sympathoadrenal development in perinatally addicted rats. *Addict. Dis. Int. J.*, **2**, 243–305.

Slotkin, T.A., Weigle, S.J., Whitmore, W.L. and Seidler, F.J. (1982) Maternal methadone administration: deficient in development of alpha-noradrenergic responses in developing rat brain as assessed by norepinephrine stimulation of ^{33}Pi incorporation into phospholipids in vivo. *Biochem. Pharmacol.*, **31**, 1899–902.

Smiley, P.L., Johnson, M., Bush, L. *et al.* (1990) Effects of cocaine on extrapyramidal and limbic dynorphin systems. *J. Pharmacol. Exp. Ther.*, **253**, 938–43.

Smith, R.F., Mattran, K.M., Kurkjian, M.F. and Kurtz, S.L. (1989) Alterations in offspring behavior induced by chronic prenatal cocaine dosing. *Neurotoxicol. Teratol.*, **11**, 35–8.

Snyder, S.H. (1984) Characterization of opiate binding sites. *Science*, **224**, 22–5.

Spain, J.W., Roth, B.L. and Coscia, C.J. (1985) Differential ontogeny of multiple opioid receptors (mu, delta, and kappa). *J. Neurosci.*, **5**, 584–8.

Spear, L.P., Kirstein, C.L. and Frambes, N.A. (1989) Cocaine effects on the developing central nervous system: behavioral, psychopharmacological and neurochemical studies. *Ann. NY Acad. Sci.*, **562**, 290–307.

Steece, K.A., DeLeon-Jones, F.A., Meyerson, L.R. *et al.* (1986) In vivo downregulation of rat striatal opioid receptors by chronic enkephalin. *Brain Res. Bull.*, **17**, 255–7.

Steketee, J.D., Striplin, C.D., Murray, T.F. and

Kalivas, P.W. (1991) Possible role for G-proteins in behavioral sensitization to cocaine. *Brain Res.*, **545**, 287–91.

Sternwies, P.C. (1986) The purified alpha subunits of G_o and G_i from bovine brain require beta gamma for association with phospholipid vesicles. *J. Biol. Chem.*, **261**, 631–7.

Strauss, M.E., Andresko, M., Stryker, J.C. *et al.* (1974) Methadone maintenance during pregnancy: pregnancy, birth and neonate characteristics. *Am. J. Obstet. Gynecol.*, **120**, 895–900.

Stryer, L. (1986) Cyclic GMP cascade of vision. *Annu. Rev. Neurosci.*, **9**, 87–119.

Tang, A.H. and Collins, R.J. (1978) Enhanced analgesic effects of morphine after chronic administration of naloxone in the rat. *Eur. J. Pharmacol.*, **47**, 473–4.

Tao, P.L., Law, P.Y. and Loh, H.H. (1987) Decrease in delta and mu opioid receptor binding capacity in rat brain after chronic etorphine treatment. *J. Pharmacol. Exp. Ther.*, **240**, 809–16.

Tao, P.L., Chang, L.R., Law, P.Y. and Loh, H. H. (1988) Decrease in delta-receptor density in rat brain after chronic (D-Ala^2, D-Leu^5) enkephalin treatment. *Brain Res.*, **462**, 313–20.

Tavani, A., Robson, L. and Kosterlitz, H.W. (1985) Differential postnatal development of mu, delta, and kappa opioid binding sites in mouse brain. *Dev. Brain Res.*, **23**, 306–9.

Tempel, A. (1991) Visualization of μ opiate receptor downregulation following morphine treatment in neonatal rat brain. *Dev. Brain Res.*, **64**, 19–26.

Tempel, A., Gardner, E.L. and Zukin, R.S. (1984) Visualization of opiate receptor upregulation by light microscopy autoradiography. *Proc. Natl. Acad. Sci. USA*, **81**, 3893–7.

Tempel, A., Gardner, E.L. and Zukin, R.S. (1985) Neurochemical and functional correlates of naltrexone-induced opiate receptor up-regulation. *J. Pharmacol. Exp. Ther.*, **232**, 439–44.

Tempel, A., Crain, S.M., Peterson, E.R. *et al.* (1986) Antagonist-induced opiate receptor upregulation in cultures of fetal mouse spinal cord-ganglion explants. *Brain Res.*, **390**, 287–91.

Tempel, A., Habas J., Paredes, W. and Barr, G.A. (1988) Morphine-induced downregulation of μ-opioid receptors in neonatal rat brain. *Dev. Brain Res.*, **41**, 129–33.

Tempel, A., Kessler, J.A. and Zukin, R.S. (1990) Chronic naltrexone treatment increases expression of preproenkephalin and preprotachykinin mRNA in discrete brain regions. *J. Neurosci.*, **10**, 741–7.

Townsend, R.R., Laing, F.C. and Jeffrey, R.B. (1988) Placental abruption associated with cocaine abuse. *Am. J. Roentgenol.*, **150**, 1339–40.

Traber, J., Gullis, R. and Hamprecht, B. (1975) Influence of opiates on the levels of adenosine 3':5'-cyclic monophosphate in neuroblastoma X glioma hybrid cells. *Life Sci.*, **16**, 1863–8.

Tsang, D. and Ng, S.C. (1980) Effect of antenatal exposure to opiates on the development of opiate receptors in rat brain. *Brain Res.*, **188**, 199–206.

Tsang D., Tan, A.T., Henry, J.L. and Lal, S. (1978) Effect of opioid peptides on L-noradrenaline-stimulated cyclic AMP formation in homogenates of rat cerebral cortex and hypothalamus. *Brain Res.*, **152**, 521–7.

Uhl, G.R., Ryan, J.P. and Schwartz, J.P. (1988) Morphine alters preproenkephalin gene expression. *Brain Res.*, **459**, 391–7.

Valencia, G., McCalla, S., da Silva, M. *et al.* (1989) Epidemiology of cocaine use during pregnancy at Kings County Hospital Center (KCHC). *Pediatr. Res.*, **25**, 265A.

Van Vliet, J., Wardeh, G., Mulder, A.H. and Schoffelmeer, A.N.M. (1991) Reciprocal effects of chronic morphine administration on stimulatory and inhibitory G-protein subunits in primary cultures of rat striatal neurons. *Eur. J. Pharmacol.*, **208**, 341–2.

Watanabe, Y., Shibuya, T., Salafsky, B. and Hill, H.F. (1983) Prenatal and postnatal exposure to diazepam: effects on opioid receptor binding in rat brain cortex. *Eur. J. Pharmacol.*, **96**, 141–4.

Weber, E., Evans, E.J. and Barchas, J.D. (1983) Multiple endogenous ligands for opioid receptors. *Trends Neurosci.*, **6**, 333.

Weiner, N. (1974) Neurotoxicity of opiates during brain development, in *Drugs and the Developing Brain* (eds A. Vernadakis and N. Weiner), Plenum Press, New York, pp. 215–27.

Wesche, D., Hollt, V. and Herz, A. (1977) Radioimmunoassay of enkephalins. Regional distribution in rat brain after morphine treatment and hypophysectomy. *Naunyn-Schmiedebergs Arch. Pharmacol.*, **301**, 79–82.

Wuster, M., Costa, T. and Gramsch, C.H. (1983) Uncoupling of receptors is essential for opiate-induced desensitization (tolerance) in neuroblastoma x glioma hybrid cells NG108-15. *Life Sci.*, **33**, 341–4.

Yoburn, B.C., Goodman, R.G., Cohen, A.C. *et al.* (1985) Increased analgesic potency of morphine and increased brain opioid binding sites in the rat following chronic naltrexone treatment. *Life Sci.*, **36**, 2325–9.

Yoburn, B.C., Kreuscher, S.P., Inturrisi, C.E. and Sierra, V. (1989) Opioid receptor up-regulation and supersensitivity in mice: effect of morphine sensitivity. *Pharmacol. Biochem. Behav.*, **32**, 727–31.

Yoshikawa, K. and Sabol, S.L. (1986) Gluco-corticoids and cyclic AMP synergistically regulate the abundance of preproenkephalin messenger RNA in neuroblastoma-glioma hybrid cells. *Biochem. Biophys. Res. Commun.*, **139**, 1–10.

Young, W.S. and Kuhar, M.J. (1979) A new method for receptor autoradiography: [^{3}H] opioid receptors in rat brain. *Brain Res.*, **179**, 255–70.

Zagon, I.S. and McLaughlin, P.J. (1977) The effect of chronic morphine administration on pregnant rats and their offspring. *Pharmacology*, **15**, 302–10.

Zagon, I.S., McLaughlin, P.J., Weaver, D.J. and Zagon, E. (1982) Opiates, endorphins, and the developing organism: a comprehensive bibliography. *Neurosci. Biobehav. Rev.*, **6**, 439–79.

Zukin, R.S. and Zukin, S.R. (1988) The sigma receptor, in *The Opiate Receptors* (ed. G.W. Pasternak), Humana Press, Clifton, NJ, pp. 143–63.

Zukin, R.S., Sugarman, J.R., Fitz-Syaga, M.L. *et al.* (1982) Naltrexone-induced opiate receptor supersensitivity. *Brain Res.*, **245**, 285–92.

Zukin, R.S., Eghbali, M., Olive, D. *et al.* (1988) Characterization and visualization of rat and guinea-pig kappa opioid receptors: evidence for kappa$_1$ and kappa$_2$ receptors. *Proc. Natl. Acad. Sci. USA*, **85**, 4061–5.

INDEX